EXS 66

Nonselective Cation Channels

Channels

Pharmacology, Physiology and Biophysics

Edited by D. Siemen
J. Hescheler

Birkhäuser Verlag
Basel · Boston · Berlin

Editors:

Professor Dr. Detlef Siemen
Institut für Zoologie
Universität Regensburg
Universitätsstrasse 31
D-93040 Regensburg

PD Dr. Jürgen Hescheler
Institut für Pharmakologie
Freie Universität Berlin
Thielallee 69–73
D-14195 Berlin

This book was made possible by generous financial support from Boehringer Ingelheim KG, Ingelheim, FRG and Hoechst AG, Frankfurt, FRG.

Library of Congress Cataloging-in-Publication Data
Nonselective cation channels: pharmacology, physiology and biophysics
  / edited by D. Siemen, J. Hescheler.
     p.   cm.—(EXS; 66)
  Includes bibliographical references and index.
  ISBN 3-7643-2888-6—ISBN 0-8176-2888-6
  1. Ion channels. 2. Cations—Physiological transport.
I. Siemen, D. (Detlef), 1950–. II. Hescheler, J. (Jürgen
Karl-Josef), 1959– III. Series.
  [DNLM: 1. Ion Channels—drug effects. 2. Cations—metabolism.
3. Cations—pharmacology.   W1 E65 v. 66 1993 / QH 603.I54 N814 1993]
QH603.I54N66 1993
591'.87'5—dc20

Deutsche Bibliothek Cataloging-in-Publication Data
Nonselective cation channels: pharmacology, physiology and
biophysics / ed. by D. Siemen; J. Hescheler.—Basel;
Boston; Berlin: Birkhäuser, 1993
  (EXS; 66)
  ISBN 3-7643-2888-6 (Basel . . .)
  ISBN 0-8176-2888-6 (Boston)
NE: Siemen, Detlef [Hrsg.]; GT

© 1993 Birkhäuser Verlag, P.O. Box 133, CH-4010 Basel, Switzerland
Printed on acid-free paper produced from chlorine-free pulp
Printed in Germany

ISBN 3-7643-2888-6
ISBN 0-8176-2888-6

Contents

vi

Gap Junction Channels

Cyclic Nucleotide-Activated Nonselective Cation Channels

Nonselective Cation Channels Activated by Intracellular Calcium

Nonselective Cation Channels as Regulatory Components of Cells from Various Tissues

Foreword

It can be argued that nonselective cation channels were the first sort of ion channel to be described, though the word *channel* was not used at the time. Their existence was implied by Fatt and Katz in 1952, when they described the action of acetylcholine at the muscle endplate as producing "a large nonselective increase of ion permeability, i.e. a short circuit". Shortly afterwards, in 1956, Katz referred to "aqueous channels through which small ions can pass . . . " (del Castillo and Katz, *Prog. Biophysics and Biophys. Chem.* 6, 121–170).

Now, more than thirty years later, it has become clear that there are far more types of nonselective cation channels than anyone could have imagined a few years ago, and that they are found in a vast range of tissues. One has, of course, become quite accustomed to such diversity in, for example, $GABA_A$ receptors, but this is not quite the same thing. In the case of $GABA_A$ receptor we are talking about a fairly narrow range of structural diversity (resulting largely from differences in subunit composition) within a single type of channel with more-or-less well-defined function. In the case of nonselective cation channels the function is often not known, and relatively few have been cloned. It seems certain though, that they encompass a wide range of quite different structural types.

After the nonselective channels activated by excitatory neurotransmitters, the function of which is clear, the next group to be discovered arose from experiments on heart muscle. Kass *et al.* (1978*a,b*) suggested that a current that was involved in oscillatory activity in the heart was carried by a nonselective cation channel that was activated by intracellular calcium ions (rather than by an extracellular neurotransmitter). Since then single channel methods have demonstrated the existence of such channels in many other sorts of cell too.

More recently, nonselective cation channels have been found to be involved in vision, olfaction and taste. Many other such channels, with unknown function, have been observed too. The casual observer might be forgiven for feeling that the panoply of channels that have been discovered has resulted in a certain degree of chaos. In this book the authors describe the current state of knowledge of a wide range of nonselective cation channels, and their reviews should certainly help to restore some order to this large, and growing, field of work. Work on the biophysical and pharmacological properties of nonselective cation channels is proceeding apace. Information on molecular structure is also beginning to emerge now, so we can look forward to the time, before very long, when many of these channels will be understood at the level that has been achieved, so far, for only a few, such as the nicotinic acetylcholine receptor and sodium channels.

D. Colquhoun
University College London
May 1993

List of Contributors

Barnstable, C. J., Department of Ophthalmology, Yale University School of Medicine, 333 Cedar Street, New Haven, CT 06510, USA

Chen, S., Department of Pharmacology, Faculty of Medicine, Kyushu University, Fukuoka 812, Japan

Dani, J. A., Department of Physiology and Biophysics, Baylor College of Medicine, One Baylor Plaza, Houston, TX 77030, USA

Dermietzel, R., Institut für Anatomie, Universität Regensburg, Universitätsstr. 31, D-93040 Regensburg, FRG

Desmedt, L., Laboratory for Physiology, K.U. Leuven, Campus Gasthuisberg, B-3000 Leuven, Belgium

De Smet, P., Laboratory for Physiology, K.U. Leuven, Campus Gasthuisberg, B-3000 Leuven, Belgium

Dotzler, Elisabeth, Institut für Zoologie, Universität Regensburg, Universitätsstr. 31, D-93040 Regensburg, FRG

Droogmans, D., K.U. Leuven, Campus Gasthuisberg, Department of Physiology, B-3000 Leuven, Belgium

Englert, H. C., Hoechst AG, G 838, D-65926 Frankfurt/Main, FRG

Frace, A. M., Department of Biophysics and Physiology, University of California, Irvine, CA 92717, USA

Gargus, J. J., Department of Physiology and Section of Medical Genetics, Emory University School of Medicine, Atlanta, Georgia 30322, USA

Geiger, J., Klinische Forschergruppe, Medizinische Uniklinik Würzburg, Josef-Schneider Str. 2, D-97078 Würzburg, FRG

Gericke, M., K.U. Leuven, Campus Gasthuisberg, Department of Physiology, B-3000 Leuven, Belgium

Gögelein, H., Hoechst AG, H 821, D-65926 Frankfurt, FRG

Hescheler, J., Pharmakologisches Institut der Freien Universität Berlin, Thielallee 69–73, D-14195 Berlin, FRG

Hoyer, J., Dr., Klinikum Steglitz, Hindenburgdamm 30, D-12202 Berlin, FRG

Inoue, R., Department of Pharmacology, Faculty of Medicine, Kyushu University, Fukuoka 812, Japan

Isenberg, G., Institut für Vegetative Physiologie der Universität zu Köln, Robert-Koch-Str. 39, D-50931 Köln, FRG

Jonas, P., Max-Planck-Institut für Medizinische Forschung, Abteilung Zellphysiologie, Jahnstr. 29, D-69120 Heidelberg, FRG

Jung, F., Department of Physiology and Section of Medical Genetics, Emory University School of Medicine, Atlanta, Georgia 30322, USA

Kleppisch, T., Institut für Physiologie, Humboldt-Universität zu Berlin, D-10115 Berlin, FRG

Koivisto, A., Wenner-Gren-Institute, Arrheniuslab. F3, University of Stockholm, S-10691 Stockholm, Sweden

Korbmacher, C., Zentrum der Physiologie, Klinikum der Universität, Theodor-Stern Kai 7, D-60596 Frankfurt, FRG

Lang, H.-J., Hoechst AG, Institut für Pharmakologie, D-65926 Frankfurt, FRG

Nedergaard, J., Wenner-Gren-Institute, Arrheniuslab. F3, University of Stockholm, S-10691 Stockholm, Sweden

Nilius, B., K.U. Leuven, Campus Gasthuisberg, Department of Physiology, B-3000 Leuven, Belgium

Partridge, L. D., School of Medicine, Department of Physiology, The University of New Mexico, Albuquerque NM 87131, USA

Petersen, O. H., The Physiological Laboratory, Crown Street, P.O. Box 147, University of Liverpool, Liverpool L69 3BX, Great Britain

Popp, R., Max-Planck-Institut für Biophysik, Kennedyallee 70, D-60596 Frankfurt, FRG

Ruß, U., Institut für Zoologie, Universität Regensburg, Universitätsstr. 31, D-93040 Regensburg, FRG

Sachs, F., Department of Biophysical Sciences, 120 Cary Hall, State University of New York, Buffalo, NY 14214, USA

Schild, D., Physiologisches Institut, Universität Göttingen, Humboldtallee 23, D-37073 Göttingen, FRG

Schultz, G., Pharmakologisches Institut der Freien Universität Berlin, Thielallee 69–73, D-14195 Berlin FRG

Schwarz, G., K.U. Leuven, Campus Gasthuisberg, Department of Physiology, B-3000 Leuven, Belgium

Siemen, D., Institut für Zoologie, Universität Regensburg, Universitätsstr. 31, D-93040 Regensburg, FRG

Siemer, Christiane, Max-Planck-Institut für Biophysik, Kennedyallee 70, D-60596 Frankfurt, FRG

Simaels, Jeannine, Laboratory for Physiology, K.U. Leuven, Campus Gasthuisberg, B-3000 Leuven, Belgium

Swandulla, D. Institut für Experimentelle und Klinische Pharmakologie, Universität Erlangen, D-91054 Erlangen, FRG

Thorn, P., The Physiological Laboratory, Crown Street, P.O. Box 147, University of Liverpool, Liverpol L69 3BX, Great Britain

Van Driessche, W., Laboratory for Physiology, K.U. Leuven, Campus Gasthuisberg, B-3000 Leuven, Belgium

Walter, U., Klinische Forschergruppe, Medizinische Uniklinik Würzburg, Josef-Schneider Str. 2, D-97078 Würzburg, FRG

Weiser, T., Boehringer Ingelheim KG, Zentralnervensystem
  Pharmakologie, Bingerstrasse, D-55216 Ingelheim, FRG
Wellner, Marie-Cécile, Institut für Vegetative Physiologie der
  Universität zu Köln, Robert-Koch-Str. 39, D-50931 Köln, FRG
Wobus, Anna M., Institut für Pflanzengenetik und
  Kulturpflanzenforschung, D-06466 Gatersleben, FRG
Yang, X., American Cyanamid Company, Medical Research Division,
  Lederle Laboratories, Pearl River, NY 10965, USA
Zufall, F., Section of Neurobiology, Yale University, School of
  Medicine New Haven, CT 06510, USA

Introduction

Nonselective Cation Channels: Pharmacology, Physiology and Biophysics
ed. by D. Siemen & J. Hescheler
© 1993 Birkhäuser Verlag Basel/Switzerland

Nonselective Cation Channels

Detlef Siemen

"Assuming that acetylcholine produces a large non-selective increase of ion permeability, i.e. a short-circuit, . . ." wrote Fatt and Katz in 1951. That was a year before Hodgkin and Huxley (1952) published their pioneering papers in which they suggested selective changes of nerve membrane permeability for Na^+-ions and K^+-ions as an explanation for the characteristic shape of action potentials. Selectivity for different ions is an important criterion for classification of ion channels. We can now state that acetylcholine-sensitive ion channels are indeed nonselective for monovalent cations, and that there are even many more types of ion channels characterized by different degrees of nonselectivity. With this book the authors try to demonstrate the current state of research in this field.

I. Variety of Nonselective Cation Channels

Nonselective cation channels are a largely heterogeneous group. An extreme example would be the probably best described ion channel, the nicotinic acetylcholine receptor from the neuromuscular endplate (Dani, this volume). In the broadest sense, we may also include the gap junction channel (Dermietzel, this volume), bacterial porins (Benz, 1988), or even, under some conditions, the voltage-dependent anion channel (VDAC), although these channels are clearly distinct concerning their molecular structure. A relatively well-defined subgroup within the nonselective cation channels encompasses those which are activated by intracellular Ca^{2+}, that do not discriminate very much between Na^+- and K^+-ions, and are more or less impermeable to divalents. This subgroup is referred to as *NSC-channels* in the following, and it is introduced separately in the second part of this chapter.

Ligand-Gated and Mechanosensitive Nonselective Cation Channels

The *nicotinic acetylcholine receptor*, a classical member of the huge family of ion channels, is a large protein which has a molecular weight of more than 250 kD and consists of five subunits, each with at least four membrane-spanning α-helices (cf. Dani, this volume). It must not be confused with the muscarinic acetylcholine receptor, which does not include a pore and which is instead connected by a G-protein to a potassium channel.

Glutamate receptors have attracted a great deal of attention due to their importance in the central nervous system, where glutamate is the predominant excitatory neurotransmitter. But in arthropods, glutamate is also found to be a transmitter in the neuromuscular junction (Dudel et al., 1990). In general, there are two classes, distinguished by their ability to respond to NMDA (N-Methyl-D-aspartate); molecular biology has shown several subclasses. In all cases the channel molecule carries the receptor site for glutamate. NMDA-receptors are, by far, better permeable to Ca^{2+} than to monovalent cations, thus they are considered only peripherally. They are voltage-dependently blocked by Mg^{2+} (Mayer and Westbrook, 1987). Details of the non-NMDR type are reviewed by Jonas (this volume).

The number of known *stretch-activated channels* (SA-channels), has rapidly increased over recent years as has the number of preparations in which they have been found. As several of them are not only activated by stretching the cells, but by volume increase or decrease as well, mechanosensitive channels should be the preferred name of this group. A review is given by Yang and Sachs (this volume). Additional results are presented by Isenberg, Popp et al., and Wellner and Isenberg (all in this volume).

Cyclic Nucleotide Gated-Channels in Visual and Odor Reception

The cGMP-gated channel of the vertebrate retinal *rod photo receptor* (reviewed by Barnstable, this volume) is, next to the nicotinic acetylcholine receptor, probably the second best known nonselective cation channel. Following Sillmann et al. (1969) it was believed earlier that mainly Na^+ would be able to pass the photoreceptor channel, thus carrying the "dark current". Later, Yau and Nakatani (1984) and Fesenko et al. (1985) found a nonselective cation channel. Fesenko et al. had shown that cGMP was the internal regulator that opened the channel for a few milliseconds. Later, Kaupp et al. (1989) deduced the amino-acid sequence of the bovine photoreceptor channel by cloning its DNA. The channel consists of one or several units of a 63 kD protein. As expressed in *Xenopus* oocytes, a mole weight of 79 601 D was deduced from the amino acid sequence (690 aminoacids and probably four or six transmembrane segments). Single channel conductance is 20 pS (100 mM KCl) and the selectivity sequence given by

Kaupp et al. is

$$NH_4^+ > K^+ \approx Na^+ > Li^+ > Rb^+ > Cs^+ \quad (2.9: 1.0: 1.0: 0.6: 0.6: 0.4).$$

In contrast to this sequence, Fesenko et al. (1985) found in retinal rods of *Rana temporaria*: $Na^+ > Li^+$, $K^+ > Rb^+$, $Cs^+ \gg Cl^-$ and a single-channel conductance of 3 pS (it may be as low as 100 fS; Hanke et al., 1988; Bodoia and Detwiler, 1985). The differences in single-channel conductance are due to different Ca^{2+} concentrations. cGMP-induced current decreases by 20 to 30%, if internal Ca^{2+} is raised from 10^{-8} to 10^{-3} M (Fesenko et al., 1985). While the order of the selectivity sequences resembles that of the NSC of adipocytes (Koivisto et al., this volume) the relative permeability values differ. The photoreceptor channel seems to be perfectly nonselective to the physiological ions Na^+ and K^+, but in contrast to the NSC, the photoreceptor channel is highly permeable to divalent cations (Hodgkin et al., 1985).

A similar cation selective channel exists in *vertebrate cones*. It has a single-channel conductance of 45–50 pS, but is also activated by cGMP (Haynes and Yau, 1990) and blocked by divalents and *l*-cis-diltiazem. (For comparison of the different kinds of cGMP-gated channels, see Kaupp, 1991.)

Initial events of the *vertebrate olfactory-transduction* process include a G_{olf}-protein-dependent rise of cAMP-concentration in the receptor neurons eliciting a generator conductance via activation of a nonselective cation channel ($P_K/P_{Na} = 0{,}82$). This channel shares 57% identity of the amino acid sequence with the cGMP-gated channel of vertebrate photoreceptors (Dhallan et al., 1990). Many kinds of odorants are able to stimulate different receptors in the plasma membrane. These receptors in turn activate the adenylyl cyclase and cAMP causes a direct stimulation of the cAMP-dependent channel. Details are given by Zufall (this volume).

Several nonselective cation channels are also involved in *olfactory signal transduction of insects*. Although the precise sequence of events remains elusive, it is assumed that pheromone-dependent cation channel opening in the olfactory receptor neurons could lead to opening of directly as well as indirectly (protein kinase C) activated cation channels, which in turn cause depolarization. Both types of nonselective cation channels occur in these cells with similar properties as the pheromone dependent cation channels. The latter have been found in cell-attached recordings from soma membranes of cultured neurons, or extruded dendrites after stimulation with the species-specific pheromone. Their reversal potential is around 0 mV, they show substates and about equal permeability for Na^+ and K^+. Burst length and opening probability depend on the pheromone concentration. The long delay time of the response suggests involvement of second messenger systems.

Sensitivity to Ca^{2+} and sometimes also to cGMP makes it likely that different populations of nonselective channels contribute (Stengl et al., 1992). Since no internal Ca^{2+} stores are apparent in outer dendrites, it is assumed that IP_3-dependent Ca^{2+}-influx starts the transduction cascade (Breer et al., 1990).

Involvement of Non-Selective Channels in Taste Reception

A different cation channel is responsible for *salt perception of vertebrate* taste receptors. It causes a long-lasting sodium influx which depolarizes the cell to threshold. At least in the frog it seems to be related to epithelial amiloride-sensitive channels, but it also exhibits differences e.g., in amiloride sensitivity with an inhibition constant of $0.3 \mu M$ (Avenet and Lindemann, 1988, 1989). Because amiloride does not block all of these cation channels, the existence of at least two populations is suggested. Single-channel conductance is very low ($<2 pS$) and the selectivity sequence is: $K^+ > Na^+ > Rb^+ > Li^+ > Cs^+ > $ N-methyl-D-glutamine. The epithelial amiloride-sensitive channel, however, is less selective: $Li^+ > Na^+ \gg K^+, Rb^+$ and has a larger single-channel conductance (5 pS), (Palmer, 1987). Nevertheless, binding of antibodies against amiloride-sensitive sodium channels of kidney cells to the apical membranes of taste pores from dogs was observed (Simon et al., 1989). In rodent taste receptor cells the corresponding ion channel more resembles the amiloride-sensitive sodium conductance (5 pS) of frog skin and kidney (Avenet and Kinnamon, 1991). The function of these channels would thus be to depolarize the plasma membrane of salty-taste receptor cells as soon as the outside concentration of Na^+ raises. Also, in undefined lingual taste cells from the mud puppy (*Necturus*) nonselective cation channels are present. They seem to be involved in a long-lasting after-depolarization following the current pulse elicited action potential (McBride and Roper, 1991).

In transduction of the *taste umami* (i.e., aminoacids), another nonselective channel ($P_{Na} \approx P_K$) is involved. In catfish it shows a single-channel conductance of $40-50 pS$ and opens after stimulation by L-arginin (maximal increase of activity at $200 \mu M$), or L-proline ($2-4 mM$). It seems to also be permeable to divalents ($P_{Ca}/P_{Na} \approx 1$, $P_{Ba}/P_{Na} = 0.1$), while permeability to anions is negligible (Kinnamon and Cummings, 1992; Teeter et al., 1990). Reconstitution experiments make it likely that the L-arg channel is the first ligand-gated channel involved in signal transduction of taste. Recently, it has been partially purified and reconstituted from a catfish preparation by Kalinoski et al. (1992). The prominent protein turned out to be a doublet of M_r 55 000 accompanied by four additional proteins.

Nonselective Channels and Ca^{2+}-Signaling

Due to the important role calcium plays as a second messenger, considerable effort has been focused on understanding regulatory pathways for control of cytosolic calcium. In addition to IP_3-mediated Ca^{2+}-release from internal stores and Ca^{2+}-influx through voltage-gated Ca^{2+}-channels, there was recently a third major pathway for Ca^{2+}-entry discovered: *receptor mediated calcium entry* (RMCE). Two agonists may stimulate two independent RMCEs: External ATP triggers a rapid and short-lived increase of the internal Ca^{2+}-concentration, while external bradykinin elicits a more sustained rise of Ca^{2+}_i. During the bradykinin response, only the plateau phase seems to be due to RMCE. Both ATP- and bradykinin response may cause membrane depolarization, due to nonselective channels permeable to both Ca^{2+} and Na^+. RMCE was described in many cells, e.g., in human platelets, neutrophils, endothelial cells, and in the neurosecretory cell line PC12 (Fasolato et al., 1990; Merritt et al., 1990). In smooth muscle Benham and Tsien (1987) found a 3:1 selectivity for Ca^{2+} over Na^+ and a 5 pS-single-channel conductance (0.1 mM Ca^{2+}) of an RMCE, activated directly by ATP.

Another 4–5 pS nonselective cation channel was found in neutrophiles (Krautwurst et al., 1992). It is better permeable to Na^+ than to Ca^{2+} and demonstrates a possible close relation to the NCS-channels summarized in the second part of this chapter (see also Hescheler and Schultz, this volume).

In several different types of endothelial cells changes of the intracellular Ca^{2+}-concentration occur as oscillations (Jacob, 1991). The underlying channel events were found by Lückhoff and Clapham (1992). In inside-out patches from endothelial cells they described a 2.5 pS channel activated by 10–100 μM internal IP_4 or by internal Ca^{2+}. In addition to Ca^{2+}, this channel is equally permeable to Mn^{2+} and Ba^{2+}. Na^+ may permeate, too ($P_{Ca}/P_{Na} > 2$:1), but 1–5 mM Ni^{2+} blocks it. In mast cells Penner et al. (1988) found a 50 pS nonselective cation channel that becomes active after secretagogue stimulation. This channel shows Ca^{2+}-permeability ($\gamma = 16$ pS in isotonic barium solution) and seems to contribute to a sustained increase in intracellular Ca^{2+} enhancing secretion. Very likely, however, it is not the primary source of hyperpolarization-driven calcium influx into the cell and it should not be confused with the highly Ca^{2+}-selective I_{CRAC} (Hoth and Penner, 1992; Nilius et al., this volume).

VDAC, Bacterial Porins and Gap Junction Channels

The *voltage-dependent anion channel* (VDAC) of outer mitochondrial membranes (reviewed by Mannella, 1992; Colombini, 1989) is formed by a 31 kDa polypeptide which appears in hexamers (Thomas et al. 1991). Like for bacterial porins the secondary structure of the VDAC consists to a great extent of a cylindrical β-sheet (Weiss et al., 1990). Some VDACs have already been cloned and sequenced and show no significant sequence homology with bacterial porins (Forte, Guy and Manella, 1987). At 0–10 mV, VDACs may have two open states for which the conductances (in the range of nS) differ by a factor of more than two. The lower-conductance state is preferred at higher potentials, or under the influence of macromolecular effectors. High pH or micromolecular concentrations of Al^{3+} inhibit closure (Mangan and Colombini, 1987; Dill et al., 1987; cf. also Guo and Mannella, 1992). Voltage-dependent closure of the VDAC occurs at both positive and negative potentials. It was shown that under the influence of succinic anhydride (other substances like polyanion or dextransulfate may also change channel conformation) VDACs may transit from a 480 pS (0,1 M KCl) partly closed state to a 300 pS partly closed state in which it is cation selective. Selectivity changes from 2:1 Cl^- over K^+ to 8:1 K^+ over Cl^-, which is why we include it among the nonselective cation channels. Voltage dependence is altered under these conditions (Colombini, 1989; Benz and Brdiczka, 1992). Conductance of a 37 kD mitochondrial porin from *Paramecium* is slightly smaller (260 pS, 0,1 M KCl) but showed cation selectivity in the low conductance state, too ($P_K/P_{Cl} \geq 10$; Ludwig et al., 1989).

Recently, another large-conductance channel from outer mitochondrial membranes was described by Henry et al. (1989), which they designated a cation channel. It opens in two steps of 220 pS (measured in 150 mM NaCl) and it can be blocked by 100 μM peptide M, a 13 residue peptide with the sequence of the N-terminal part of the cytochrome C oxidase subunit VI. The sequence is known to direct the protein into the mitochondrial matrix (cf. Henry et al., 1989). The authors did not discuss, however, the possibility that VDACs can also show selectivity for cations.

A 36 kD polypeptide that is 70% identical with the VDAC of human B lymphocytes has been isolated together with $GABA_A$-receptors from mammalian brain membrane preparations (Bureau et al., 1992). Since we find VDAC-like channels in patches from plasma membranes of cultured rat astrocytes, it seems possible that such channels are more widely distributed than expected (Marrero et al., 1991; Dermietzel et al., submitted).

Another group of proteins forming large cation-selective pores are the *gap junction channels* which are reviewed by Dermietzel (this volume). They enable cells to communicate by signal coupling in several tissues, and are probably best known in the vertebrate heart. The quaternary structure consists of six subunits forming one half-channel. The other half-channel is contributed by the neighboring cell. In neurons, they form together an (formerly called) electrical synapse.

Major intrinsic polypeptides (MIP) also form large nonselective channel like pores. They are found in lense fiber cell membranes and show a single-channel conductance of ≈ 360 pS (100 mM KCl) when reconstituted into planar lipid bilayers. Voltage dependence is similar to that of gap junctions channels, i.e., the channel closes when the membrane potential exceeds a value of ± 30 mV. In vivo the channel is phosphorylated. If it is reconstituted into bilayers, however, it loses voltage dependence in the dephosphorylated form (Ehring et al., 1991). Their relationship to the gap junction channels is debated.

Porins of Gram-negative bacteria form aqueous channels, too, allowing diffusion of small hydrophilic molecules down their electrochemical gradient across the outermost of the two bacterial membranes. They are constructed by 16-stranded anti-parallel beta-barrels which occur in trimers. Each subunit contains a pore. Matrix porin (OmpF) and osmoporin (OmpC) of *E. coli* are weakly cation-selective and are regulated by osmotic pressure and temperature (cf. Cowan et al., 1992; for review see Benz, 1988, 1985). By site-directed mutagenesis cation selectivity may be induced in the originally anion-selective porin PhoE by replacing Lys-125 with glutamic acid (Bauer et al., 1989).

Inside the beta-barrel exists a narrower loop of the molecule, dubbed L3, which restricts its smallest cross-section to 7×11 Å, i.e., much larger than the 6.5×6.5 Å of the endplate channel (Dwyer et al., 1980). Immediately beyond the L3-segment the diameter increases to 15×22 Å. The porin is thus large enough to allow a hydrated molecule of glucose to pass through (Cowan et al., 1992; Schindler and Rosenbusch, 1978). Nikaido (1992) points out that there are two types of channels in bacterial outer membranes: nonspecific and specific channels. Specific channels contain stereospecific binding sites which accelerate diffusion at low concentration of the solute and slow down diffusion at higher concentrations. As a result, the channel shows saturation of the current, which may be described by Michaelis-Menten enzyme kinetics. Nonspecific porins do not contain such ligand-binding sites and show a more linear increase of current with increasing solute concentration. In order to avoid confusion, Nikaido recommends to use the word "porin" only for the non-specific type of channel. The two mentioned porins of *E. coli* satisfy this criterion.

Other Nonselective Cation Channels

Finally, there are channels described which are definitely nonselective to cations, but which do not really fit into one or the other group. Mostly, the reports are too brief to really classify the channels. As an example, a 69 pS nonselective cation channel has been found in peptidergic nerve terminals of a land crab. It is permeable to Na^+ and K^+, but not to

Cs^+ or Cl^-. Interestingly, it is activated by intracellular Na^+, but not by internal Ca^{2+} (Lemos et al., 1986). Another channel from boar sperm plasma membranes was described in planar lipid bilayers (Cox et al., 1991). Keratinocytes show a 14 pS nonselective cation channel that is permeable to Ca^{2+} (Mauro et al., 1993).

Even *cell organelles* may show an ion channel nonselective to monovalent cations. In the *sarcoplasmic reticulum* of skeletal muscle and ventricular muscle it has studied extensively (cf. Miller, 1978; Hill et al., 1989). In heart, this channel has two open states: a main state with a single-channel conductance of 35 pS and a subconductance state with 20 pS. Selectivity calculated by comparison of the reversal potentials under biionic conditions revealed a selectivity sequence:

$$NH_4^+ > Cs^+ > Rb^+ \geq K^+ > Na^+ > Li^+ (1.7:\ 1.25:\ 1.0:\ 1.0:\ 0.5:\ 0.2).$$

Ca^{2+} was not measurably permeant. Interestingly, this channel does not saturate dependent on ion activity and it shows a prominent anomalous mole fraction behavior (Cs^+ with K^+; for theory cf. Hille, 1992). It seems to be a multi-ion pore, differing from skeletal sarcoplasmic reticulum channel in biophysical properties. This may demonstrate that, within a few years, the group of nonselective cation channels has grown to a size than can hardly be reviewed completely.

II. Nonselective Cation Channels Activated by Intracellular Ca^{2+}

History

An early hint of the presence of nonselective ion channels in pancreatic acinar cells came from Nishiyama and Petersen (1975), who compared simultaneous changes in the permeabilities for Na^+- and K^+-ions after ACh-stimulation with the events following stimulation of the motor endplate. Petersen and Iwatsuki (1978) found that an increase in intracellular Ca^{2+}-concentration caused the increase in membrane conductance and the resulting depolarization with a reversal potential near $-15\,mV$.

In the 1970s it was already known that a complete understanding of the oscillatory electrical activity of Purkinje fibers of the vertebrate heart required knowledge of an additional transient inward current; it becomes prominent after strophantidin treatment (Lederer and Tsien 1976). Kass et al. (1978a, b) studied this cation current in more detail and found that it was carried mainly by Na^+-ions. Tetrodotoxin (TTX), however, had no blocking effect. The remainder of the observed current was carried by an ion with a negative reversal potential, either Cl^- or K^+. As Cl^- could be experimentally excluded, it was assumed that the channel is permeable to Na^+ and K^+. Kass et al. compared this permeability with the nicotinic AChR and pointed at a major difference: Ca^{2+} acts as an internal "transmitter." The last sentence of their paper is: "Future tests... would be greatly aided by the discovery of pharmacological agents which specifically inhibit the various pathways for Na entry". This is still true today.

In 1981, Colquhoun et al. reported single-channel events of an ion channel in cultured rat ventricular muscle cells exhibiting K^+ – as well as Na^+ – permeability. It turned out to be a Ca^{2+}-activated channel nonselective for monovalent cations (abreviated NSC). In the following year three more papers appeared that proved the presence of NSC-channels in neuroblastoma and in pancreatic acinar cells of rat and mice (Yellen, 1982; Maruyama and Petersen, 1982a, b). These channels had a single-channel conductance of $22-30$ pS ($22-25°C$) and a selectivity for Na^+ and K^+ over Cs^+, Li^+ and Rb^+. Ca^{2+} seemed to be crucial; today, we know that it may pass a few of the channels (v. Tscharner et al., 1986; Nilius, 1990), others are not affected, or only to very small extent (Weber and Siemen, 1989; Cook et al., 1990). The anion permeability was below the threshold for detection.

Hints for a neuronal NSC in bursting neurons of *Helix pomatia* also came in 1981 from Hofmeier and Lux. In a subsequent paper, Swandulla and Lux (1985) described macroscopic currents through a NSC-like channel, which is permeable also to cations as large as choline, TEA or Tris. Later, those channels, which show Ca-sensitivity and nonselectivity for cations, were called CAN-channels (for Ca-activated nonspecific).

So far, concepts of excitable membrane function have been based on the known presence of two types of ion channels: those activated by neurotransmitters and those activated by voltage (Colquhoun et al., 1981). With ion channels that were activated by intracellular messengers an additional class was found. Due to heterogeneity of the channels and difficulties to understand their function little attention was given them at first. In a cultured insulin-secreting cell line the 25 pS-channel appeared not earlier than at passage number 77. After passage number 70, however, several types of K^+-channels observed earlier disappeared and insulin secretion ceased at the same time (Light et al., 1987). From this observation, one could have thought of the NSC as a kind of degenerative K^+-channel. The ambiguous name "nonspecific" channels, which, unfortunately, was still used in TiPS Receptor Nomenclature Supplement (Watson and Abbot, 1992) could be regarded as an expression of these early doubts.

In 1984, Bevan et al., Maruyama and Petersen, Marty et al., and Gallacher, et al. found NSC channels in rat glial cells and in different gland cells. Yellen (1982) had already collected first data in neuroblastoma concerning selectivity for specific ions other than Na^+ and K^+.

Colquhoun et al. (1981) described temperature dependence of the cardiac NSC. They calculated a Q_{10} of 2.3 for the current amplitudes, which seems too large for the data shown. The Q_{10}-value for the NSC channel of brown adipocytes was found to be 1.4 (Siemen and Reuhl, 1987). This is well in the range of aqueous diffusion as shown for several ion channels (Hille, 1992). But the most exciting question was that for the internal Ca^{2+}-concentration necessary to activate the channels: Colquhoun et al. (1981) gave a threshold of 10^{-6} M, a value later confirmed by Yellen (1982) for neuroblastoma and by Maruyama et al. (1985) for thyroid follicular cells of rats. Also, Matsunaga et al. (1991) found the NSC-channel of cultured rat mesangial cells active at 10^{-6} M internal Ca^{2+}, but not at 10^{-8} M. Several of the early papers state dependence of the open probability on intracellular Ca^{2+}, but do not give the concentrations. Some preparations show a higher threshold. Sturgess et al. (1987a) gave a lower limit of 0.1 mM internal Ca^{2+} in the rat insulinoma cell line which is close to what was found in excised patches from brown and white adipocytes (Koivisto et al., this volume) and even 10 times higher than the value from NSCs of lacrimal glands (Marty et al., 1984). Finally, Cook et al. (1990) measured a dose-response relation with a K_D of 1.8 mM. The Hill coefficient for Ca^{2+}-binding was 1.2 in their experiments, while Ehara et al. (1988) found a Hill coefficient of 3.0 for a small NSC in ventricular muscle cells. Quite often, internal Ca^{2+} seems not to be the only activator of NSC-channels. But it is not always easy to find out whether the complete activating pathway includes a final Ca^{2+}_i-raising step. In rat mesangial cells, which show similarities with smooth muscle cells, angiotensin II as well as vasopressin stimulate an IP_3-induced increase in internal Ca^{2+} by opening a 25 pS channel (Matsunaga et al., 1991).

Unfortunately, several data concerning concentration dependence of the 25 pS channel activation on cytosolic Ca^{2+} may have to be reinvestigated, since Thorn and Petersen (1992) found that in mouse pancreatic acinar cells it is not possible to get reliable data from excised patches without adding ATP to the bath before excising. The continuous presence of ATP (2 mM) seems to keep the nonselective channel in an Ca^{2+}-sensitive state. Surprisingly, ATP (2 mM) may also block the channels, however, only after excising the patches into an ATP-free solution (Thorn and Petersen, 1992). The underlying mechanism is far from being understood. If the phenomenon observed by Thorn and Petersen occurs in other preparations, too, it can be expected that several of the threshold Ca^{2+}_i-concentrations given in Table 1 are much too high. Candidates for a revision could be channels showing a characteristic fast run down occurring within a few minutes that was shown to be ATP-sensitive (Thorn and Petersen, 1992). Adipocytes are one of the preparations in which the threshold Ca^{2+}_i-concentration is surprisingly high (Koivisto et al., this volume). Sasaki and Gallacher

(1992) show that also extracellular ATP may augment the channel response. This happens via a G-protein and cAMP-mediated phosphorylation of the channel or of some regulatory component. cAMP stimulation of a 25 pS-channel was also seen by Lechleiter et al. (1988).

In 1988, Partridge and Swandulla gave a first review about NSC-channels. This class of ion channels seems to be more unified than it first appeared. In several cases, if one criterion does not fit, another may be found that also does not match. As examples Fichtner et al. (1987) found a nonselective cation channel in endothelial cells of bovine aorta showing a single-channel conductance of 12 pS. This channel is neither activated by internal Ca^{2+} nor does it open upon depolarization – it closes instead. In neutrophiles, a nonselective channel with a slightly smaller single-channel conductance of 22 pS (24°C) exists. This channel is equally permeable to Na^+, K^+, but also to Ca^{2+} (v. Tscharner et al., 1986). Epithelial Na^+-channels are a group of ion channels that has been suspected to be related to the NSC-channels. They are blocked, however, by the diuretic amiloride with apparent inhibition constants of 1 μM or less. They usually have P_{Na}/P_K-ratios far higher than those of the NSCs, and their single-channel conductance (around 5 pS) very often does not match the typical NSC values of 25–30 pS (review by Palmer, 1987; cf. Van Driessche, this volume). In contrast, the NSC of the inner medullary collecting duct of rats seems to be the only NSC that can be blocked by amiloride.

Inhibition

To date, only a small number of blocking substances has been found for NSCs (Hescheler and Schultz, this volume). Sturgess et al. (1986; 1987a, b) were able to block an 18 pS NSC from a human insulinoma cell line by internal AMP and another 25 pS channel from a rat insulinoma cell line by quinine (10 μM–1 mM), quinidine (10–400 μM), different adenine derivatives, as well as internal 4-aminopyridine (4AP, 2–10 mM). They found also that the blocking potency of the adenine derivatives decreased in the order AMP > ADP > ATP > adenosine and that 100 μM AMP blocked the channel completely. Also, in brown adipocytes ATP, ADP, cAMP and cGMP were able to block the NSC. Thorn and Petersen (1992), however, pointed out that in mouse pancreatic acinar cells ATP-block depends on the history of the patch (see above). Light et al. (1989, 1988) used cGMP (the second messenger of atrial natriuretic peptide) and amiloride to block a nonselective cation channel in renal collecting duct cells. But as there is a large variety of amiloride sensitive Na^+-channels existing with different single-channel conductances and different selectivities, it is not clear whether the renal channel really belongs to the family of NSC-channels.

Table 1. NSC-channels

Preparation	γ (pS)	Selectivity	Activ. (μmol/l)	Vol. dep.	Block[1]	Authors
1. Excitable membranes						
Purkinje cells (heart); calf	∅	Na, K	Ca^{2+}_i	∅	∅	Kass et al., 1978b[2]
Adult ventr. myocytes; guinea pig	15 (20–25°C)	Na ≈ K ≈ Li ≈ Cs; Cl:-	0.3–10 Ca^{2+}_i (TC_{50} = 1.2)	—	∅	Ehara et al., 1988
Ventr.; guinea pig	28 (20–25°C)	Na ≈ K	1000 Ca^{2+}_i	↑	∅	Ehara et al., 1988
Neonatal ventricular myocytes; rat	30 (25°C)	Na ≈ K, anions:-	>1 Ca^{2+}_i	—	∅	Colquhoun et al., 1981
Jejunum, smooth muscle; rabbit	∅	Na, K	(Ca^{2+}_i facilit.)	↑	∅	Benham et al., 1985
Neuroblastoma; mouse	22 (24°C)	N ≈ K > Cs, Li; Ca, Cl:-	Ca^{2+}_i	—	∅	Yellen, 1982
Neuroblastoma-glioma; mouse	∅	Na ≈ K	Ca^{2+}_i[2]	∅	∅	Higashida and Brown, 1986
2. Endocrine cells						
Thyroid follicular; rat	35 (22°C)	Na ≈ K	>1 Ca^{2+}_i	—	∅	Maruyama et al., 1985
Insulinoma; human	18 (22°C)	Na ≈ K	Ca^{2+}_i	∅	AMP	Sturgess et al., 1987b
Insulinoma; rat	25 (23°C)	Na ≈ K; Cl:-	>100 Ca^{2+}_i	↑	quinine, ATP, 4AP, 0.1 AMP > ADP > ATP > adenosine	Sturgess et al., 1987a Sturgess et al., 1986
3. Exocrine cells						
Pancreas acini; mouse	33 (22°C)	Na ≈ K ≈ Rb	>5 Ca^{2+}_i	∅	∅	Gallacher et al., 1984
Pancreas acini; mouse, rat	30 (22°C)	Na ≈ K; Cl:-	Ca^{2+}_i	—	∅	Maruyama and Petersen, 1982a, b
Pancreas acini; mouse	26 (∅°C)	—	≪1 Ca^{2+}_i		1 ATP	Maruyama and Petersen, 1984
Pancreas acini; guinea pig	27 (room)		Ca^{2+}_i (100 SITS,)		DCDPC,(DPC,, NPPB, ATP, ADP, Mg^{2+}, incr. pH	Suzuki and Petersen, 1988
Pancreas acini; rat	25 (21–23°C)	Na ≈ K	1 Ca^{2+}_i	↑	0.1 AMP	Gögelein and Pfannmüller, 1989
Pancreatic duct; rat						Gray and Argent, 1990
ST₈₈₅ (salivary epithel); mouse	25 (20°C)	NH₄ > K > Li ≈ Na > Rb ≫ Mg > Ca	Ca^{2+}_i (K_D = 1.8 mM)	↑	0.1 DPC₀ (slow) 0.1 quinine, (interm.) 1–5 4AP (slow), 0.1 SITS	Cook et al., 1990 Poronnik et al., 1991
Lacrimal-gland; rat	25 (22°C)	Na ≈ K; Cl:-	10 Ca^{2+}_i	↑	∅	Marty et al., 1984
Lacrimal acinar cells; mouse	30 (23°C)	Na ≈ K > Ca	Ca^{2+}_i, 1 mM ATP	—		Sasaki and Gallacher, 1990
Thyroid follicular all; rat	35 (22°C)	Na ≈ K	100 Ca^{2+}_i	—		Maruyama et al., 1985
Gastric glands, parietal cells; rabbit	22 (23–26°C)	Na ≈ K	—	↑		Sakai et al., 1989

4. Kidney

Cell / tissue	Temp	Ion selectivity	Ca²⁺	Activity	Blocking substances	References
Renal mesangial cells; rat	25	Na ≈ K	1 Ca²⁺ᵢ	—	∅	Matsunaga et al., 1991
Ascend. limb of Henle's loop; mouse	25–30 (∅°C)		0.1 Ca²⁺ᵢ		0.02 ATP, 0.02 ADP, 0.002 AMP (50%)	Paulais and Teulon, 1989
Princ. cells, cortical coll. duct; rabbit	28.4 (22–23°C)	Na ≈ K ≫ Cl⁻	—	↑	flufenamic acid	Ling et al., 1991
M-1 cell-line (cort. coll. duct); mouse	34 (18–37°C)	Na ≈ K ≫ Cl⁻	Ca²⁺ᵢ	↑	DPC	Korbmacher and Barnst., this vol.
Renal tubulus; rabbit	28 (35°C)	Na > K; Cl ≈ 0.5	Ca²⁺ᵢ	—	0.0005 amiloride; DPC:-	Gögelein and Greger, 1986
Inner medullary coll. duct; rat	27.5 (∅°C)	Na ≈ K ≈ NH₄, Cl:-	Ca²⁺ᵢ:?			Light et al., 1988

5. Other

Cell / tissue	Temp	Ion selectivity	Ca²⁺	Activity	Blocking substances	References
Schwann cell; rat	32 (20°C)	Na ≈ K; Cl <0.2, Ca <0.05	100 Ca²⁺ᵢ	↑	∅	Bevan et al., 1984
Brown adipocytes; rat, Djung. Hamster	30 (25°C)	Na ≈ K; Cl:-		↑	∅	Siemen and Reuhl, 1987
Brown adipocytes; rat		NH₄ > Na > Li > K > Rb ≈ Cs Ca, Ba, Cl:-		↑	∅	Weber and Siemen, 1989
White-, brown adipocytes; rat	30 (25°C)	Na ≈ K ≈ Cl	10 Ca²⁺ᵢ	↑	mef. acid, aden. nucleot.	Koivisto et al., this volume
Oocyte; mouse	30 (∅°C)	Na ≈ K ≈ Ca	∅	↑	∅	Hunter and Main, 1987
Neutrophile; human	22.5 (24°C)	Na ≈ K ≈ Ca	Ca²⁺ᵢ	↑	∅	v. Tscharner et al., 1986
Cornea endothelial; rabbit	22 (21–22°C)	Na ≈ K; Cl, Ca:-	>100 Ca²⁺ᵢ	∅	ATP, ADP, AMP	Rae et al., 1990
Aorta, endothelial; bovine	12 (20°C)	Na ≈ K	∅	↓	∅	Fichtner et al., 1987
Unbilic. vein endothelial; human	20 (20–23°C)		1 Ca²⁺ᵢ	↓	Naᵢ	Bregestovski et al., 1988
Vascular endothelial; human	26 (∅°C)	K > Na > Ca	histamine, Ca²⁺ᵢ:?	↓	∅	Nilius, 1990
Vascular endothelial; rat, pig	23 (20°C)	Na > K	∅	∅	amiloride, phenamil	Vigne et al., 1989
Distal colon; rat	29 (35°C)	Na ≈ K	Ca²⁺ᵢ	↑	1 quinine, 1 DCDPC	Gögelein and Capek, 1990
Colonic cell lines T₈₄, HT29D₄; human	19 (20–24°C)	Na > K; Cl:-	>100 Ca²⁺ᵢ		1 ATP; DPC, DCDPC	Champingny et al., 1991
Nasal epithelium; human	20–21 (room)	Na ≈ K > Cs	100–1000 Ca²⁺ᵢ	↑	amiloride	Jorissen et al., 1990
Alveolar epithelium; rat	23 (20°C)	Na > K; Cl:-	Ca²⁺ᵢ		∅	Orser et al., 1991
Vestibular dark cells, gerbil	28 (20–22°C)	NH₄ > Na ≈ K ≈ Li ≈ Rb ≈ Cs	10 Ca²⁺ᵢ		serum deprivation	Marcus et al., 1992
Fibroblast; mouse	28 (∅°C)	Na ≈ K ≈ Cs; Cl, Ca:-	—			Frace and Gargus, 1989

6. Invertebrates

Cell / tissue	Temp	Ion selectivity	Ca²⁺	Activity	Blocking substances	References
Neuron; *Helix pomatia*	30 (room)	Na, K ≫ Tris ≈ TEA; Cl:-	>0.1 Ca²⁺ᵢ	—	∅	Partridge and Swandulla, 1987; Swandulla and Lux, 1985
Pept., nerve ending; *crustacea*	213 (∅)³⁾	Na ≈ K ≈ Cs	>1 Ca²⁺ᵢ	∅	∅	Lemos et al., 1986

Signs for charges of ions are omitted for brevity. γ: single-channel conductance; —: negative or very little; ∅: no details in cited refs.; ↑: increased channel activity during depolarizing potentials; ↓: increased channel activity during hyperpolarizing potentials. ¹⁾ mostly direct; ²⁾ no single-channel data; ³⁾ measured in 310 mM KCl. Abbreviations for blocking substances explained throughout this volume (see index).

Another amiloride-sensitive channel with a single-channel conductance of 23 pS and a permeability ratio of $P_{Na}/P_K = 1.5$ was found by Vigne et al. (1989) in the endothelial cells of rat brain vessels.

Gögelein and Greger (1986), Gögelein and Pfannmüller (1989), and Gögelein et al. (1990) showed that NSCs in cells from kidney, pancreas and colon may be blocked by the Cl-channel blockers DPC and related compounds like DCDPC, as well as by some antiinflammatory drugs (for details see Popp et al., this volume). These drugs are non-specific and act by increasing the mean channel closed time (slow block), while 100 μM SITS, DIDS or DNDS in the presence of 1.3 mM internal Ca^{2+} increased the open probability (Gögelein and Pfannmüller, 1989). Thus, there are no specific blockers for NSCs at present. Much of the data currently available on NSCs are summarized in Table 1.

The Current View

Thus far, NSC-channels have not been cloned or sequenced. Probably, the most detailed reports on the electrophysiology and biophysics of these channels appeared for brown adipocytes (Weber and Siemen, 1989) and for the apical membrane of the secretory epithelial cell line ST_{885} (Cook et al., 1990). Whether the 15 pS-ventricular channel found by Ehara et al. (1988) belongs to the same type is not absolutely clear. On the one hand, it shows clear differences to the 30 pS-ventricular channel found by Colquhoun et al. (1981); on the other hand, a larger channel has been found by Ehara et al., too, but was not studied in detail (see Isenberg, this volume). Table 2 gives a comparison showing the small channel of Ehara et al. (1988) with those NSC-channels described earlier.

Experiments by Ehara et al. (1988) were done on adult ventricular muscle, while experiments by Colquhoun et al. (1981) were carried out on cultured neonatal ventricular cells. It would be interesting to find out if the difference between the channels could at least partly be due to different developmental stages as described for the nAChR (Mishina et al., 1986), or if it represents a species difference.

Table 2. Two guinea pig ventricular channels in comparison with other NSC-channels

Preparation	γ (pS)	Vol. dep.	$Ca^{2+}{}_i$	Ref.
Ventr. muscle; rat	30 (25°C)	—	10^{-6}	Colquhoun et al., 1981
Ventr. muscle; guinea pig	15 (22.5°C)	—	10^{-6}	Ehara et al., 1988
Ventr. muscle; guinea pig	28 (22.5°C)	↑	10^{-3}	Ehara et al., 1988
Brown adipocyte; rat	30 (25°C)	↑	10^{-4}	Siemen and Reuhl; 1987 Koivisto et al., this vol.

γ: single-channel conductance; vol. dep: voltage dependence; ↑: increasing with depolarization.

While the single-channel conductance seems to be about the same in both brown adipocytes and ST_{885} cells, there is difference in the selectivity sequences concerning the K^+-ion. It is slightly better permeant than Na^+-ions in ST_{885} cells. In brown fat cells the permeability ratios for the two ions are almost reversed (Table 3).

Unfortunately, both series of experiments were carried out under different conditions (biionic in brown adipocytes, ion mixtures outside ST_{885} cell membranes) which means that differing results could also be due to anomalous mole fraction behavior (Hille, 1992). Ca^{2+}-permeability of the NSC in brown adipocytes is very low, as in several other NSCs (e.g., Cook et al., 1990). In vascular endothelial cells, however, it is much higher. Weber and Siemen (1989) as well as Nilius (1990) calculated a simple two-barrier one-site model with which the currents were fitted. It turned out that the well energies for the Ca^{2+}-binding site within the middle of the channels' electrical field were slightly less negative for brown adipocytes (-7.2 kJ/M) than for endothelial cells (-9.5 kJ/M). The difference of the barrier energies for Na^+ (26.4 kJ/M for brown adipocytes vs. 24.3 kJ/M for endothelial cells), however, was about three times smaller than the difference of the barrier energies for Ca^{2+} (36.1 kJ/M vs. 29.5 kJ/M. This illustrates that the energy barrier a Ca^{2+}-ion has to pass when entering the pore is considerably higher in brown adipocytes (relative permeability for Ca^{2+}: <0.02) as compared with vascular endothelial cells (relative permeability for Ca^{2+}: 0.02, but cf. Yamamoto et al., 1992).

For understanding many different cells, it is a central question whether opening of NSC-channels is able to explain increases in cytosolic Ca^{2+}-concentration (Poronnik et al., 1991). Even in the well-studied glandular acinar cells it seems not absolutely clear whether Ca^{2+} entry is mediated by NSC-channels (Petersen, 1992). Obviously, this question has to be tested separately for every single cell type in which NSCs occur. Due to the steep concentration gradient across the cell membrane and the high Ca^{2+} sensitivity of several cell proteins, already a small relative permeability is able to contribute considerably to the control mechanisms of the cell. Ca^{2+} entry through nonselective channels has been discussed or observed in glandular acinar cells (Petersen and Maruyama, 1983; Sasaki and Gallacher, 1990), neutrophils (von Tscharner et al., 1986), different epithelial cells of the eye (Cooper et al., 1986; Rae et al., 1990), and vascular endothelial cells (Nilius, 1990). The

Table 3. Selectivity sequences from NSCs of brown fat cells and a secretory cell line

	NH_4^+	Na^+	Li^+	K^+	Rb^+	Cs^+	Mg^{2+}	Ca^{2+}	Ba^{2+}
Brown fat	1.6	1.0	0.9	0.8	0.8	0.78	—	<0.02	<0.02
ST_{885}	1.9	1.0	1.0	1.1	0.8	—	0.1	0.002	—

same problem is also relevant for stretch-activated or other nonselective channels (Bear, 1990). How easily very small amounts of entering Ca^{2+} may escape detection is pointed out by Petersen and Maruyama (1983) who calculated a 0.5 mV shift in reversal potential at physiological salt concentrations, even at a permeability ratio P_{Ca}/P_{Na} of 1!

Kinetics of the NSC are complex. Both open and closed time distribution histograms are best described by two exponentials (Table 4). In brown adipocytes temperature dependence of the slow components of on- and off-time seems to be steeper than that of the fast components. In general, it can be concluded that kinetics of the nonselective channels are rather complicated. A possible explanation could be that binding of internal Ca^{2+} (and of other activators) and gating of the channel are separate steps with different kinetics, requiring at least a three-state model for description.

Function

For many NSCs activated by intracellular Ca^{2+} the function is not yet known. A relatively well-founded picture already exists for heart cells. In sheep cardiac Purkinje fibers oscillatory inward currents (I_{TI}) are elicited by brief depolarizing pulses in K^+-free external solution, or in raised intracellular Ca^{2+}. This I_{TI} is mediated by channels that are permeable to Na^+, K^+ and Ca^{2+} (Cannell and Lederer, 1986). But NSCs may also contribute to the plateau-phase of the cardiac action potential. The rise in intracellular Ca^{2+} seems to be large enough to justify this suggestion (see Isenberg, this volume). Different types of nonselective channels are involved in excitation-contraction coupling. In smooth muscle cells muscarinic acetylcholine receptor activation, for example, switches on a NSC-channel via a stimulatory G-Protein. This response is facilitated by internal Ca^{2+}. The variety of nonselective channels of heart and smooth muscle is reviewed by Isenberg (this volume).

Little is known about function of the NSCs in nerve cells. In neuroblastoma (N1E115) they appear so frequently that this preparation may

Table 4. Open-time and closed-time constants of different NSC-channels

Preparation	τ_{on} (ms)		τ_{off} (ms)		V_H (mV)	Ca^{2+}_i (mM)	Temp. °C
ST_{885}	1	9	9	43	−60	2.7	20
Insulinoma	4	25	1	9	−40	0.5	23
Brown adipocytes	8	76	5	130	−70	1.2	26
Ventricular cells	4	140	2	15	−80	0.01	20−25
Schwann cells	0.5	15	0.3	8	−60	1.0	20

Ref. see Table 1.

be nicely used to demonstrate nonselective channels in practical courses (Yellen, 1982; personal observation). Nevertheless, their function remains obscure. Only in the soma of *Helix* burster neurons was it shown that the NSC carries the calcium-activated inward current producing the maintained depolarizing drive that generates endogenous bursts (Swandulla and Lux, 1985; Partridge and Swandulla, 1987; Swandulla, this volume).

In exocrine glands, acinar cells may respond to secretagogues with several different types of membrane potential changes. This response is controlled at least by internal Ca^{2+} and may be partly due to NSCs in the basolateral membrane. As properties and distribution of these channels may differ between different exocrine glands from different species, several hypotheses of their involvement in fluid secretion or sustained KCl secretion were developed (Marty et al., 1984; Sasaki and Gallacher, 1990, 1992; Thorn and Petersen, 1992). Gray and Argent (1990) also found NSC-channels in apical and basolateral membranes of pancreatic duct cells. Current theories are explained by Thorn and Petersen (this volume), but the complete mechanism is far from being understood.

NSC-channels are also present in different parts of the vertebrate renal tubular system (reviewed by Korbmacher and Barnstable, this volume). They occur in basolateral membranes of proximal tubular cells (Gögelein and Greger, 1986), in cortical thick ascending limb of Henle's loop (Paulais and Teulon, 1989), in inner medullary collecting duct (Light et al., 1988), and in principal cells of cortical collecting tubules (Ling et al., 1991). They may influence several steps of the transcellular NaCl transport. Membrane depolarization could change the electromotive driving forces for recycling K^+ ions as well as for absorbing Cl^- ions. Na^+ entry could reduce the activity of the Na^+-K^+-$2Cl^-$- and of the Na^+-H^+-cotransport system, but it could also activate the Na^+/K^+-ATPase. As most of these steps would occur simultaneously, it is nearly impossible to predict the resulting effect on NaCl reabsorption of the whole system of the loop of Henle (Paulais and Teulon, 1989). Availability of a specific blocker for the NSC is needed to differentiate the different current components. A key role in renal Na^+-excretion is certainly played by the atrial natriuretic peptide, which inhibits Na^+-reabsorption across the inner medullary collecting duct via its intracellular messenger cGMP. The NSC of an insulinoma cell line, endothelial cells, as well as brown adipocytes may be blocked or also down-regulated by cGMP, too. Whether it is blocked by phosphorylation has to be shown (Reale et al., 1992; Nilius et al., this volume; Koivisto et al., this volume).

In the inner medullary collecting duct the NSC could be involved in K^+ secretion as well as K^+ absorption, depending on the metabolic state. In addition, the amiloride blockable part of the Na^+ absorptive fluxes may be due to NSCs. Finally, as the channel is highly permeable for NH_4^+, it may also be involved in ammonium secretion, too (Light et al., 1988). Principal cells of the cortical collecting tubule are important for regulating total body Na^+ balance as they are a primary target for mineralocorticoids.

Distal colon crypt cells mediate salt and water secretion. In these cells NSCs are also likely to be involved in the transcellular ion transport mechanisms (Gögelein and Capek, 1990; Siemer and Gögelein, this volume).

Sturgess et al. (1987a) pointed out that in insulinoma cells (like in other cells) the open-state probability of the NSC is a complex interrelation between the internal Ca^{2+}-concentration, membrane voltage, and intracellular nucleotide concentrations. Several of the cells showing NSCs respond to metabolic signals (which in turn alter the nucleotide balance within the cell) by depolarization and an increased cytosolic Ca^{2+}-concentration. Thus, a channel sensitive to all these factors could possibly modulate the normal cell function. This statement fits to several of the NSCs listed in Table 1. It may even be complicated by a possible function in pH-regulation or NH_3-turnover. A NH_4-permeability of the NSCs, which is, for example, in brown adipocytes clearly higher than the Na^+-permeability, would be the basis. The mechanism could work as was described for stretch-activated nonselective channels of *Xenopus* oocytes (Burckhardt and Frömter, 1992).

Human neutrophils respond to activation with an increased cytosolic Ca^{2+} concentration. Ca^{2+} comes partly from the extracellular space. Ca^{2+} from internal stores is rate-limiting. Thus, it is very likely that the 22.5 pS-nonselective channel observed by von Tscharner et al. (1986) is activated by these sources and finally constitutes a positive feedback, because, in contrast to almost all the other Ca^{2+}_i-activated channels described here, it is highly Ca^{2+} permeable.

Conclusion

It is far from being clear if treating the NSCs as a separate group within the nonselective cation channels is really justified on a molecular basis. Nevertheless, most of them share some pharmacological, physiological, or biophysical characteristics and were thus separately introduced within the fast growing field of nonselective cation channels.

Acknowledgements
I would like to thank Drs Monika Stengl, J. Hescheler, U. Ruß, and W. Vogel for critically reading, as well as Magdalena Dietl for typing the manuscript.

References

Avenet P, Lindemann B (1988). Amiloride-blockable sodium currents in isolated taste receptor cells. J. Membrane Biol. 105:245–255.

Avenet P, Lindemann B (1989). Perspectives of taste reception. J. Membrane Biol. 112:1–8.

Avenet P, Kinnamon, SC (1991). Cellular basis of taste reception. Curr. Opinion Neurobiol. 1:198–203.

Bauer K, Struyvé M, Bosch D, Benz R, Tommassen J (1989). One single lysine residue is responsible for the special interaction between polyphosphate and the outer membrane porin PhoE of *Escherichia coli*. J. Biol. Chem. 248:16393–16398.

Bear CE (1990). A nonselective cation channel in rat liver cells is activated by membrane stretch. Am. J. Physiol. 258:C421–C428.

Benham CD, Bolton TB, Lang RJ, Takewaki T (1985). The mechanism of action of Ba^{2+}-activated K^+-channels in arterial and intestinal smooth muscle cell membranes. Pflügers Arch. 403:120–127.

Benham CD, Tsien RW (1987). A novel receptor-operated Ca^{2+}-permeable channel activated by ATP in smooth muscle. Nature 328:275–278.

Benz R (1985). Porins from bacterial and mitochondrial outer membranes. CRC Crit. Rev. Biochem. 19:145–190.

Benz R (1988). Structure and function of porins from gram-negative bacteria. Ann. Rev. Microbiol. 42:359–393.

Benz R, Brdiczka D (1992). The cation-selective substate of the mitochondrial outer membrane pore: Single-channel conductance and influence on intermembrane and peripheral kinases. J. Bioenerg. Biomembr. 24:33–39.

Bevan S, Gray PTA, Ritchie JM (1984). A calcium-activated cation-selective channel in rat cultured Schwann cells. Proc. R. Soc. Lond. B222:349–355.

Bodoia RD, Detwiler PB (1984). Patch-clamp recordings of the light-sensitive dark noise in retinal rods from the lizard and frog. J. Physiol. 367:183–216.

Breer H, Boekhoff I, Tareilus E (1990). Rapid kinetics of second messenger formation in olfactory transduction. Nature 345:65–68.

Bregestovski P, Bakhramov A, Danilov S, Moldobaeva A, Takeda K (1988). Histamine-induced inward currents in cultured endothelial cells from human umbilical vein. Br. J. Pharmacol. 95:429–436.

Bureau MH, Khrestchatisky M, Heeren MA, Zambrowicz EB, Kim H, Grisar TM, Colombini M, Tobin AJ, Olsen RW (1991). Isolation and cloning of a voltage-dependent anion channel-like M_r 36 000 polypeptide from mammalian brain. J. Biol. Chem. 267:8679–8684.

Burckhardt BC, Frömter E (1992). Pathways of NH_3/NH_4^+ permeation across *Xenopus laevis* oocyte cell membrane. Pflügers Arch. 420:83–86.

Cannell MB, Lederer WJ (1986). The arrhythmogenic current I_{TI} in the absence of electrogenic sodium-calcium exchange in sheep cardiac Purkinje fibers. J. Physiol. 374:201–219.

Champigny G, Verrier B, Lazdunski M (1991). A voltage, calcium, and ATP sensitive non selective cation channel in human colonic tumor cells. Biochem. Biophys. Res. Comm. 176:1196–1203.

Colombini M (1989). Voltage gating in the mitochondrial channel, VDAC. J. Membrane Biol. 111:103–111.

Colquhoun D, Neher E., Reuter H., Stevens CF (1981). Inward current channels activated by intracellular Ca in cultured cardiac cells. Nature 294:752–754.

Cook DI, Poronnik P, Young JA (1990). Characterization of a 25-pS nonselective cation channel in a cultured secretory epithelial cell line. J. Membrane Biol. 114:37–52.

Cooper KE, Tang JM, Rae JL, Eisenberg RS (1986). A cation channel in frog lens epithelia responsive to pressure and calcium. J. Membrane Biol. 93:259–269.

Cowan SW, Schirmer T, Rummel G, Steiert M, Ghosh R, Pauptit RA, Jansonius JN, Rosenbusch JP (1992). Crystal structures explain functional properties of two *E. coli* porins. Nature 358:727–733.

Cox T, Campbell P, Peterson RN (1991). Ion channels in boar sperm plasma membranes: Characterization of a cation selective channel. Mol. Reprod. Devel. 30:135–147.

Dermietzel R, Hwang T-K, Buettner R, Hofer A, Dotzler E, Kremer M, Thinnes FP, Siemen D (1993). Cloning and *in situ* localisation of a brain-derived porin (BR1-VDAC) that constitutes a large conductance anion channel in astrocytic plasma membranes. Submitted.

Dhallan RS, Yau KW, Schrader KA, Reed RR (1990). Primary structure and functional expression of a cyclic nucleotide-activated channel from olfactory neurons. Nature 347:184–187.

Dill ET, Holden MJ, Colombini M (1987). Voltage gating in VDAC is markedly inhibited by micromolar quantities of aluminium. J. Membrane Biol. 99:187–196.

Dudel J, Franke C, Hatt H (1990). A family of glutamatergic, excitatory channel types at the crayfish neuromuscular junction. J. Comp. Physiol. A 166:757–768.

Dwyer TM, Adams DJ, Hille B (1980). The permeability of the endplate channel to organic cations in frog muscle. J. Gen. Physiol. 75:469–492.

Ehara T, Noma A, Ono N (1988). Calcium-activated non-selective cation channel in ventricular cells isolated from adult guinea-pig hearts. J. Physiol. 403:117–133.

Ehring GR, Lagos N, Zampighi GA, Hall JE (1991). Phosphorylation modulates the voltage dependence of channels reconstituted from the major intrinsic protein of lens fiber membranes. J. Membrane Biol. 126:75–88.

Fasolato C, Pizzo P, Pozzan T (1990). Receptor-mediated calcium influx in PC12 cells. J. Biol. Chem. 265:20351–20355.

Fatt P, Katz B (1951). An analysis of the endplatte potential recorded with an intracellular electrode. J. Physiol. 115:320–370.

Fesenko EE, Kolesnikov SS, Lyubarsky AL (1985). Induction by cyclic GMP of cationic conductance in plasma membrane of retinal rod outer segment. Nature 313:310–313.

Fichtner H, Fröbe U, Busse R, Kohlhardt M (1987). Single nonselective cation channels and Ca^{2+}-activated K^+ channels in aortic endothelial cells. J. Membrane Biol. 98:125–133.

Forte M, Guy HR, Mannella CA (1987). Molecular genetics of the VDAC ion channel: Structural model and sequence analysis. J. Bioenerg. Biomembr. 19:341–350.

Frace AM, Gargus JJ (1989). Activation of single-channel currents in mouse fibroblasts by platelet-derived growth factor. Proc. Natl. Acad. Sci. USA 86:2511–2515.

Gallacher DV, Maruyama Y, Petersen OH (1984). Patch-clamp study of rubidium and potassium conductances in single cation channels from mammalian exocrine acini. Pflügers Arch. 401:361–367.

Gögelein H, Greger R (1986). A voltage-dependent ionic channel in the basolateral membrane of late proximal tubules of the rabbit kidney. Pflügers Arch. 407:S142–S148.

Gögelein H, Pfannmüller B (1989). The nonselective cation channel in the basolateral membrane of rat exocrine pancreas. Pflügers Arch. 413:287–298.

Gögelein H, Capek K (1990). Quinine inhibits chloride and nonselective cation channels in isolated rat distal colon cells. Biochim. Biophys. Acta 1027:191–198.

Gögelein H, Dahlem D, Englert HC, Lang HJ (1990). Flufenamic acid, mefenamic acid and niflumic acid inhibit single nonselective cation channels at the rat exocrine pancreas. FEBS Lett. 268:79–82.

Gray MA, Argent BE (1990). Non-selective cation channel on pancreatic duct cells. Biochim. Biophys. Acta 1029:33–42.

Guo XW, Mannella CA (1992). Classification of projection images of crystalline arrays of the mitochondrial, voltage-dependent anion-selective channel embedded in aurothioglucose. Biophys. J. 63:418–427.

Hanke W, Cook NJ, Kaupp UB (1988). cGMP-dependent channel protein from photoreceptor membranes: single-channel activity of the purified and reconstituted protein. Proc. Natl. Acad. Sci. USA 85:94–98.

Haynes LW, Yau KW (1990). Single-channel measurement from the cyclic GMP-activated conductance of catfish retinal cones. J. Physiol. 429:451–481.

Henry JP, Chich JF, Goldschmidt D, Thieffry M (1989). Blockade of a mitochondrial cationic channel by an addressing peptide: An electrophysiological study. J. Membrane Biol. 112:139–147.

Higashida H, Brown DA (1986). Membrane current responses to intracellular injections of inositol 1,3,4,5-tetrakisphosphate and inositol 1,3,4-trisphosphate in NG108-15 hybrid cells. FEBS Lett. 208:283–286.

Hill JA, Coronado R, Strauss HC (1989). Potassium channel of cardiac sarcoplasmic reticulum is a multi-ion channel. Biophys. J. 55:35–45.

Hille B (1992). Ionic Channels of Excitable Membranes. 2nd ed. Sinauer Associates, Sunderland, Mass.

Hodgkin AL, Huxley AF (1952). A quantitative description of membrane current and its application to conduction and excitation in nerve. J. Physiol. 117:500–544.

Hodgkin AL, McNaughton PA, Nunn BJ (1985). The ionic selectivity and calcium dependence of the light-sensitive pathway in toad rods. J. Physiol. 358:447–468.

Hofmeier G, Lux D (1981). The time courses of intracellular free calcium and related electrical effects after injection of $CaCl_2$ into neurons of the snail, *Helix pomatia*. Pflügers Arch. 391:242–251.

Hoth M, Penner R (1992). Depletion of intracellular calcium stores activates a calcium current in mast cells. Nature 355:353–356.

Hunter MJ, Main KE (1987). Excision-activated non-selective ion channels in mouse oocytes. J. Physiol. 386:80P.

Jacob R (1991). Calcium oscillations in endothelial cells. Cell Calcium 12:127–134.

Kalinoski DL, Teeter JH, Spielman AI, Brand JG (1992). Partial purification of an L-arginine receptor from catfish taste epithelium. Abstr. of Xth ECRO Congress, München, 1992, p 67.

Kass RS, Lederer WJ, Tsien RW, Weingart R (1978a). Role of calcium ions in transient inward currents and aftercontractions induced by strophanthidin in cardiac Purkinje fibres. J. Physiol. 281:187–208.

Kass RS, Tsien RW, Weingart R (1978b). Ionic basis of transient inward current induced by strophanthidin in cardiac Purkinje fibres. J. Physiol. 281:209–226.

Kaupp UB, Niidome T, Tanabe T, Terada S, Bönigk W, Stühmer W, Cook NJ, Kangawa K, Matsuo H, Hirose T, Miyata T, Numa S (1989). Primary structure and functional expression from complementary DNA of the rod photoreceptor cyclic GMP-gated channel. Nature 342:762–766.

Kaupp UB (1991). The cyclic nucleotide-gated channel of vertebrate photoreceptors and olfactory epithelium. TINS 14:150–157.

Kinnamon SC, Cummings TA (1992). Chemosensory transduction mechanisms in taste. Annu. Rev. Physiol. 54:715–731.

Krautwurst D, Seifert R, Hescheler J, Schultz G (1992). Formyl peptides and ATP stimulate Ca^{2+} and Na^+ inward currents through non-selective cation channels via G-proteins in dibutyryl cyclic AMP-differentiated HL-60 cells. Biochem. J. 288:1025–1035.

Lechleiter JD, Dartt DA, Brehm P (1988). Vasoactive intestinal peptide activates Ca^{2+}-dependent K^+ channels through a cAMP pathway in mouse lacrimal cells. Neuron 1:227–235.

Lederer WJ, Tsien RW (1976). Transient inward current underlying arrhythmogenic effects of cardiotonic steroids in Purkinje fibres. J. Physiol. 263:73–100.

Lemos JR, Nordmann JJ, Cooke IM, Stuenkel EL (1986). Single channels and ionic currents in peptidergic nerve terminals. Nature 319:410–412.

Light DB, Van Eenenaam DP, Sorenson RL, Levitt DG (1987). Potassium-selective ion channels in a transformed insulin-secreting cell line. J. Membr. Biol. 95:63–72.

Light DB, McCann F, Keller TM, Stanton BA (1988). Amiloride-sensitive cation channel in apical membrane of inner medullary collecting duct. Am. J. Physiol. 255:F278–F286.

Light DB, Schwiebert EM, Karlson KH, Stanton BA (1989). Atrial natriuretic peptide inhibits a cation channel in renal inner medullary collecting duct cells. Science 243:383–385.

Ling BN, Hinton CF, Eaton DC (1991). Amiloride-sensitive sodium channels in rabbit cortical collecting tubule primary cultures. Am. J. Physiol. 261:F933–F944.

Ludwig O, Benz R, Schultz JE (1989). Porin of *Paramecium* mitochondria isolation, characterization and ion selectivity of the closed state. Biochim. Biophys. Acta 978:319–327.

Lückhoff A, Clapham DE (1992). Inositol 1,3,4,5-tetrakisphosphate activates an endothelial Ca^{2+}-permeable channel. Nature 355:356–358.

Mangan PS, Colombini M (1987). Ultrasteep voltage dependence in a membrane channel. Proc. Natl. Acad. Sci. 84:4896–4900.

Marrero H, Orkand PM, Kettenmann H, Orkand RK (1991). Single channel recording from glial cells on the untreated surface of the frog optic nerve. Eur. J. Neurosci. 3:813–819.

Mannella CA (1992). The "ins" and "outs" of mitochondrial membrane channels. TIBS 17:315–320.

Marcus DC, Takeuchi S, Wangemann P (1992). Ca^{2+}-activated nonselective cation channel in apical membrane of vestibular dark cells. Am. J. Physiol. 262:C1423–C1429.

Marty A, Tan YP, Trautmann A (1984). Three types of calcium-dependent channel in rat lacrimal glands. J. Physiol. 357:293–325.

Maruyama Y, Petersen OH (1982a). Single-channel currents in isolated patches of plasma membrane from basal surface of pancreatic acini. Nature 299:159–161.

Maruyama Y, Petersen OH (1982b). Cholecystokinin activation of single-channel currents is mediated by internal messenger in pancreatic acinar cells. Nature 300:61–63.

24

Maruyama Y, Petersen OH (1984). Single calcium-dependent cation channels in mouse pancreatic acinar cells. J. Membrane Biol. 81:83–87.

Maruyama Y, Moore D, Petersen OH (1985). Calcium-activated cation channel in rat thyroid follicular cells. Biochim. Biophys. Acta 821:229–232.

Matsunaga H, Yamashita N, Miyajima Y, Okuda T, Chang H, Ogata E, Kurokawa K (1991). Ion channel activities of cultured rat mesangial cells. Am. J. Physiol. 261:F808–F814.

Mauro TM, Isseroff RR, Lasarow R, Pappone PA (1993). Ion channels are linked to differentiation in keratinocytes. J. Membrane Biol. 132:201–209.

Mayer ML, Westbroook GL (1987). Permeation and block of N-methyl-D-aspartic acid receptor channels by divalent cations in mouse cultured central neurones. J. Physiol. 394:501–527.

McBride DW, Roper SD (1991). Ca^{2+}-dependent chloride conductance in *Necturus* taste cells. J. Membrane Biol. 124:85–93.

Merritt JE, Armstrong WP, Benham CD, Hallam TJ, Jacob R, Jaxa-Chamiec A, Leigh BK, McCarthy SA, Moores KE, Rink TJ (1990). SK&F 96365, a novel inhibitor of receptor-mediated calcium entry. Biochem. J. 271:515–522.

Miller C (1978). Voltage-gated cation conductance channel from fragmented sarcoplasmic reticulum: steady-state electrical properties. J. Membrane Biol. 40:1–23.

Mishina M, Takai T, Imoto K, Noda M, Takahashi T, Numa S, Methfessel C, Sakmann B (1986). Molecular distinction between fetal and adult forms of muscle acetylcholine receptor. Nature 321:406–411.

Nikaido H (1992). Porins and specific channels of bacterial outer membranes. Mol. Microbiol. 6(4):435–442.

Nilius B (1990). Permeation properties of a non-selective cation channel in human vascular endothelial cells. Pflügers Arch. 416:609–611.

Nishiyama A, Petersen OH (1975). Pancreatic acinar cells: ionic dependence of acetylcholine-induced membrane potential and resistance change. J. Physiol. 244:431–465.

Orser BA, Bertlik M, Fedorko L, O'Brodovich H (1991) Cation selective channel in fetal alveolar type II epithelium. Biochim. Biophys. Acta 1094:19–26.

Palmer LG (1987). Ion selectivity of epithelial Na channels. J. Membrane Biol. 96:97–106.

Partridge LD, Swandulla D (1987). Single Ca-activated cation channels in bursting neurons of *Helix*. Pflügers Arch. 410:627–631.

Partridge LD, Swandulla D (1988). Calcium activated non-specific cation channels. Trends Neurosci. 11:69–72.

Paulais M, Teulon J (1989). A cation channel in the thick ascending limb of Henle's loop of the mouse kidney: Inhibition by adenine nucleotides. J. Physiol. 413:315–327.

Penner R, Matthews G, Neher E (1988). Regulation of calcium influx by second messengers in rat mast cells. Nature 334:499–504.

Petersen OH, Iwatsuki N (1978). The role of calcium in pancreatic acinar cell stimulus-secretion coupling: an electrophysiological approach. Ann. NY Acad. Sci. 307:599–617.

Petersen OH, Maruyama Y (1983). What is the mechanism of the calcium influx to pancreatic acinar cells evoked by secretagogues. Pflügers Arch. 396:82–84.

Petersen OH (1992). Stimulus-secretion coupling: cytoplasmic calcium signals and the control of ion channels in exocrine acinar cells. J. Physiol. 448:1–51.

Poronnik P, Cook DI, Allen DG, Young JA (1991). Diphenylamine-2-carboxylate (DPC) reduces calcium influx in a mouse mandibular cell line (ST_{885}). Cell Calcium 12:441–447.

Rae JL (1985). The application of patch clamp methods to ocular epithelia. Current Eye Research 4.4:409–420.

Rae JL, Dewey J, Cooper K, Gates P (1990). A non-selective cation channel in rabbit corneal endothelium activated by internal calcium and inhibited by internal ATP. Exp. Eye Res. 50:373–384.

Reale V, Hales CN, Ashford MLJ (1992). Cyclic AMP regulates a calcium-activated non-selective cation channel in a rat insulinoma cell line. J. Physiol. 446:312P.

Sakai H, Okada Y, Morii M, Takeguchi N (1989). Anion and cation channels in the basolateral membrane of rabbit parietal cells. Pflügers Arch. 414:185–192.

Sasaki T, Gallacher DV (1990). Extracellular ATP activates receptor-operated cation channels in mouse lacrimal acinar cells to promote calcium influx in the absence of phosphoinositide metabolism. FEBS Lett. 264:130–134.

Sasaki T, Gallacher DV (1992). The ATP-induced inward current in mouse lacrimal acinar cells is potentiated by isoprenaline and GTP. J. Physiol. 447:103–118.

Schindler H, Rosenbusch JP (1978). Matrix protein from *Escherichia coli* outer membranes forms voltage-controlled channels in lipid bilayers. Proc. Natl. Acad. Sci. USA 75:3751–3755.

Siemen D, Reuhl T (1987). Non-selective cationic channel in primary cultured cells of brown adipose tissue. Pflügers Arch. 408:534–536.

Sillman AJ, Ito H, Tomita T (1969). Studies on the mass receptor potential of the isolated frog retina II. On the basis of the ionic mechanism. Vision Res. 9:1443–1451.

Simon SA, Holland VF, Benos DJ (1989). XI-th Annual Meeting Association for Chemoreceptive Sciences. Abstr. 202.

Stengl M, Hatt H, Breer H (1992). Peripheral processes in insect olfaction. Annu. Rev. Physiol. 54:665–681.

Sturgess NC, Hales CN, Ashford MLJ (1986). Inhibition of a calcium-activated, non-selective cation channel, in a rat insulinoma cell line, by adenine derivates. FEBS Lett. 208:397–400.

Sturgess NC, Hales CN, Ashford MLJ (1987a). Calcium and ATP regulate the activity of a non-selective cation channel in a rat insulinoma cell line. Pflügers Arch. 409:607–615.

Sturgess NC, Carrington CA, Hales CN, Ashford MLJ (1987b). Nucleotide-sensitive ion channels in human insulin producing tumour cells. Pflügers Arch. 410:169–172.

Suzuki K, Petersen OH (1988). Patch-clamp study of single-channel and whole-cell K^+ currents in guinea pig pancreatic acinar cells. Am. J. Physiol. 255:G275–G285.

Swandulla D, Lux HD (1985). Activation of a nonspecific cation conductance by intracellular Ca^{2+} elevation in bursting pacemaker neurons of *Helix pomatia*. J. Neurophysiol. 54:1430–1443.

Teeter JH, Brand JG, Kumazawa T (1990). A stimulus-activated conductance in isolated taste epithelial membranes. Biophys. J. 58:253–259.

Thomas L, Kocsis E, Colombini M, Erbe E, Trus BL, Steven AC (1991). Surface topography and molecular stoichiometry of the mitochondrial channel, VDAC, in crystalline arrays. J. Structural Biol. 106:161–171.

Thorn P, Petersen OH (1992). Activation of nonselective cation channels by physiological cholecystokinin concentrations in mouse pancreatic acinar cells. J. Gen. Physiol. 100:11–25.

Tscharner von V, Prod'hom B, Baggioloni M, Reuter H (1986). Ion channels in human neutrophils activated by a rise in free cytosolic calcium concentration. Nature 324:369–372.

Vigne P, Champigny G, Marsault R, Barbry P, Frelin C, Lazdunski M (1989). A new type of amiloride-sensitive cationic channel in endothelial cells of brain microvessels. J. Biol. Chem. 264:7663–7668.

Watson S, Abbott A (1992). TIPS Receptor Nomenclature Supplement pp 1–35.

Weber A, Siemen D (1989). Permeability of the non-selective channel in brown adipocytes to small cations. Pflügers Arch. 414:564–570.

Weiss MS, Wacker T, Weckesser J, Welte W, Schulz GE (1990). The three-dimensional structure of porin from *Rhodobacter capsulatus* at 3 Å resolution. FEBS Lett. 267:268–272.

Yamamoto Y, Chen G, Miwa K, Suzuki H (1992). Permeability and Mg^{2+} blockade of histamine-operated cation channel in endothelial cells of rat intrapulmonary artery. J. Physiol. 450:395–408.

Yau KW, Nakatani K (1984). Cation selectivity of light-sensitive conductance in retinal rods. Nature 309:352–354.

Yellen G (1982). Single Ca^{2+}-activated nonselective cation channels in neuroblastoma. Nature 296:357–359.

Nonselective Cation Channels: Pharmacology, Physiology and Biophysics
ed. by D. Siemen & J. Hescheler
© 1993 Birkhäuser Verlag Basel/Switzerland

Nonselective Cation Channels: Physiological and Pharmacological Modulations of Channel Activity

J. Hescheler and G. Schultz

Pharmakologisches Institut, Freie Universität Berlin, Thielallee 69-73, D-14195 Berlin 33, FRG

Summary
Cation channels play a major role in fast and sustained cellular responses to hormones and neurotransmitters. They contribute to depolarization of the membrane and – in most cases – to an increase in the intracellular Ca^{2+} concentration. Nonselective cation channels presumably form a large family of diverse channels which are modulated by various extracellular and intracellular signals. Structure and regulation of ligand-operated and cyclic nucleotide-activated nonselective cation channels found in synapses and sensory receptor cells, respectively, are well documented; none of the structures of other cation channels are known. Except for ligand-operated and stretch-activated channels, G-proteins form the link between the involved receptors and signalling cascades stimulating nonselective cation channels. Observed in numerous cellular systems is hormonal activation of cation channels by hormones or neurotransmitters interacting with heptahelical receptors inducing a phosphoinositide breakdown (PI response); several pathways stimulated within the PI response may generate signals involved in cation channel activation. Pharmacological modifications of nonselective cation channels by inorganic and organic blockers are so far extremely limited; various blockers have been described but unfortunately lack high specificity for these channels.

Introduction

In contrast to the ionic channels, which are characterized according to their selectivity as Na^+, K^+, Ca^{2+} or Cl^- channels, there is growing evidence for the occurrence of channels which partially or completely lack cation selectivity. These nonselective cation channels presumably form a new class of channels which exert specific cellular functions including last and long term activation of cellular responses as well as proliferation and differentiation (Jung et al., 1992; Felder et al., 1991). Although the ion selectivity can be described by the Goldman-Hodgkin-Katz formalism (Hille, 1991), there is still no correlate for a lack of ion selectivity on the molecular level. Recent site-directed mutagenesis experiments demonstrated that just the exchange of a single amino acid is enough to switch ion selectivity. For example, Heinemann et al. (1992) exchanged lysine (position 1,422 in repeat III) or alanine (position 1,714 in repeat IV) of the Na^+ channel by glutamic acid and found a change of ion-selectivity from Na^+ to Ca^{2+}. Site-directed mutagenesis of the nicotine-gated cation channel in the MII region, possibly facing the lumen of the channel's pore, induced anion selectivity (Galzi et al.,

1992). Sorbera and Morad (1990) reported that atrionatriuretic peptide changes the selectivity of voltage-dependent Na^+ channels.

Up to now, cDNAs of only two types of nonselective cation channels have been cloned, i.e. of ligand-operated channels and of cyclic nucleotide-gated channels. These channels apparently form two families of channels with each family showing a high degree of amino acid identity. All other types of nonselective cation channels, however, which have not yet been biochemically characterized nor cloned, exhibit a huge diversity with respect to their biophysical properties (Siemen, this volume), pharmacological modulations and the mechanisms involved in the channel regulation. Some pharmacological and biochemical similarities to the selective channels may even suggest that nonselective channels are variants of the same proteins lacking amino acid residues responsible for ion selectivity.

In view of the large diversity of nonselective cation channels, a simple classification of the selective channels appears to be impossible, particularly as there are no specific blockers available which are suitable for classification. The biochemical pathways involved in the regulation of nonselective cation channel activity vary considerably from cell type to cell type. Therefore, this synopsis has to be considered rather preliminary and descriptive.

Differentiation of Nonselective Cation Channels According to Regulations by Extra- and Intracellular signals

Although nonselective cation channels form a large heterogenous family of pore-forming proteins whose structure is only partly known, what most of them have in common is that their activity is only insignificantly affected by the membrane potential but rather stimulated by hormones and neurotransmitters interacting with membrane-bound receptors.

Ligand-Operated Cation Channels

Best-studied are ligand- or receptor-operated cation channels (Fig. 1). Nicotinic receptors for acetylcholine, ionotropic NMDA-, AMPA- and kainate-subtypes of glutamate receptors and possibly other receptors for neurotransmitters are located on an N-terminal domain of transmembrane proteins with four hydrophobic α-helical domains and N- and C-termini facing the extracellular space (Jan and Jan, 1992). These receptor-operated cation channels are generally composed of five homologous subunits forming the cation-permeable pore which is opened upon neurotransmitter binding to the N-terminal receptor domain. It is likely that some purinergic receptors belong to the same family, e.g. P_{2x}

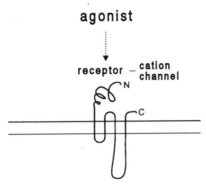

agonist

receptor – cation channel

Figure 1. Receptor-operated cation channels. Numerous receptors for neurotransmitters (nicotinic acetylcholine receptors, NMDA, AMPA and kainate types of glutamate receptors) have been cloned and sequenced; they are intramolecularly coupled to transmembrane proteins with four transmembrane domains. P_{2x} and other ATP and ADP receptors may have a similar structure. The receptor/cation complex is generally formed of five homologous subunits. These receptor-operated cation channels are opened upon neurotransmitter binding and are mostly expressed in neuronal but also in neuroendocrine and skeletal muscle cells.

and possibly P_{2z} ATP receptors occurring in neuronal and other cells (Benham and Tsien 1987; O'Connor et al., 1991; Inoue and Nakazawa, 1992) and related ADP receptors occurring in platelets (Geiger and Walter, this volume). In general, these receptor-operated cation channels allow very fast generation of intracellular signals, i.e. increases in cytoplasmic Na^+ and Ca^{2+} concentrations within the millisecond range, and thereby induce very fast reactions of neuronal, muscular, neuroendocrine and other cells to the stimulus.

Cyclic Nucleotide-activated Cation Channels

Other unselective cation channels with known structures are those stimulated by cyclic nucleotides (Fig. 2, Barnstable, this volume). These cation channels belong to the family of hexahelical transmembrane proteins (a structure related to that of potassium channels) with cytoplasmic N- and C-termini (Jan and Jan, 1992). These channels have a homooligomeric structure. Binding of cyclic GMP (in retinal sensory cells) or cyclic AMP (in olfactory cilia) to a C-terminal cyclic nucleotide-binding domain causes opening of these channels and membrane depolarisation (Matthews, 1991). Cyclic GMP-stimulated cation channels may occur in cells other than the sensory system and may contribute to cyclic GMP effects in various cells.

The only other cation-permeable channels with known structures are those located in intracellular organelles which serve as release channels for Ca^{2+} (ryanodine- and IP_3-sensitive channels, Mikoshiba et al., 1991;

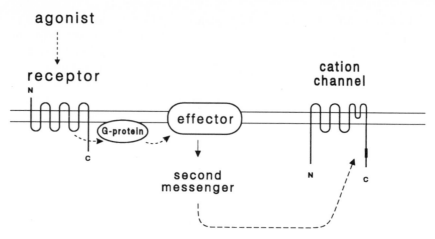

Figure 2. Cyclic nucleotide-stimulated cation channels. In sensory cells (retina, olfactory cilia) and probably other cells, cation channels are expressed that are homooligomers of hexahelical transmembrane proteins with a cytoplasmic C-terminal cyclic nucleotide-binding domain. Increased cyclic GMP and cyclic AMP concentrations cause opening and membrane depolarization in retina and olfactory cells, respectively.

Meissner et al., 1991). Our knowledge about all other cation channels is significantly more descriptive and based on limited electrophysiological studies, ^{45}Ca flux measurements or cytoplasmic Ca^{2+} determinations by the Fura II method.

Other Second Messenger-Gated Nonselective Cation Channels

The I_f current activated by hyperpolarisation is found in various cells including atrial pacemaker (DiFrancesco, 1986) and neuronal (McCormick and Pape, 1990) cells. It was shown by DiFrancesco and coworkers that this type of nonselective channel is directly activated by cAMP (DiFrancesco and Tortora, 1991) so that it may be structurally related to the cyclic nucleotide-stimulated cation channels observed in sensory cells. Another type of nonselective cation channel belonging to the group of second messenger-gated channels is the Ca^{2+}-activated nonselective cation channel (for review see Partidge and Swandulla, 1988). These channels are found in many cells including neutrophils (Von Tscharner et al., 1986), cardiomyocytes (Ehara et al., 1988), neurons (Yellen, 1982), endocrine- and exocrine cells (Sturgess et al., 1987; Maruyama and Peterson, 1984), and epithelial cells (Gögelein and Greger, 1986). The mechanism of activation is in most cases linked to phosphoinositide breakdown (PI response) (see below). Hormonally stimulated formation of inositol 1,3,4-trisphosphate (IP_3) causes a Ca^{2+} release from internal stores, and the increased intracellular Ca^{2+} concentration stimulates these cation channels, which in turn may lead to an additional influx of Ca^{2+} from the extracellular space into the cytoplasm. Hence, Ca^{2+}-activated nonselective cation channels are supposed to amplify the Ca^{2+} signal after induction of PI response. It is tempting to speculate that Ca^{2+}-activated nonselective cation channels may have some at least functional similarity to the ryanodine receptors, identified as release channels responsible for the Ca^{2+}-dependent Ca^{2+} release from internal stores (Meissner et al., 1991).

Phosphorylation-Gated Nonselective Cation Channels

There are many reports on selective channels modulated by phosphorylation/dephosphorylation. One of the best studied examples includes voltage-dependent Ca^{2+} channels of the heart, which are stimulated by cAMP-dependent protein kinase (Trautwein and Hescheler, 1990). It is assumed that the kinase phosphorylates the channel protein itself or a protein closely related to the channel (α_1- and β-subunits). A very similar modulatory cascade for nonselective cation channels was detected in mesodermal cells (MES-1 cell line) which represent precursors of cardiomyocytes (see Kleppisch et al., this volume). As denoted in Fig. 3, β-adroceptors via a G-protein (G_s) stimulate the effector enzyme adenylyl cyclase generating cAMP as a second messenger. Cyclic-AMP activates the cAMP-dependent protein kinase, which presumably increases the open probability of the nonselective cation channel by phosphorylating the channel protein itself or an associated protein.

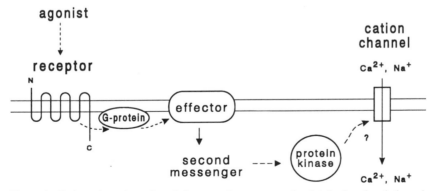

Figure 3. Cation channels activated by second messenger-stimulated phosphorylation. A cation channel with unknown structure, phosphorylated and opened by cAMP-dependent protein kinase has been observed in MES-1 cells. Whether or not cation channels stimulated by phosphorylation via second messenger-dependent proteinkinases occur in other systems and this modulation represents a general regulatory principle for cation channel regulation, needs to be elucidated.

Nonselective Cation Channels Related to PI Response and IP₃-Induced Calcium Release

Many hormonal factors interacting with heptahelical membrane receptors have been shown to stimulate ^{45}Ca influx or cation channel activity in various cellular systems (Meldolesi and Pozzan, 1987; Merritt and Rink, 1987; Putney, 1990; Merritt et al., 1989; Hallam and Rink 1989; Sage et al., 1989; Jacob, 1990; Meldolesi et al., 1991). These hormones and neurotransmitters both induce a biphasic increase in cytoplasmic Ca^{2+} concentration, with a rapid and transient increase observed within a few seconds, independent of the presence of extracellular Ca^{2+}, and a second, more slowly occurring and longer lasting phase of increased cytosolic Ca^{2+} depending on the presence of extracellular Ca^{2+}. The rapid increase in the cytosolic Ca^{2+} concentration is caused by Ca^{2+} release from intracellular storage sites within the PI response these hormonal factor induce (Berridge, 1987); hormonally activated heptahelical receptors interact with G-proteins, in most systems belonging to the pertussis toxin-insensitive group of G_q and G_{11} (Strathmann and Simon, 1990), whose activation causes stimulation of phospholipase C-β1 (Fig. 4). This enzyme catalyzes degradation of phosphatidylinosi-

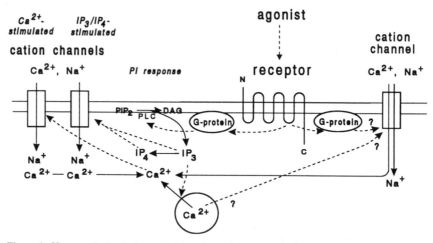

Figure 4. Hormonal stimulations of cation channels connected with stimulated phosphatidylinositol 4,5-bisphosphate degradation and inositol 1,4,5-trisphosphate-induced release of intracellular calcium. In many cellular systems, hormonal factors inducing a PI response cause opening of cation channels. The underlying control mechanisms for cation channel activity are not clear and are apparently caused by several factors. IP₃- and IP₃ plus IP₄-stimulated cation channels have been observed in some cellular systems; Ca^{2+}-stimulated cation channels, first observed in neutrophiles, appear to occur in various cells. Even more ubiquitously observed is opening of cation channels in connection to empty IP₃-sensitive Ca^{2+} pools, although the signalling mechanism between these Ca^{2+} stores and the channels is obscure. At least in some systems, the agonists that induce a PI response may independently cause opening of cation channels by a membrane-defined mechanism; in most systems, this stimulation may involve pertussis toxin-insensitive G-proteins.

tol 4,5-bisphosphate to diacylglycerol and inositol 1,3,4-trisphosphate (IP_3), which causes release of Ca^{2+} from intracellular, non-mitochondrial Ca^{2+} stores. IP_3 may additionally cause influx of Ca^{2+} through nonselective cation channels from the extracellular space and, thereby, a slower and longer lasting increase in the cytoplasmic Ca^{2+} concentration; IP_3- stimulated cation channels have been observed in T-cells, mast cells and A 431 epithelial cells (Kuno and Gardner, 1987; Penner et al., 1988; Chapron et al., 1989). In *Xenopus laevis* oocytes and exocrine cells (Irine and Moor, 1987; Morris et al., 1987), cation channels have been observed in electrophysiological studies that are activated by IP_3 plus inositol 1,3,4,5-tetrakisphosphate (IP_4). Whether or not this channel is different from the IP_3-stimulated cation channel is not clear, as IP_3 can be converted to IP_4 in most cellular systems. Elevation of the cytoplasmic Ca^{2+} concentration induced by the PI response may also serve as a signal to induce opening of nonselective cation channels; Ca^{2+}-stimulated cation channels were first reported to occur in neutrophils but have since been observed in various cellular systems (see above).

In many cells, e.g. neutrophils, mast cells, platelets, vascular muscles, hepatocytes and neuronal cells, stimulation of nonselective cation currents has been observed, caused by agonists inducing a PI response via heptahelical receptors and pertussis toxin-insensitive G-proteins (Meldolesi and Pozzan, 1987; Strathmann and Simon, 1990). In most instances, it is not known whether opening of these channels occurs as a consequence of the PI response or as a parallel, more direct event (with the same or different pertussis toxin-insensitive G-proteins functionally coupling the receptor to phospholipase C-β and to the cation channel). This stimulation of cation channels in connection with the release of intracellularly stored Ca^{2+} may have two roles, i.e. allowing a longer lasting cellular response and a refill of the IP_3-sensitive Ca^{2+} stores.

Not only hormonal factors inducing a PI response but also thapsigargin, an inhibitor of the calcium ATPase pumping Ca^{2+} from the cytoplasm into the IP_3-sensitive Ca^{2+} pool, and related drugs causing emptying of these Ca^{2+} pools have been shown to cause opening of plasma membrane cation channels (Thastrup, 1990; Thastrup et al., 1990; Mason et al., 1991; Dolar et al., 1992). The molecular basis of communication between the intracellular IP_3-sensitive Ca^{2+} pools to the plasma membrane cation channels is not known and is the subject of continuous speculation. In some cellular systems, hormonal factors interacting with heptahelical G-protein-interacting receptors cause fast effects on cation channels which can, so far at least in a few systems, be observed in isolated patches where effects of intracellular organelles or second messengers on the ion channel activity cannot occur. This relatively direct effect comes closest to the classical textbook assumption regarding the occurrence of "receptor-operated Ca^{2+} channels".

No cation channels stimulated in connection with a PI response have been purified or cloned, and no reconsitution experiments with other regulatory components have been performed; the signaling sequences described above are hypothetical and yet to be proved. Whether G-proteins identical to those inducing phospholipase C stimulation or other pertussis toxin-insensitive G-proteins are involved in these cation channel regulations needs to be elucidated. Both thapsigargin-activated cation channels and rapidly hormonally activated cation channels observed in isolated patch configurations can be blocked by some of the cation channel blockers described below. Therefore, Ca^{2+} release-activated cation channels and more directly G-protein-activated cation channels may represent the same entity of channels or may belong to a group of closely related cation channels. Whether or not Ca^{2+}-, IP_3- or $IP_3 + IP_4$-stimulated cation channels can also be blocked by these drugs and may belong to the same family of related cation channels will become clear in the near future.

It is interesting to see that most cells express a related but Ca^{2+}-selective ion channel that is also activated by the release of Ca^{2+} from intercellular storage sites (CRAC, Hoth and Penner, 1992); this channel may be an additional member of the proposed family of related, more or less selective cation channels.

Pharmacological Modulations of Nonselective Cation Channels

Inorganic Compounds

The permeation of ions through a channel is generally described by energy profiles, assuming that the charged amino acids work as binding sites for ions with various affinities. This model also provides a theoretical basis to explain the fact that ions do not only penetrate the channel but also work as blockers, particularly if their affinity is high enough to hinder the passage of other ions (Hille, 1991). Inorganic blockers of cation channels are divalent or trivalent cations and exceed the molecular weight of Na^+. Probably the most efficient ion to block nonselective cation channels is gadolinium (Gd^{3+}), a high atomic mass lanthanide with an ionic radius (0.938 Å) close to that of Na^+ (o.97 Å) and Ca^{2+} (0.99 Å). In various cell types including *Xenopus* oocytes (Yang and Sachs, 1989), vascular smooth muscle cells (Krautwurst, unpublished results) and neutrophils (Krautwurst et al., 1993) it blocks stretch- as well as receptor-activated nonselective cation channels at concentrations ranging from 50 nM to 10 μM. Other lanthanides, lanthanum (La^{3+}, 1.061 Å) and lutetium (0.85 Å), also block nonselective cation channels but at higher concentrations. Interestingly, these ions are also quite powerful blockers of voltage-dependent Ca^{2+} channels, suggesting a similarity in the permeation pore. A rather complete comparative analysis of inorganic compounds blocking agonist-stimulated Ca^{2+} entry through nonselective cation channels and depolarization-induced Ca^{2+} entry through voltage-dependent Ca^{2+} channels was performed for vascular smooth muscle cells (Rüegg et al., 1989; Wallnöfer et al., 1989). The agonist-stimulated nonselective cation channels were blocked by multivalent cations with the rank order of potency $La^{3+} > Cd^{2+} > Mn^{2+} > Co^{2+} > Ni^{2+} > Mg^{2+}$. Voltage-dependent Ca^{2+} channels were significantly more sensitive to Cd^{2+} and significantly less sensitive to La^{3+} than nonselective cation channels. Ni^{2+} shows a higher affinity to T-type voltage-dependent Ca^{2+} channels.

Organic Compounds

There are numerous compounds including amiloride, quinine, 4 aminopyridine, 4-acetamido-4'-isothiocyanatostilbene-2,2'-disulfonic acid (SITS) and 4,4'-diisothiocyanatostilbene-2,2'-disulfonic acid (DIDS) which in the range of 0.1 to 10 mM unspecifically inhibit a variety of membraneous ionic channels and exchangers. These compounds were also found to be blockers of nonselective cation channels (for review see Siemen, this volume). Substances which inhibit nonselective cation channels in the micromolar range are summarized in Table 1. They are rather unspecific and especially crossreact with selective channels, most frequently voltage-sependent Ca^{2+} channels. In the following section we will describe the most commonly used compounds in more detail.

l-cis-Diltiazem

The benzothiazepine diltiazem is well known as a blocker of voltage-dependent Ca^{2+} channels but was also found to affect cGMP-activated nonselective cation channels in rod photoreceptor cells (Stern et al., 1986). The ED_{50} of l-cis diltiazem was $10 \mu M$ for cGMP-activated nonselective cation channels, d-cis-diltiazem was ineffective. Other types of nonselective cation channels have not been reported to be blocked by l-cis diltiazem.

Flufenamic Acid, Mefenamic Acid

Flufenamic acid and mefenamic acid are derivatives of 3',5'-dichlorodiphenylamine-2-carboxylic acid (DCDPC, blocker of Cl^- channels) and were originated as non-steroidal anti-inflammatory drugs (for more details see Gögelein et al., this volume). With a half-maximally active concentration of about $10 \mu M$, they inhibit Ca^{2+}-activated nonselective cation channels in the basolateral membrane of rat ex-

Table 1. Organic compounds affecting the activity of nonselective cation channels. The table summarizes recent literature, giving informations on the biological systems used and the cellelar response affected by the cation channel blocker or opener. Whether or how Ca^{2+} entry was measured, is not indicated.

A. BLOCKERS:

Flufenamic acid:
Human erythrocyte band 3 anion transport protein (Knauf et al.; Am-J-Physiol. 1989; 257:C277-89
Rat isolated myocardium (Northover; Br-J-Pharmacol. 1990; 100:477-82)
Rat exocrine pancreas (Gögelein et al.; FEBS-Lett. 1990; 268:79-82)
Xenopus oocytes; Ca^{2+}-activated Cl^- channels (White et al.; Mol-Pharmacol. 1990; 37:720-4)
Rat distal colon crypt cell basolateral membrane (Siemer & Gögelein; Pflügers-Arch. 1992; 420:319-28)
Mammalian airway; chloride secretion (Chao & Mochizuki; Life-Sci. 1992; 51:1453-71)
Guinea-pig parietal cells; H^+ production (Beil et al.; Clin-Exp-Pharmacol-Physiol. 1992; 19:555-61)
Mouse L-M(TK$^-$) fibroblasts; cell proliferation (Jung et al.; Am-J-Physiol. 1992; 262:C1464-70)

SK&F 96365:
PC-12 cells (Fasolato et al.; J-Biol-Chem. 1990; 265:20351-5)
Human platelets; neutrophils; endothelial cells (Merritt et al.; Biochem-J. 1990; 271:515-22)
Human platelets (Sage et al.; Biochem-J. 1990; 265:675-80)
Polyoma middle T anitigen-transformed NIH-3T3 fibroblasts (Okano et al.; Biochem-Biophys-Res-Commun. 1991; 176:813-9)
NG108-15 cells (Chan & Greenberg; Biochem-Biophys-Res-Commun. 1991; 177:1141-6)
Rat lymphocytes (Mason et al.; J-Biol-Chem. 1991; 266:10872-9)
Pig aortic microsomes (Blayney et al.; Biochem-J. 1991; 273:803-6)
Human endothelial cells; histamine-induced formation of endothelium-derived relaxing factor (Graier et al.; Biochem-Biophys-Res-Commun. 1992; 186:1539-45)
Human platelets (Sage et al.; Biochem-J. 1992; 285:341-4)
Vascular endothelial cells (Schilling et al.; Biochem-J. 1992; 284:521-30)
C6 glioma cells; RIN insulimoma cells (Soergel et al.; Mol-Pharmacol. 1992; 41:487-93)
Pig aortic microsomes (Blayney et al.; Biochem-J. 1992; 282:81-4)
HL-60 cells (Demaurex et al.; J-Biol-Chem. 1992; 267:2318-24)
Human neutrophils (Davies et al.; FEBS-Lett. 1992; 313:121-5)
Human leukemia cells; mitosis-arrest (Nordstrom et al.; Exp-Cell-Res. 1992; 202:487-94)

SC38249:
Cerebellar neurons, glial and pheochromocytoma PC-12 cells (Ceardo & Meldolesi; Eur-J-Pharmacol. 1990; 188:417-21)
NIH-3T3 fibroblasts (Magni et al.; J-Biol-Chem. 1991; 266:6329-35)
PC-12 cells (Clementi et al.; J-Biol-Chem. 1992; 267:2164-72)

L-651.582:
Madin-Darby Bovine kidney (MDBK) cells; proliferation (Hupe et al.; J-Cell-Physiol. 1990; 144:457-66; Hupe et al.; J-Pharmacol-Exp-Ther. 1991; 256:462-7)
Ovarian cancer; proliferation; tumor cell adhesion (Kohn & Liotta; J-Natl-Cancer-Inst. 1990; 82:54-60)
Human melanoma and ovarian cancer; calcium; arachidonate; inositol phosphates (Kohn et al.; Cancer-Res. 1992; 52:3208-12)
Chinese hamster ovary cells; receptor-stimulated calcium influx and arachidonic acid release (Felder et al.; J-Pharmacol-Exp-Ther. 1991; 257:697-714)
Rat polymorphonuclear leukocytes, receptor-mediated and voltage-dependent calcium entry (Hupe et al.; J-Biol-Chem. 1991; 266:10136-42)

R56865:
Pithed normotensive rats; alpha-adrenoceptors (Koch et al.; Arch-Int-Pharmacodyn-Ther. 1989; 299:134-43)

Table 1 (continued)

Rat heart isolated atria and ventricles; ouabain-induced intoxication (Finet et al.; Eur-J-Pharmacol. 1989; 164:555–63)

Ca^{2+} displacement from phosphatidylserine monolayers (Vogelgesang & Scheufler; Eur-J-Pharmacol. 1990; 188:17–22)

Rat heart; antiarrhythmic effects (Garner et al.; J-Cardiovasc-Pharmacol. 1990; 16:468–79)

Guinea-pig cardiac ventricle; Na$^+$-activated K$^+$ current (Luk & Carmeliet; Pflügers-Arch. 1990; 416:766–8; Rodrigo & Chapman; Exp-Physiol. 1990; 75:839–42)

Ventricular cardiomyocytes (Himmel et al.; Eur-J-Pharmacol. 1990; 187:235–40)

Cardiac glycoside toxicity and myocardial ischemia (Damiano et al.; J-Cardiovasc-Pharmacol. 1991; 18:415–28)

Rat and rabbit cardiomyocytes; myocardial protection (Ver-Donck & Borgers; Am-Physiol. 1991; 261:H1828–35)

Cardiac Purkinje cells; veratridine-induced, non-inactivating Na$^+$ current (Verdonck et al.; Eur-J-Pharmacol. 1991; 203:371–8)

Ventricular cardiomyocytes; Na- and L-type Ca channels (Wilhelm et al.; Br-J-Pharmacol. 1991; 104:483–9)

Rabbit heart; ischemic damage (Vandeplassche et al.; Eur-J-Pharmacol. 1991; 202:259–68)

Rat cardiac trabeculae; triggered propagated contractions (Daniels et al.; J-Cardiovasc-Pharmacol. 1992; 20:187–96)

Hippocampus of the rat (Scheufler et al.; Neuropharmacology. 1992; 31:481–6)

Lysophosphatidylcholine-induced Ca^{2+}-overload in isolated cardiomyocytes (Ver-Donck et al.; J-Mol-Cell-Cardiol. 1992; 24:977–88)

Isolated perfused rabbit heart; ouabain intoxication (Tegtmeier et al.; J-Cardiovasc-Pharmacol. 1992; 20:421–8)

Guinea-pig ventricular single cells; Na$^+$ channel (Carmeliet et al.; Eur-J-Pharmacol. 1991; 196:53–60)

Guinea-pig ventricular myocytes; transient inward current (Leyssens & Carmeliet; Eur-J-Pharmacol. 1991; 196:43–51)

B. OPENERS

Maitotoxin:

PC-12 and NCB-20 cells; phosphoinositide breakdown (Gusovsky et al.; Mol-Pharmacol. 1989; 36:44–53; FEBS-Lett. 1989; 243:307–12)

Bovine parathyroid cells; inhibition of parathyroid hormone release (Fitzpatrick et al.; Endocrinology 1989; 124:97–103)

PC-12 cells (Meucci et al.; Pharmacol-Res. 1989; 21 Suppl 1:1–2)

Liver; cell death (Kutty et al.; Toxicol-Appl-pharmacol. 1989; 101:1–10)

Rat tuberoinfundibular neurons; dopamine release (Ohmichi et al.; Neuroendocrinology. 1989; 50:481–7)

Myocardium; cell injury (Santostasi et al.; Toxicol-Appl-Pharmacol. 1990; 102:164–73)

Parathyroid C-cells; calcitonin secretion (Nishiyama et al.; Horm-Metab-Res. 1990; 22:258–9)

Purified rat brain synaptosomes (Taglialatela et al.; Biochim-Biophys-Acta. 1990; 1026: 126–32)

Cerebellar neurons; phosphoinositide turnover; glutamatergic and muscarinic cholinergic receptor function (Lin et al.; J.Neurochem. 1990; 55:1563–8)

Insulinoma HIT T15 cells; insulin release (Soergel et al.; J-Pharmacol-Exp-Ther. 1990; 255:1360–5)

MMQ pituitary cells (Login et al.; Cell-Calcium. 1990; 11:525–30)

PC-12 cells (Meucci et al.; Pharmacol-Res. 1990; 22 Suppl 1:75–6)

Rat anterior pituitary cells (Sortino et al.; Pharmacol-Res. 1990; 22 Suppl 1:73–4)

Rat ventricular myocytes; calcium current and background inward current (Faivre et al.; Toxicon. 1990; 28:925–37)

HL-60 cells; phosphoiositide breakdown (Gusovsky et al.; J-Pharmacol-Exp-Ther. 1990; 252:466–73)

Sea urchin egg; Ca^{2+} permeabilities (Pesando et al.; Biol-Cell. 1991; 72:269–73)

Insulinoma HIT T15 cells and rat glioma C6 cells (Murata et al.; Toxicon. 1991; 29:1085–96)

Rat anterior pituitary cells; inositol phosphates (Sortino et al.; J-Mol-Endocrinol. 1991; 6:95–9)

ocrine pancreatic cells (Gögelein et al., 1990). It is very likely that they bind directly to nonselective cation channels, since the blocking effect was also identified in inside-out patches. Other non-steroidal anti-inflammatory drugs, such as indomethacin, acetylsalicyl acid, and ibuprofen were without effect. In mouse L-M (TK$^-$) fibroblasts, a 28 pS nonselective cation channel is activated by platelet-derived growth factor (PDGF) within seconds, suggesting that this channel is involved as an early component in the PDGF-induced signalling cascade (Jung et al., 1992). Flufenamic acid and mefenamic acid rapidly and reversibly blocked this nonselective cation channel, and inhibited the cloning effciency and growth rate of these cells, substantiating the hypothesis that nonselective cation channel activation forms a neccessary component in the transduction of the mitogenic signal from PDGF receptors.

SK&F 96365

1-$\{\beta$-[3-(4-methoxyphenyl)propoxy]-4-methoxyphenethyl$\}$-1H-imidazole hydrochloride has been used as a blocker of nonselective cation channels in various cell types, including human platelets, neutrophils and endothelial cells. As demonstrated in Ca^{2+}-fluorimetric measurements, SK&F 96365 did not affect the receptor-mediated Ca^{2+} release but blocked the Ca^{2+} entry through nonselective cation channels (Merritt et al., 1990; Howson et al., 1990; Krautwurst et al., 1992; Graier et al., 1992). The half-maximal inhibitory concentration amounted to approximately 10 μM. In patch clamp studies on neutrophil HL60 cells (Krautwurst et al., 1992), SK&F 96365 blocked currents through nonselective cation channels activated by ATP or fMLP with similar IC_{50} values. In artery smooth muscle cells of the rabbit ear, SK&F 96365 did not block a ATP-gated Ca^{2+} permeable nonselective cation channel but significantly reduced the Ca^{2+} entry through voltage-gated channels, suggesting that (i) SK&F 96365 has little selectivity for voltage-gated Ca^{2+} entry and nonselective cation channels, but (ii) may discriminate between different types of nonselective cation channels.

SC 38249

The imidazole derivate SC 38249 is an analog to SK&F 96365 and has also been described as a blocker of voltage-gated L-type channels and nonselective cation channels which open following activation of receptors. SC 38249 has been tested in mouse NIH-3T3 fibroblasts overexpressing epidermal growth factor (EGF) receptors. It has been demonstrated that SC 38249 (10 μM) markedly inhibits the EGF-induced growth of cells, suggesting that the sustained Ca^{2+} influx via nonselective cation channels plays a role in the activity of EGF (Magni et al., 1991).

L-651.582

5-amino-[4-(4-chlorobenzoyl)-3,5-dichlorobenzyl-1,2,3-triazole-4-car-bozamide is an antiproliferative and antiparasitic agent which has been shown to affect $^{45}Ca^{2+}$ uptake into mammalian cells. The leukotriene B_4 production in rat polymorphonuclear leukocytes induced by fMet-Leu-Phe (fMLP) was blocked by L-651.582 (IC_{50} 0.5 μg/ml) probably due to blockage of fMLP-induced Ca^{2+} entry through nonselective cation channels, since L-651.582 did not inhibit calmodulin or enzymes cata-lyzing arachidonate metabolism (Hupe et al., 1991b). However L-651.582 is not specific for nonselective cation channels. IC_{50} values of 0.2 μg/ml and 1.4 μg/ml were reported for L-type and T-type Ca^{2+} channels, respectively (Hupe et al., 1991a). The importance of extracel-lular Ca^{2+} in the regulation of cell proliferation in culture and the derangement of these signal transduction events in transformed cells is well recognized and has been the subject of a large body of experimental work. This has led to the use of L-651.582 as a possible tool to inhibit malignant cell growth. Indeed L-651.582 could be demonstrated to reduce anchorage-dependent and -independent growth in a large series of human cancer cell lines. L-651.582 pretreatment of HT-29 human colon cancer and ras-transfected rat embryo fibroblast cells inhibited the formation and growth of experimental pulmonary metastases in nude mice. Oral administration of L-651.582 arrested growth and metastasis of transplanted human melanoma and ovarian cancer xeno-grafts (Kohn et al., 1992).

In another study, L-651.582 blocked muscarinic m5 receptor-stimu-lated Ca^{2+} influx and release of arachidonic acid at low micromolar concentrations. Muscarinic receptor-stimulated release of arachidonic acid was previously shown to be dependent on Ca^{2+} influx but not on intracellular Ca^{2+} release, suggesting L-651.582 may be useful as Ca^{2+} channel blocker (Hupe et al., 1990; Felder et al., 1991; Hupe et al., 1991a,b).

R56865

The benzothiazolamine derivative R56865 was first demonstrated as a tool against digitalis toxicity, i.e. the delayed after-depolarizations and arrhythmias in cardiomyocytes. It was later found that it blocked the transient inward (TI) current (apparent IC_{50} 7.5×10^{-8} M). This is at least partially due to a cation current through nonselective cation channels activated by intracellular Ca^{2+}. The blockage appeared to occur at the level of Ca^{2+}-release from the sarcoplasmic reticulum (Leyssens et al., 1991).

Maitotoxin, a Possible Activator of Nonselective Cation Channels

Maitotoxin, isolated from the marine dinoflagellate Gambierdiscus, is a water-soluble polyether with a molecular weight of 3424 as disodium salt (Yokoyama et al., 1988). As a highly polar substance, maitotoxin is not expected to cross membrane lipid bilayers. At concentrations ranging between 100 pM and 30 nM, it has been reported to stimulate the Ca^{2+} uptake, neurotransmitter release and phosphoinositide breakdown, contraction of smooth and skeletal muscle, and to have stimulatory effects on the heart. All effects of maitotoxin appear to be dependent on extracellular Ca^{2+} (Choi et al., 1990; Gusovsky and Daly, 1990; Soergel et al., 1992).

In neuroblastoma-glioma hybrid cells and certain pituitary tumor cells, dihydropyridines block the influx of $^{45}Ca^{2+}$ elicited by maitotoxin, whereas in the muscular cells BC3H1, aortic myocytes and synaptosomes, these L-channel blockers have no effect on maitotoxin-elicited Ca^{2+} influx (Yoshii et al., 1987; Choi et al., 1990; Gusovsky and Daly, 1990; Soergel et al., 1992; see Tab. 1). Maitotoxin also causes markable stimulations on phosphoinositide breakdown resulting in the generation of IP_3. Toxic effects presumably occur due to the elevated intracellular Ca^{2+} concentration (Gusovsky et al., 1990).

Concluding Remarks

Due to the diversity of the regulatory mechanisms, the understanding of the pathways and signals involved in stimulations of nonselective cation channels is very limited. Elucidating the structures of the channel-forming and possible connected proteins will allow expression of these channels in defined systems and will lead to better understanding of the control mechanisms; this will also allow the development of more selective blockers of specific types of cation channels than presently available.

References

Benham CD, Tsien RW (1987). A novel receptor-operated Ca^{2+}-permeable channel activated by ATP in smooth muscle. Nature 328:275–278.
Berridge MJ (1987). Inositol trisphosphate and diacylglycerol: two interacting second messengers. Ann. Rev. Biochem. 56:159–193

Chapron Y, Cochet C, Crouzy S, Jullien T, Keramidas M, Verdi J (1989). Tyrosine protein kinase activity of the EGF receptor is required to induce activation of receptor-operated calcium channels. Biochem. Biophys. Res. Commun. 158:527–533.

Choi OH, Padgett WL, Nishizawa Y, Gusovsky F, Yasumoto T, Daly JW (1990). Maitotoxin: Effects on calcium channels, phosphoinositide breakdown, and arachidonate release in pheochromocytoma PC12 cells. Molec. Pharmacol. 37:222–230

DiFrancesco D (1986). Characterization of single pacemaker channels in cardiac sino-atrial node cells. Nature 324:470–473.

DiFrancesco D, Tortora P (1991). Direct activation of cardiac "pacemaker" (i_f) channels by intracellular cAMP. Nature 351:145–147.

Dolar RJ, Hurwitz LM, Mirza Z, Strauss HC, Whorton R (1992) Regulation of extracellular calcium entry in endothelial cells: role of intracellular calcium pool. Am. J. Physiol. 262:C171–C181.

Ehara T, Noma A, Ono K (1988). Calcium activated non-selective cation channels in ventricular cells isolated from adult guinea-pig hearts. J. Physiol. 403:117–133.

Felder CC, Ma AL, Liotta LA, Krohn EC (1991). The antiproliferative and antimetastatic compound L651582 inhibits muscarinic acetylcholine receptor-stimulated calcium influx and arachidonic acid release. J. Pharmacol. Exp. Therapeutics 257:967–971.

Galzi JL, Devillers-Thiery A, Hussy N, Bertrand S, Changeux JP, Bertrand D (1992). Mutations in the channel domain of a neuronal nicotinic receptor convert ion selectivity from cationic to anionic. Nature 359:500–505.

Gögelein H, Dahlem D, Englert HC, Lang HJ (1990). Flufenamic acid, mefenamic acid and niflumic acid inhibit single nonselective cation channels in the rat exocrine pancreas. FEBS Lett. 268:79–82.

Gögelein H, Greger R (1986). A voltage-dependent ionic channel in the basolateral membrane of late proximal tubules of the rabbit kidney. Pflügers Arch. 407.2:142–148.

Graier WF, Groschner K, Schmidt K, Kukovetz WR (1992). SK&F 96365 inhibits histamine-induced formation of endothelium-derived relaxing factor in human endothelial cells. Biochem. Biophys. Res. Commun. 186:1539–1545.

Gusovsky F, Bitran JA, Yasumoto T, Daly JW (1990). Mechanism of maitotoxin-stimulated phosphoinositide breakdown in HL-60 cells. J. Pharmacol. Exp. Therapeutics 252:466–473.

Gusovsky F, Daly JW (1990). Maitotoxin: a unique pharmacological tool for research on calcium-dependent mechanisms. Biochem. Pharmacol. 39:1633–1639.

Hallam TJ, Rink TJ (1989). Receptor-mediated Ca^{2+} entry: diversity of function and mechanism. Trends Pharmacol. Sci. 10:8–10.

Heinemann SH, Terlau H, Stuhmer W, Imoto K, Numa S (1992). Calcium channel characteristics conferred on the sodium channel by single mutations. Nature 356:441–443.

Hille B (1991). Ionic channels of excitable membranes. Sinauer Associates Inc, Sunderland.

Hoth M, Penner R (1992). Depletion of intracellular calcium stores activates a calcium current in mast cells. Nature 355:353–356.

Howson W, Armstrong WP, Cassidy K, Novelli R, Tchorzewska MA, Jaxa-Chamiec A, Dolle RE, Hallam TJ, Leigh BK, Merritt JE, Moores KE, Rink TJ (1990). Design and synthesis of a series of glycerol-derived receptor mediated calcium entry (RMCE) blockers. Eur. J. Med. Chem. 25:595–602.

Hupe DJ, Behrens ND, Boltz R (1990). Antiproliferative activity of L651,582 correlates with calcium mediated regulation of nucleotide metabolism at phosphoribosyl pyrophosphate synthetase. J. Cell. Physiol. 144:457–466.

Hupe DJ, Boltz R, Cohen CJ, Felix J, Ham E, Miller D, Soderman D, Van Skiver D (1991a). The inhibition of receptor-mediated and voltage-dependent calcium entry by the antiproliferative L-651,582. J. Biol. Chem. 266:10136–10142.

Hupe DJ, Pfefferkorn ER, Behrens ND, Peters K (1991b). L-651,582 Inhibition of intracellular parasitic protozoal growth correlates with host-cell directed effects. J. Pharmacol. Exp. Therapeutics 156:462–467.

Inoue K, Nakazawa K (1992). ATP receptor-operated Ca^{2+} influx and catecholamine release from neuronal cells. NIPS 7:56–59.

Irvine RF, Moor RM (1987). Inositol (1,3,4,5) tetrakisphosphate-induced activation of sea urchin eggs requires the presence of inositol trisphosphate. Biochem. Biophys. Res. Commun. 146:284–290.

Jacob R (1990). Agonist-stimulated divalent cation entry into single cultured human umbilical vein endothelial cells. J. Physiol. 421:55–77.

Jan LY, Jan YN (1992). Tracing the roots of ion channels. Cell 69:715–718.

Jung F, Selvaraj S, Gargus JJ (1992). Blockers of platelet-derived growth factor-activated nonselective cation channel inhibit cell proliferation. Am. J. Physiol. 262:C1464–C1470.

Kohn EC, Sandeen MA, Liotta LA (1992). In vivo efficacy of a novel inhibitor of selected signal transduction pathways including calcium, arachidonate and inositol phosphates. Cancer Research 52:3208–3212.

Krautwurst D, Hescheler J, Arndts D, Lösel W, Hammer R, Schultz G (1993). A novel potent inhibitor of receptor-activated nonselective cation current in HL-60 cells. Mol. Pharm. 43(5):655–659.

Krautwurst D, Seiffert R, Hescheler J, Schultz G (1992) Formyl peptides and ATP stimulate Ca^{2+} and Na^+ inward currents through non-selective cation channels via G-proteins in dibutyryl cyclic AMP-differentiated HL-60 cells. Biochem. J. 288:1025–1035.

Kuno M, Gardner P (1987). Ion channels activated by inositol 1,4,5-trisphosphate in plasma membrane of human T-lymphocytes. Nature 326:301–304.

Leyssens A, Carmeliet E (1991). Block of the transient inward current by R56865 in guinea-pig ventricular myocytes. Eur. J. Pharmacol. 196:43–51.

Magni M, Meldolesi J, Pandiella A (1991). Ionic events induced by epidermal growth factor. J. Biol. Chem. 266:6329–6335.

Maruyama Y, Peterson OH (1984). Single calcium-dependent cation channels in mouse pancreatic acinar cells. J. Membrane Biol. 81:83–87.

Mason MJ, Garcia-Rodriguez C, Grinstein S (1991). Coupling between intracellular Ca^{2+} stores and the Ca^{2+} permeability of the plasma membrane. J. Biol. Chem. 266:20856–20862.

Matthews G (1991). Ion channels that are directly activated by cyclic nucleotides. Trends Pharmacol. Sci. 12:245–247.

McCormick DA, Pape HC (1990). Properties of a hyperpolarization-activated cation current and its role in rhythmic oscillation in thalamic relay neurones. J. Physiol. Lond. 431:291–318.

Meissner G, Lai FA, Anderson K, Xu L, Liu QY, Herrmann-Frank A, Rousseau E, Jones RV, Lee HB (1991). Purification and reconstitution of the ryanodine- and caffeine-sensitive Ca^{2+} release channel complex from muscle sarcoplasmic reticulum. Adv. Exp. Med. Biol. 304:241–256.

Meldolesi J, Clementi E, Fasolato C, Zacchetti D, Pozzan T (1991). Ca^{2+} influx following receptor activation. Trends Pharmacol. Sci. 12:289–292.

Meldolesi J, Pozzan T (1987). Pathways of Ca^{2+} influx at the plasma membrane: Voltage-, receptor-, and second messenger-operated channels. Exp. Cell Res. 171:271–283.

Merritt JE, Armstrong WP, Benhan CD, Hallam TJ, Jacob R, Jaxa-Chamiec A, Leigh BK, McCarthy SA, Moores KE, Rink TJ (1990). SK&F 96365, a novel inhibitor of receptor-mediated calcium entry. Biochem. J. 271:515–522.

Merritt JE, Jacob R, Hallam TJ (1989). Use of manganese to discriminate between calcium influx and mobilization from internal stores in stimulated human neutrophils. J. Biol. Chem. 264:1522–1527.

Merritt JE, Rink TJ (1987). Regulation of cytosolic free calcium in fura-2-loaded rat parotid acinar cells. J. Biol. Chem. 262:17362–17369.

Mikoshiba K, Furuichi T, Maeda N, Yoshikawa S, Miyawaki A, Niinobe M, Wada K (1991). Primary structure and functional expression of the inositol 1,4,5-trisphosphate receptor. Adv. Exp. Med. Biol. 287:83–95.

Morris AP, Gallacher DV, Irvine RF, Peterson OH (1987). Synergism of inositol trisphosphate and tetrakisphosphate in activating Ca^{2+}-dependent K^+ channels. Nature 330:635–655.

O'Connor SE, Dainty IA, Leff P (1991). Futher subclassification of ATP receptors based on agonist studies. Trends Pharmacol. Sci. 12:137–141.

Partidge LD, Swandulla D (1988). Calcium activated non-specific cation channels. Trends Neurosci. 11:69–72.

Penner R, Matthews G, Neher E (1988). Regulation of calcium influx by second messengers in rat mast cells. Nature 334:499–504.

Putney JW (1990). Receptor-regulated calcium entry. Pharmacol. Ther. 48:427–434.

Rüegg UT, Wallhöfer A, Weir S, Cauvin C (1989). Receptor-operated calcium-permeable channels in vascular smooth muscle. J. Cardiovasc. Pharmacol. 14 (Suppl.) S49–S58.

Sage SO, Merrit JE, Hallam TJ, Rink TJ (1989). Receptor-mediated calcium entry in fura-2-loaded human platelets stimulated with ADP and thrombin. Dual-wavelength studies with Mn^{2+}. Biochem. J. 258:923–926.

Soergel DG, Yasumoto T, Daly JW, Gusovsky F (1992). Maitotoxin effects are blocked by SK&F 96365, an inhibitor of receptor-mediated calcium entry. Molec. Pharmacol. 41:487–493.

Sorbera LA, Morad M (1990). Atrionatriuretic peptide transforms cardiac sodium channels into calcium-coduction channels. Science 247:969–973.

Stern JH, Kaupp UB, MacLeish PR (1986). Control of the light-regulated current in rod photoreceptors by cyclic GMP, calcium and 1-cis-diltiazem. Proc. Natl. Acad. Sci. USA 83:1163–1167.

Strathmann M, Simon MI (1990). G protein diversity: a distinct class of alpha subunits is present in vertebrates and invertebrates. Proc. Natl. Acad. Sci. USA 87:9113–9117.

Sturgess NC, Hales CN, Ashford MLJ (1987). Calcium and ATP regulate the activity of a non-selective cation channel in a rat insulinoma cell line. Pflügers Arch. 409:607–615.

Thastrup O (1990). Role of Ca^{2+}-ATPases in regulation of cellular Ca^{2+} signalling, as studied with the selective microsomal Ca^{2+}-ATPase inhibitor, thapsigargin. Agents and Actions 29:8–23.

Thastrup O, Cullen PJ, Drobak B K, Hanley MR, Dawson AP (1990). Thapsigargin, a tumor promoter, discharges intracellular Ca^{2+} stores by specific inhibition of the endoplasmic reticulum Ca^{2+}-ATPase. Proc. Natl. Acad. Sci. USA 87:2466–2470.

Trautwein W, Hescheler J (1990). Regulation of cardiac L-type calcium current by phosphorylation and G proteins. Annu. Rev. Physiol. 52:257–274.

von Tscharner V, Prod'hom B, Bagglioni M, Reuter H (1986). Ion channels in human neutrophils activated by a rise in free cytosolic calcium concentration. Nature 324:369–372.

Wallnöfer A, Cauvin C, Lategan TW, Rüegg UT (1989). Differential blockade of agonist- and depolarization-induced $^{45}Ca^{2+}$ influx in smooth muscle cells. Am. J. Physiol. 257:C607–C611.

Yang XC, Sachs F (1989). Block of stretch-activated ion channels in Xenopus oocytes by gadolinium and calcium ions. Science 243:1068–1071.

Yellen G (1982). Single Ca^{2+}-activated nonselective cation channels in neuroblastoma. Nature 296:357–359.

Yokoyama A, Murata M, Oshima Y, Iwashita T, Yasumoto A (1988). Some chemical properties of maitotoxin, a putative calcium channel agonist isolated from a marine dinoflagellate. J. Biochem. (Tokyo) 104:184–187.

Yoshii M, Tsunoo A, Kuroda Y, Wu CH, Narahashi T (1987). Maitotoxin-induced membrane currents in neruoblastoma cells. Brain Res. 424:119–125.

Receptor-Activated Nonselective
Cation Channels

Nonselective Cation Channels: Pharmacology, Physiology and Biophysics
ed. by D. Siemen & J. Hescheler

Structure, Diversity, and Ionic Permeability of Neuronal and Muscle Acetylcholine Receptors

John A. Dani

Department of Physiology and Biophysics, Baylor College of Medicine, Houston, TX 77030, USA

Nicotinic acetylcholine receptors (nAChRs) form a family of ligand-gated, cation-selective channels that are concentrated at cholinergic synapses on vertebrate neurons and muscle cells. At the neuromuscular endplate, muscle nAChRs bind acetylcholine released by the persynaptic motor neuron. The receptors then undergo a conformational change that opens their ion channels. Cations move passively through the water-filled pores down their electrochemical gradients, completing synaptic transmission by depolarizing the postsynaptic muscle. The channel only weakly discriminates among permeant cations, which include all monovalent and divalent cations that are small enough to fit through the narrowest cross section. The membrane-spanning region of the pore is lined by uncharged domains that are bracketed by residues with net negative charge. The pore has large entrance vestibules, especially facing extracellularly. The narrowest cross-section is located near the cytoplasmic end of the membrane-spanning region, and this short narrow region probably provides the main cation binding site that is directly in the permeation pathway. Neuronal nAChRs share many of the properties of muscle nAChRs, but the neuronal receptor subtypes are more heterogenous genetically, pharmacologically, and functionally. There are especially important functional differences between muscle and neuronal nAChRs. For example, neuronal nAChRs are more highly permeable to Ca^{2+} and physiological levels of Ca^{2+} very potently modulate neuronal nicotinic currents. This variety of nAChRs suggests that these receptor/channels serve many roles in the excitable tissues of vertebrates.

Introduction

Nicotinic acetylcholine receptors (nAChRs) are ligand-gated cation-selective ion channels. It is likely that nAChRs are composed of five subunits (Raftery et al., 1980; Cooper et al., 1991). Although various subunit combinations can produce many different nAChR subtypes, the receptor/channel family can be classified into three main functional categories: muscle nAChRs, neuronal nAChRs, and neuronal α-bungarotoxin (α-BGT) binding proteins.

Muscle nAChRs mediate the high efficiency synaptic transmission at the vertebrate neuromuscular junction (review: Katz, 1966). In muscle, the five subunits that form the receptor are two α_1 and one each of β_1, δ, and either γ or ε. During development the muscle nAChR contains a γ-subunit, but in the adult form at the neuromuscular endplate the γ-subunit is replaced by an ε-subunit (Mishina et al., 1986). An important pharmacological characteristic of muscle nAChRs is that they are potently inhibited by α-BGT.

Neuronal nAChRs are a more diverse second category of nicotinic receptors that are located both presynaptically and postsynaptically in the central and peripheral nervous system (reviews: Aquilonius and Gillberg, 1990; Steriade and Biesold, 1990; Patrick et al., 1992; Sargent, 1992). Many neuronal nAChR subunits have been cloned. Based on their homology to muscle subunits, seven neuronal subunits have been classified as α_{2-8} and three as β (non-α)$_{2-4}$. Unlike muscle nAChRs, functional neuronal channels have been expressed in *Xenopus* oocytes from some pairs of α and β subunits, and α_7 can form a homo-oligomeric channel (Boulter et al, 1986, 1987; Goldman et al., 1987; Schoepfer et al., 1988; Wada et al., 1988; Duvoisin et al., 1989). Genetic diversity in the subunit composition of the neuronal receptor subtypes probably underlies the functional and pharmacological diversities that have been observed (Mathie et al., 1987; Moss et al., 1989; Papke et al., 1989; Luetje et al., 1990; Mulle and Changeux, 1990; Luetje and Patrick, 1991). Although the homo-oligomer receptors formed by a α_7 in oocytes are inhibited by α-BGT (Couturier et al., 1990; Séguéla et al., 1992), the more traditional neuronal nAChRs are not inhibited by α-BGT, but some subtypes are inhibited by neuronal bungarotoxin (Ascher et al., 1979; Chiappinelli, 1983; Lipton et al., 1987; Boulter et al., 1987; Luetje et al., 1990; review: Loring and Zigmond, 1988).

A third functional category of receptors is the α-BGT binding proteins that are homologous to nAChRs (Carbonetto et al., 1978; Conti-Tronconi et al., 1985). These proteins are present in neuronal tissues, but they are distinct from the traditional neuronal nAChRs (Jacob and Berg, 1983; Clarke et al., 1985; Schulz et al., 1991). The function and complete composition of native neuronal α-BGT receptors is still unknown, but the homo-oligomeric channels formed by α_7 in oocytes are inhibited by α-BGT (Schoepfer et al., 1990; Couturier et al., 1990; Séguéla et al., 1992). Therefore, it is possible that high-affinity neuronal α-BGT receptors also could function as ion channels under some circumstances. For example, Listerud et al. (1991) showed that α-BGT sensitive receptors were formed in sympathetic neurons, but only after the usual composition of the nAChRs had been disrupted by deletion of α_3 by an antisense oligonucleotide.

Ionic Permeability Properties

Permeability Ratios

The nAChR provides an aqueous pore that selects against anions, but passes many nonelectrolytes and cations (Takeuchi and Takeuchi, 1960; Huang et al., 1978; Dwyer et al., 1980). The weak selectivity of the muscle nAChR has been extensively characterized by calculating permeability ratios from reversal potential measurements. The selectivity sequence for the alkali metals obtained from permeability ratio determinations is $Cs^+ > Rb^+ > K^+ > Na^+ > Li^+$ (Gage and Van Helden, 1979; Adams et al., 1980; Lewis and Stevens, 1983; Dani and Eisenman, 1987). For the alkaline earth metals, the sequence obtained from permeability ratio determinations is $Mg^{++} > Ca^{++} > Ba^{++} > Sr^{++}$ (Adams et al., 1980; Lewis and Stevens, 1983; Dani and Eisenman, 1987). For organic monovalent cations, the permeability ratio generally varies inversely with ionic size as if frictional drag decreases permeability (Dwyer et al., 1980; Adams et al., 1981; Sanchez et al., 1986). For some large cations, however, size is not the dominating factor. The hydrophobicity of the cation and interactions with specific residues within the pore can have major influences (Dwyer et al., 1980; Cohen et al., 1992).

Occupancy of the Pore

Cations bind within the pore and, therefore, do not move through the channel independently (Lewis, 1979; Dwyer et al., 1980; Adams et al., 1981; Sanchez et al., 1986). Various experiments showed that the muscle nAChR behaves as though there is only single occupancy by a permeant cation at a site (or sites) that is directly in the permeation pathway. The concentration dependence of the single-channel conductance levels off at higher concentrations as if a single-site is becoming saturated (Dani and Eisenman, 1987). The mole-fraction dependence of the single-channel conductance in mixtures of Na^+ and organic cations is a monotonic function (Sanchez et al., 1986). The monotonic fall in conductance theoretically indicates there is single occupancy of sites directly along the permeation pathway (see Levitt, 1986). The conductance of a multi-occupied channel would have an anomalous mole-fraction relationship: a maximum in the conductance (Neher, 1975). The strongest evidence about the number of binding sites along the permeation pathway is obtained from the concentration dependence of the permeability ratio (see Levitt, 1986). The permeability ratio of the muscle nAChR is independent of concentration over a wide range, again indicating that there is single occupancy (Dani, 1989a).

Open-Channel Blockers

Many organic cations block the nAChR channel (reviews: Changeux and Revah, 1987; Lester, 1992). Although other binding sites may exist (Papke and Oswald, 1989), voltage-dependent noncompetitive inhibitors often exert their primary cation by blocking the open pore. Neher and Steinbach (1978) provided the first single-channel records showing open-channel block of nAChRs. At low concentrations, the quaternary ammonium derivative of lidocaine, QX-222, caused single-channel openings to be shorter and to occur in bursts containing brief closures. The voltage dependence of the block indicated that QX-222 entered deeply into the pore. As is the case for QX-222, complete permeation by most open-channel blockers is prevented because they are too large to pass through the narrowest cross-section of the pore.

General Structure of the Pore

Structure of the Pore

Figure 1 schematically illustrates the general structure of the nAChR. The nAChR is composed of five polypeptide subunits arranged around the pore like staves of a barrel (Raftery, 1980; Cooper et al., 1991). Structural analysis of dense arrays of nAChRs indicated that the receptor is about 14 nm long and 8 nm wide (Brisson and Unwin, 1985; Toyoshima and Unwin, 1988, 1990; Mitra et al., 1989). The structural studies directly showed that the external entrance vestibule and opening are larger than the internal vestibule and opening. This finding had been suggested by the asymmetric action of open-channel blockers (del Castillo and Katz, 1957; Aguayo et al., 1981; Farley and Narahashi, 1983). The external opening of the pore is about 3 nm wide and runs at a constant diameter up to the membrane-spanning region. At the membrane, the pore tapers to a narrowest diameter of about 0.8–0.9 nm. The narrowest cross-section was determined by finding the largest permeant cation that could fit through the opening (Huang et al., 1978; Dwyer et al., 1980; Dani, 1989a; Cohen et al., 1992). The voltage dependence of small open-channel blockers suggested that the narrowest region is near the cytoplasmic end of the membrane spanning region (Neger and Steinbach, 1978; Sine and Steinbach, 1984). Streaming potential measurements directly showed that the narrowest cross-

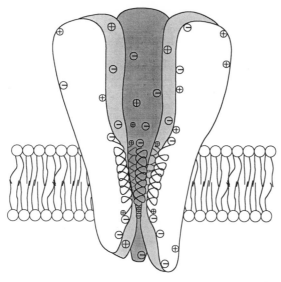

Figure 1. A schematic representation of the open nAChR channel. Three of the 5 subunits are shown is this cut-away view. The channel is drawn to scale with the opening to the outer vestibule at the top of the figure. M2 domains (dark) line the pore, and M1 domains (light) are exposed to the pore near the outer vestibule at the interstices of the M2 domains. Possible locations of charged amino acids are shown as circles containing pluses or minuses. The number and exact location of the charged residues are not intended to be percisely accurate. Detailed features such as glycosylation. phosphorylation, and connections to the cytoskeleton are not represented (taken from Dani, 1989b).

section is very short (Dani, 1989a). Probably one or two sets of amino acid side chains compose the narrowest region. Outward from the narrowest region, the membrane spanning regions may spread out, exposing a second set of transmembrane domains (see Figure 1).

The Subunits and Mutational Studies

All of the muscle subunits and many neuronal subunits for nAChRs have been cloned (reviews: Lindstrom et al., 1987; Claudio, 1989; Steinbach and Ifune, 1989; Luetje et al., 1990; Patrick et al., 1992). All of the subunits share sequence homology and have similar hydropathy profiles, including four hydrophobic domains (M1, M2, M3, M4) that have been predicted to be transmembrane. The proposed extracellular domain of the subunits has a highly conserved region containing two conserved cysteine residues separated by 13 amino acids. A disulfide bond between the two conserved cysteines may provide the structural integrity necessary for ligand binding in the extracellular domain of the receptors.

Much evidence indicates that the uncharged M2 domain, possibly as an α-helix, lines the transmembrane region of the pore (Figure 2;

```
                                          -1'  2'  6'  10'
                                           |   |   |   |
α   RI   PLYFVVNVIIPCLLFSFLTGLVFYLPT   DSG.EK   MTLSISVLLSLTVFLLVIV   ELIPSTSSAVPLIGK   YMLFT
β   RK   PLFYIVYTIIPCILISILAILVFYLPP   DAG.EK   MSLSISALLAVTVFLLLLA   DKVPETSLSVPIIIR   YLMFI
γ   RK   PLFYIINIIAPCVLISSLVVLVYFLPA   QAGGQK   CTLSISVLLAQTIFLFLIA   QKVPETSLNVPLIGK   YLIFV
δ   RK   PLFYVINFITPCVLISFLASLAFYLPA   ESG.EK   MSTAISVLLAQAVFLLLTS   QRLPETALAVPLIGK   YLMFI
         <───────M1──────────>        INNER    <──────M2──────>      OUTER            <─M3─
```

Figure 2. Partial amino acid sequences for *Torpedo californica* nAChR subunits. The locations of the transmembrane domains M1, M2, and part of M3 are indicated. The amino acids connecting the transmembrane regions are labelled as being in the inner or the outer vestibule of the channel. The M2 transmembrane domain is thought to be an α-helix, and therefore, only the amino acids on one side of the cylinder formed by the helix face into the pore. Several positions thought to be facing the pore are numbered with respect to the inner end of M2 (see Dani, 1989b).

review: Dani, 1989b). Noncompetitive antagonists that are thought to be open-channel blockers were able to label by photoaffinity the M2 domain, indicating that M2 is exposed within the pore (Giraudat et al., 1986; Hucho et al., 1986; Revah et al., 1990). Also, studies with chimeras between bovine and *Torpedo* δ-subunits showed that M2 and the amino acids connecting M2 and M3 influence ion permeation (Imoto et al., 1986).

The relationship between structure and function was further defined by comparing wild-type and mutated nAChRs expressed in *Xenopus* oocytes. Single-channel currents with mutated receptors showed that the charged residues bracketing M2 are important determinants of ion permeation (Imoto et al., 1988) as had been suggested by studies of ion permeation (Dani and Eisenman, 1987). Mutations in M2 also influenced the residence time of the open-channel blocker, QX-222. Polar residues (6') stabilized the charged end of QX-222 deeply in the pore while the hydrophobic end was stabilized by nonpolar residues (10') one turn outward on the M2 α-helix (Figure 2; Leonard et al., 1988; Charnet et al., 1990). Conductance and permeability ratios were most influenced by mutations at positions located at the inner end of M2 (Figure 2, columns -1' and 2'; Imoto et al., 1988; Charnet et al., 1990; Konno et al., 1991; Cohen et al., 1992; Villarroel and Sakmann, 1992). The voltage-dependence of open-channel blockers (Neher and Steinbach, 1978; Sine and Steinbach, 1984) indicated that this may be the narrowest cross-section, and streaming potentials indicated that this narrowest region is very short (Dani, 1989a). Together, the evidence suggests that turns -1' and 2' of the M2 α-helix form the main binding site at the narrow region directly in the permeation pathway.

Evidence also indicates that the M1 may be exposed within the pore (reviews: Dani, 1989b; Karlin, 1991). A noncompetitive antagonist, quinacrine, affinity labels the M1 domain when the channel is open (DiPaola et al., 1990). Because the narrowest region is very short, the M2 helices may spread apart as the pore diameter increases near the outer vestibule (Figure 1). Therefore, M1 domains may be exposed to the pore at the interstices of the M2 domains.

Models of Ion Permeation

Some permeation data can be described by very simple models, but they have severe limitations. The Goldman-Hodgkin-Katz (GHK) equations are often used, especially to calculate permeability ratios. This theory, however, has never completely described any ion channel because the GHK model does not allow binding of permeant ions within the pore. Much work has shown that the GHK equations do not accurately describe the nAChR (Lewis, 1979; Adams et al., 1981; Dwyer and Farley, 1984; Dani and Eisenman, 1987). A simple 2-barrier, 1-site (2B1S) reaction rate model describes the data better because it does account for the main ion binding site (Lewis, 1979; Lewis and Stevens, 1979; Marchais and Marty, 1979). Aside from inherent limitations of reaction rate theory for channels (review: Dani and Levitt, 1990), the 2B1S model still is not adequate. It cannot describe reversal potentials in mixtures of monovalent and divalent cations (Lewis and Stevens, 1979). Also, it cannot explain both conductance and current block (Adams et al., 1981; Dwyer and Farley, 1984). Finally, it cannot describe the concentration dependence of currents (Sanchez et al., 1986; Dani and Eisenman, 1987).

Although no easily tractable model can be expected to account for all the complexities of ion permeation, a simple extension of the 2B1S model has been very successful in describing the permeation data available for the muscle nAChR (Dani, 1986; Dani and Eisenman, 1987). A 2B1S reaction rate model describes the short narrow region that contains the single main binding site. The model also accounts for net negative charge bracketing the membrane spanning region. The negative charge attracts an atmosphere of cations into a geometry given by the size and shape of the entrance vestibules. The details of the model and the calculations are given elsewhere (Dani, 1986). The model has adequately described much data (Dani and Eisenman, 1987; Dani, 1988; Decker and Dani, 1990; Konno et al., 1991) and has quantitatively predicted the Ca^{2+} flux through nAChRs (Decker and Dani, 1990) before it was directly measured (Vernino and Dani, 1993). Using similar notions of the permeation process, Levitt (1991) more accurately described the vestibules as non-equilibrium diffusion regions, but the model was not used to describe sets of experimental data.

Functional Differences Between Muscle and Neuronal nAChRs

Because muscle and neuronal nAChRs are composed of different subunits, they have pharmacological and functional differences (Mathie et al., 1987; Moss et al., 1989; Papke et al., 1989; Luetje et al., 1990; Mulle and Changeux, 1990; Luetje and Patrick, 1991; reviews: Deneris et al.,

1991; Sargent, 1992; Patrick et al., 1992). Two of the most important functional differences are that neuronal nAChRs have a higher Ca^{2+} permeability and that they are directly modulated by external Ca^{2+} in a dose-dependent manner.

The higher Ca^{2+} permeability of traditional neuronal nAChRs is supported by strong evidence. Currents through neuronal nAChRs activate Ca^{2+}-dependent conductances, indicating that Ca^{2+} carries a significant amount of the current (Tokimasa and North, 1984; Fuchs and Morrow, 1992; Mulle et al, 1992a; Vernino et al., 1992). Permeability ratio (P_r) measurements also indicate that Ca^{2+} passes through neuronal nAChRs better than muscle nAChRs. The permeability ratio of Ca^{2+} to Na^+ or Cs^+ is much less than 1 for muscle nAChRs ($P_r \approx 0.2$; Adams et al., 1980; Decker and Dani, 1990; Vernino et al., 1992), but neuronal nAChRs have a higher relative Ca^{2+} permeability ($P_r \approx 1.5$; Fieber and Adams, 1991; Sands and Barish, 1991; Vernino et al., 1992). Even more surprising was the recent result obtained with homo-oligomeric α-BGT-sensitive nAChRs composed of α_7 subunits expressed in oocytes. These channels had a much greater relative Ca^{2+} permeability ($P_r \approx 20$; Séguéla et al., 1992), which is significantly greater than the P_r of 5 found for the NMDA subtype of glutamate receptors (Mayer and Westbrook, 1987). This high Ca^{2+} permeability for α_7 channels carries special significance because it suggests that α-BGT receptors in the central nervous system may be sites of Ca^{2+} influx (see Vijayaraghavan et al., 1992). Recently, a direct measure of the Ca^{2+} influx was made using a new approach that simultaneously measures membrane current and intracellular Ca^{2+}. Near the resting potential of a cell, Ca^{2+} carries 2.4% of the net current through muscle nAChRs and 5.3% of the net current through traditional neuronal nAChRs (Vernino and Dani, 1993). The results indicate that cholinergic synaptic activity can produce intracellular Ca^{2+} signals that have important consequences.

Ca^{2+} also directly modulates muscle and neuronal nAChRs, but in much different ways. ACh-induced currents through muscle nAChRs decrease as the extracellular Ca^{2+} concentration increases, which was explained by single-channel measurements (Decker and Dani, 1990). Ca^{2+} competes with Na^+ and K^+ for occupancy of the channel. Then, Ca^{2+} moves through the channel more slowly than the monovalent cations. Therefore, Ca^{2+} decreases the macroscopic currents through the muscle nAChRs by decreasing the underlying single-channel conductance. In contrast, currents through traditional neuronal nAChRs are enhanced by extracellular Ca^{2+} even though the single-channel conductance decreases (Vernino et al., 1992). Neuronal nAChRs respond to external Ca^{2+} in a dose-dependent manner (Mulle et al., 1992b; Vernino et al., 1992). The results suggest that external Ca^{2+} may directly modulate neuronal nAChRs. An intracellular Ca^{2+}-dependent

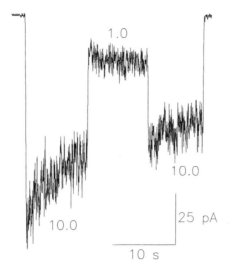

Figure 3. Modulation of neuronal nAChRs by rapid changes in external Ca^{2+}. Agonist-induced nAChR currents in an adrenal chromaffin cell rapidly and repeatedly respond to the extracellular concentration of Ca^{2+}. In this record, the cell was held at -70 mV and the concentration of applied agonist remained constant, but the concentration of external Ca^{2+} was rapidly changed between 10 mM and 1 mM as labeled next to the record. Rapid responses to changes in external Ca^{2+} were seen for as long as whole-cell seals lasted. The cell was internally perfused by a pipette containing mostly CsMethanesulfonate with 10 mM BAPTA, no ATP, and no enzymes. Therefore, Ca^{2+}-activated conductances were inhibited by BAPTA, and enzyme systems were disrupted (taken from Vernino et al., 1992).

enzyme cascade is a very unlikely mechanism for several reasons. The response to changes in external Ca^{2+} are very rapid and repeatable even for cells perfused with Ca^{2+} chelators and no enzymes (Figure 3; Vernino et al., 1992). Also, Ca^{2+} influx is strongly voltage dependent, but the modulation is not. Since Ca^{2+} binds to nAChRs (Chang and Neumann, 1976), that binding may directly modulate neuronal nAChRs.

Biological Roles of nAChRs

The primary role of many nAChRs is to provide a ligand-gated channel for cations to induce a depolarization that completes synaptic transmission. The roles of neuronal nAChRs may be much more varied, especially because they have both a presynaptic and postsynaptic location (Brown et al., 1984; McCormick and Prince, 1987). Depending on the nAChR subtypes expressed and post-translational modifications, an incredible spectrum of pharmacological and functional responses are possible. The high Ca^{2+} permeability of these receptors suggests that

56

nicotinic cholinergic synapses participate in processes of neuronal plasticity vital for development and learning.

Acknowledgements
The work in my laboratory has been supported by the Muscular Dystrophy Association, by the Whitaker Foundation, and by the National Institutes of Health.

References

Adams, DJ, Dwyer, TM, Hille, B (1980). The permeability of endplate channels to monovalent and divalent metal cations. J. Gen. Physiol. 75:493–510.

Adams DJ, Nonner W, Dwyer TM, Hille B (1981). Block of endplate channels by permeant cations in frog skeletal muscle. J. Gen. Physiol. 78:593–615.

Aguayo LG, Pazhenchevsky B, Daly JW, Albuquerque EX (1981). The ionic channel of the acetylcholine receptor: regulation by sites outside and inside the cell membrane which are sensitive to quaternary ligands. Molec. Pharm. 20:345–355.

Aquilonius S-M, Gillberg P-G, editors (1990). Cholinergic neurotransmission: functional and clinical aspects. Amsterdam: Elsevier.

Ascher P, Large WA, Rang HP (1979). Studies on the mechanism of action of acetylcholine antagonists on rat parasympathetic ganglion cells. J. Physiol. 295:139.

Boulter J, Connolly J, Deneris ES, Goldman D, Heinemann S, Patrick J (1987). Functional expression of two neuronal nicotinic acetylcholine receptors from cDNA clones identifies a gene family. Proc. Natl. Acad. Sci. 84:7763–7767.

Boulter J, Evans K, Goldman D, Martin G, Treco D, Heinemann S, Patrick J (1986). Isolation of a cDNA clone coding for a possible neural nicotinic acetylcholine receptor α-subunit. Nature 319:368–374.

Brown DA, Docherty RJ, Halliwell JV (1984). The action of cholinomimetic substances on impulse conduction in the habenulointerpeduncular pathway of the rat. J. Physiol. 353:101–109.

Carbonetto ST, Fambrough DM, Muller KJ (1978). Nonequivalence of α-bungarotoxin receptors and acetylcholine receptors in chick sympathetic neurons. Proc. Natl. Acad. Sci. USA 75:1016–1020.

Chang HW, Neumann E (1976). Dynamic properties of isolated acetylcholine receptor proteins: release of calcium ions caused by ACh binding. Proc. Natl. Acad. Sci. USA 73:3364–3368.

Changeux J-P, Revah, F (1987). The acetylcholine receptor molecule: allosteric sites and the ion channel. Trends Neurosci. 10:245–250.

Clarke PS, Schwartz RD, Paul SM, Pert CB, Pert A (1985). Nicotinic binding in rat brain: autoradiographic comparison of ^{3}H-acetylcholine, ^{3}H-nicotine, and ^{125}I-α-bungarotoxin. J Neurosci. 5:1307–1315.

Charnet P, Labarca C, Leonard RJ, Vogelaar NJ, Czyzyk L, Gouin A, Davidson N, Lester HA (1990). An open-channel blocker interacts with adjacent turns of α-helices in the nicotinic acetylcholine receptor. Neuron 4:87–95.

Chiappinelli VA (1983). Kappa-bungarotoxin: a probe for the neuronal nicotinic receptor in the avian ciliary ganglion. Brain. Res. 277:9–22.

Claudio T (1989). Molecular genetics of acetylcholine receptor-channels. In: Frontiers in Molecular Biology: Molecular Neurobiology. Glover DM, Hames BD, editors. Oxford: IRL Press, pp. 63–142.

Cohen BN, Labarca C, Davidson N, Lester HA (1992). Mutations in M2 alter the selectivity of the mouse nicotinic acetylcholine receptor for organic and for alkali metal cations. J. Gen. Physiol. 100:373–400.

Conti-Tronconi BA, Dunn SMJ, Barnard EA, Dolly O.J, Lai FA, Ray N, Raftery MA (1985). Brain and muscle nicotinic acetylcholine receptors are different but homologous proteins. Proc. Natl. Acad. Sci. USA 82:5208–5212.

Cooper E, Couturier S, Ballivet M (1991). Pentameric structure and subunit stoichiometry of a neuronal actylcholine receptor. Nature 350:235–238.

Couturier S, Bertrand D, Matter J-M, Hernandez M-C, Bertrand S, Millar N, Valera S, Barkas T, Ballivet M (1990). A neuronal nicotinic acetylcholine receptor subunit ($\alpha 7$) is developmentally regulated and forms a homo-oligomeric channel blocked by α-bungarotoxin. Neuron 5:847–856.

Dani JA (1986). Ion-channel entrances influence permeation. Net charge, size, shape, and binding considerations. Biophys. J. 49:607–618.

Dani JA (1988). Ionic permeability and the open channel structure of the nicotinic acetylcholine receptor. In: Transport Through Membranes: Carriers, Channels and Pumps. Pullman A, editor. Dordrecht, Holland: D. Reidel Publishing Co., pp. 297–319.

Dani JA (1989a). Open channel structure and ion binding sites of the nicotinic acetylcholine receptor channel. J, Neurosci. 9:882–890.

Dani JA (1989b). Site-directed mutagenesis and single-channel currents define the ionic channel of the nicotinic acetylcholine receptor. Trends Neurosci. 12:125–128.

Dani JA Eisenman G (1987). Monovalent and divalent cation permeation in acetylcholine receptor channels: ion transport related to structure. J. Gen. Physiol. 89:959–983.

Dani JA, Levitt DG (1990). Diffusion and kinetic approaches to describe permeation in ion channels. J. Theoret. Biol. 146:289–301.

Decker ER, Daini JA (1990). Calcium permeability of the nicotinic acetylcholine receptor: the single-channel calcium influx is significant. J. Neurosci. 10:3413–3420.

Del Castillo J, Katz B (1957). A study of curare action with an electrical micro method. Proc. Roy. Soc. Lond. B 147:339–356.

Deneris ES, Connolly J, Rogers SW, Duvoisin R (1991). Pharmacological and functional diversity of neuronal micotinic acetylcholine receptors. Trends Pharm. Sci. 12:34–40.

DiPaola M, Kao PN, Karlin A (1990). Mapping the alpha-subunit site photolabeled by the noncompetitive inhibitor 3H-quinacrine azide in the active state of the nicotinic acetylcholine receptor. J. Biol. Chem. 265:11017–11029.

Duvoisin RM, Deneris ES, Patrick J, Heinemann S (1989). The functional diversity of the neuronal nicotinic acetylcholine receptors is increased by a novel subunit: beta 4. Neuron 3:487–496.

Dwyer TM, Adams DJ, Hille B (1980). The permeability of the endplate channel to organic cations in frog muscle. J. Gen. Physiol. 75:469–492.

Dwyer TM, Farley JM (1984). Permeability properties of chick myotube acetylcholine-activated channels. Biop. J. 45:529–539.

Farley JM, Narahashi T (1983). Effects of drugs on acetylcholine-activated ionic channels of internally perfused chick myoballs. J. Physiol. 337:753–768.

Fieber LA, Adams DJ, (1991). Acetylcholine-evoked currents in cultured neurones dissociated from rat parasympathetic cardiac ganglia. J. Physiol. 434:215–237.

Fuchs PA, Murrow BW (1992). Cholinergic inhibition of short (outer) hair cells of the chick's cochlea. J. Neurosci. 12:800–809.

Gage P.W, Van Helden DF, (1979). Effects of permeant monovalent cations on end-plate channels. J. Physiol. (Lond.) 288:509–528.

Giraudat J, Dennis M, Heidmann T, Chang J-Y, Changeux J-P (1986). Structure of the high-affinity binding site for noncompetitive blockers of the acetylcholine receptor: serine-262 of the delta subunit is labeled by [^{3}H]chlorpromazine. Proc. Natl. Acad. Sci. USA 83:2719–2723.

Goldman D, Deneris E, Luyten W, Kochhar A, Patrick J, Heinemann S (1987). Members of a nicotinic acetylcholine receptor gene family are expressed in different regions of the mammalian central nervous system. Cell 48:965–973.

Huang LM, Catterall WA, Ehrenstein G (1978). Selectivity of cations and nonelectrolytes for acetylcholine-activated channels in cultured muscle cells. J. Gen. Physiol. 71:397–410.

Hucho F, Oberthur W, Lottspeich F (1986). The ion channel of the nicotinic acetylcholine receptor is formed by the homologous helices MII of the receptor subunits. FEBS Lett. 205:137–142

Imoto K, Busch C, Sakmann B, Mishina M, Konno T, Nakai J, Bujo H, Mori Y, Fukuda K, Numa S (1988). Rings of negatively charged amino acids determine the acetylcholine receptor channel conductance. Nature 335:645–648.

Imoto K, Methfessel C, Sakmann B, Mishina M, Mori Y, Konno T, Fukuda K, Kurasaki M, Bujo H, Fujita Y, Numa S (1986). Location of the δ-subunit region determining ion transport through the acetylcholine receptor channel. Nature 324:670–674.

58

Jacob MH, Berg DK (1983). The ultrastructural localization of δ-bungarotoxin binding sites in relation to synapses on chick ciliary ganglion neurons. J. Neurosci. 3:260–271.

Karlin A (1991). Explorations of the nicotinic acetylcholine receptor. Harvey. Lect. Ser. 85:71–107.

Katz B (1966). Nerve, Muscle and Synapse. New York: McGraw-Hill.

Konno T, Busch C, Von Kitzing E, Imoto K, Wang F, Nakai J, Mishina M, Numa S, Sakmann B (1991). Rings of anionic amino acids as structural determinants of ion selectivity in the acetylcholine receptor channel. Proc. R. Soc. Lond. Biol. 244:69–79.

Leonard RJ, Labaraca CG, Charnet P, Davidson N, Lester H (1988). Evidence that the M2 membrane-spanning region lines the ion channel pore of the nicotinic receptor. Science 242:1578–1581.

Lester HA, (1992). The permeation pathway of neurotransmitter-gated ion channels. Ann. Rev. Biophys. Biomol. Struct. 21:267–292.

Levitt DG (1986). Interpretation of biology ion channel flux data: reaction-rate versus continuum theory. Ann. Rev. Biophys. Biophys. Chem. 15:29–57.

Levitt DG (1991). General continuum theory for multiion channel. II. Application to acetylcholine channel. Biophys. J. 59:278–288.

Lewis CA (1979). Ion-concentration dependence of the reversal potential and the single channel conductance of ion channels at the frog neuromuscular junction. J. Physiol. 286:417–455.

Lewis CA, Stevens CF (1979). Mechanism of ion permeation through channels in a post-synaptic membrane. In: Membrane Transport Processes. Stevens CF, Tsien RW, editors. New York: Raven Press, pp. 89–103.

Lindstrom J, Schoepfer R, Whiting P (1987). Molecular studies of the neuronal nicotinic acetylcholine receptor family. Molec. Neurobiol. 1:281–337.

Lipton SA, Aizenman E, Loring RH (1987). Neural nicotinic acetylcholine responses in solitary mammalian retinal ganglion cells. Pflügers Arch. 410:37–43.

Listerud M, Brussaard AB, Devay P, Colman DR, Role LW (1991). Functional contribution of neuronal nAChR subunits revealed by antisense oligonucleotides. Science 254:1518–1521.

Loring RH, Zigmond RE (1988). Characterization of neuronal nicotinic receptors by snake venom neurotoxins. Trends Neurosci. 11:73–78.

Luetje CW, Patrick J (1991). Both alpha and beta subunits contribute to the agonist sensitivity of neuronal nicotinic acetylcholine receptors. J. Neruosci. 11:837–845.

Luetje CW, Wada K, Rogers S, Abramson SN, Tsuji K, Heinemann S, Patrick J (1990). Neurotoxins distinguish between different neuronal nicotinic acetylcholine receptor subunit combinations. J. Neruochem. 55:632–640.

Marchais D, Marty A (1979). Interaction of permeant ions with channels activated by acetylcholine in *Aplysia* neurones. J. Physiol. 297:9–45.

Mathie A, Cull-Candy SG, Colquhoun D (1987). Single-channel and whole-cell currents evoked by acetylcholine in dissociated sympathetic neurons of the rat. Proc. R. Soc. 232:239–248.

Mayer M, Westbrook G (1987). Permeation and block of N-methyl-D-aspartic acid receptor channels by divalent cations in mouse cultured central neurons. J. Physiol. 394:501–527.

McCormick DA, Prince DA (1987). Acetylcholine causes rapid nicotinic excitation in the medial habenular nucleus of guinea pig, in vitro. J. Neurosci. 7:742–752.

Mishina M, Takai T, Imoto K, Noda M, Takahashi T, Numa S, Methfessel C, Sakmann B (1986). Molecular distinction between fetal and adult forms of muscle acetylcholine receptor. Nature 321:406–411.

Moss BL, Schuetze SM, Role LW (1989). Functional properties and developmental regulation of nicotinic acetylcholine receptors on embryonic chicken sympathetic neurons. Neuron 3:597–607.

Mulle C, Changeux JP (1990). A novel type of nicotinic receptor in the rat central nervous system characterized by patch-clamp techniques. J. Neurosci. 10:169–175.

Mulle C, Choquet D, Korn H, Changeux JP (1992a). Calcium influx through nicotinic receptor in rat central neurons: its relevance to cellular regulation. Neuron 8:135–143.

Mulle C, Lena C, Changeux JP (1992b). Potentiation of nicotinic receptor response by external calcium in rat central neurons. Neuron 8:937–945.

Neher E (1975). Ionic specificity of the gramicidin channel and the thallous ion. Biochim. Biophys. Acta 401:540–544.

Neher E, Steinbach JH (1978). Local anesthetics transiently block currents through single acetylcholine-receptor channels. J. Physiol. (Lond.) 277:153–176.

Papke RL, Boulter J, Patrick J, Heinemann S (1989). Single-channel currents of rat neuronal nicotinic acetylcholine receptors expressed in Xenopus oocytes. Neuron 3:598–596.

Papke RL, Oswald RE (1989). Mechanisms of noncompetitive inhibition of acetylcholine-induced single-channel currents. J. Gen. Physiol. 93:785–811.

Patrick J, Seguéla P, Vernino S, Amador M, Luetje C, Dani JA (1993). Functional diversity of neuronal nicotinic acetylcholine receptors. Prog. Brain. Res. In press.

Raftery MA, Hunkapiller MW, Strader CD, Hood LE (1980). Acetylcholine receptor: complex of homologous subunits. Science 208:1454–1457.

Revah F, Galzi JL, Giraudat J, Haumont PY, Lederer F, Changeux JP (1990). The noncompetitive blocker 3H:chlorpromazine labels three amino acids of the acetylcholine receptor gamma subunit: implications for the alpha-helical organization of regions MII and for the structure on the ion channel. Proc. Natl. Acad. Sci. USA 87:4675–4679.

Sanchez JA, Dani JA, Siemen D, Hille B (1986). Slow permeation of organic cations in acetylcholine receptor channels. J. Gen. Physiol. 87:985–1001.

Sands SB, Barish ME (1991). Calcium permeability of neuronal nicotinic acetylcholine receptor channels in PC-12 cells. Brain Res. 560:38–42.

Sargent PB (1993). The diversity of neuronal nicotinic acetylcholine receptors. Ann. Rev. Neurosci. 16:403–43.

Schulz DW, Loring RH, Aizenman E, Zigmond RE (1991). Autoradiographic localization of putative nicotinic receptors in the rat brain using ^{125}I-neuronal bungarotoxin. J. Neurosci. 11:287–297.

Schoepfer R, Conroy W, Whiting P, Gore M, Lindstrom J (1990). Brain α-bungarotoxin binding protein cDNAs and mAbs reveal subtypes of this branch of the ligand-gated ion channel gen superfamily. Neuron 5:35–48.

Schoepfer R, Whiting P, Esch F, Blacher R, Shimasaki S, Lindstrom J (1988). cDAN clones coding for the structural subunit of a chicken brain nicotinic acetylcholine receptor. neuron 1:241–248.

Séguéla P, Wadiche J, Dineley-Miller K, Dani JA, Patrick JW (1993). Molecular cloning, functional expression and distribution of rat brain α7: a nicotinic cation channel highly permeable to calcium. J. Neurosci. 13:596–604.

Sine SM, Steinbach JH (1984). Activation of a nicotinic acetylcholine receptor. Biophys. J. 45:175–185.

Steinbach JH, Ifune C (1989). How many kinds of nicotinic acetylcholine receptor are there? Trends Neurosci. 12:3–6.

Steriade M, Biesold D, editors (1990). Brain Cholinergic Systems. Oxford: Oxford Univ. Press.

Takeuchi A, Takeuchi N (1960). On the permeability of end-plate membrane during the action of transmitter. J. Physiol. 154:52–67.

Tokimasa T, North RA (1984). Calcium entry through acetylcholine channels can activate potassium conductance in bullfrog sympathetic neurons. Brain Res. 295:364–367.

Vernino S, Amador M, Luetje CW, Patrick J, Dani JA (1992). Calcium modulation and high calcium permeability of neuronal nicotinic acetylcholine receptors. Neuron 8:127–135.

Vernino S, Dani JA (1993). Quantitative measurement of calcium flux through muscle and neuronal nicotinic acetylcholine receptors. J. Neurosci. Submitted.

Vijayaraghavan S, Pugh PC, Zhang Z, Rathouz MM, Berg DK (1992). Nicotinic receptors that bind α-bungarotoxin on neurons raise intracellular free calcium. Neuron 8:353–362.

Villarroel A, Sakmann B (1992). Threonine in the selectivity filter of the acetylcholine receptor channel. Biophys. J. 62:196–205.

Wada K, Ballivet M, Boulter J, Connolly J, Wada E, Deneris ES, Swanson LW, Heinemann S, Patrick J (1988). Functional expression of a new pharmacological subtype of brain nicotinic acetylcholine receptor. Science 240:330–334.

Nonselective Cation Channels: Pharmacology, Physiology and Biophysics
ed. by D. Siemen & J. Hescheler

AMPA-Type Glutamate Receptors – Nonselective Cation Channels Mediating Fast Excitatory Transmission in the CNS

Peter Jonas

Max-Planck-Institut für medizinische Forschung, Abteilung Zellphysiologie, D-69120 Heidelberg, FRG

Summary
In recent years, considerable progress in our understanding of the molecular events underlying excitatory synaptic transmission has been made. This progress was mainly achieved by technical advances, among them the patch-clamp technique in brain slices (Edwards et al., 1989), fast application of agonists (Franke et al., 1987), and cloning and functional expression of GluR channels of the nonNMDA type (e.g., Hollmann et al., 1989).

A suitable model for studying excitatory postsynaptic currents (EPSCs) in the brain slice with patch-clamp techniques is the mossy fiber synapse on CA3 pyramidal cells of rat hippocampus (MF-CA3 synapse). This synapse is located close to the cell soma and should provide almost ideal space-clamp conditions. A comparison of MF-CA3 EPSCs with the currents activated by fast application of glutamate on membrane patches isolated from CA3 cell somata suggests that the concentration of glutamate in the synaptic cleft during excitatory synaptic transmission is high (about 1 mM) and that the transmitter remains in the synaptic cleft only briefly (about 1 ms). It seems likely that desensitization influences the peak amplitude of the EPSC in several ways. Brief pulses of glutamate cause desensitization, from which the glutamate receptor channels recover only slowly, and micromolar ambient gluta-mate concentrations produce desensitization at equilibrium.

From the functional properties of recombinant GluR channels, in situ hybridization data, and patch-clamp experiments on different neuronal and nonneuronal cell types, a picture of the molecular identity of native channels emerges. In neurons of the hippocampus the pharmacological features of these channels were similar to recombinant channels assembled from subunits of the AMPA/kainate subtype which are strongly expressed in these cells. The native channels are characterized by outward rectification of the steady-state I-V and low Ca permeability, similar to recombinant channels containing the GluR-B subunit. This is consis-tent with the ubiquitous expression of this subunit in hippocampal neurones. In contrast, GluR channels from cerebellar glial cells, which uniquely in the central nervous system lack the expression of GluR-B subunits, show double rectification and high Ca permeability. The results suggest that the native functional nonNMDA glutamate receptor channels in the CNS are assembled form subunits of the AMPA/kainate subtype in a cell-specific way, with the functional properties of GluR channels in neurones being dominated by the GluR-B subunit.

Glutamate: The Principal Excitatory Transmitter in the CNS

Glutamate is the major excitatory neurotransmitter in the central nervous system (CNS). At the postsynaptic membrane, glutamate directly activates nonselective cation channels (glutamate receptor (GluR) channels). Almost every neurone in the CNS carries GluR channels.

According to pharmacological features, these were subdivided into one class which is specifically activated by N-Methyl-D-aspartate (NMDA), and a second class which is not activated by NMDA, but by quisqualate, S-α-Amino-3-hydroxy-5-methyl-4-isoxazolepropionic acid (AMPA), and kainate. Originally, it was thought that within the second (the so-called nonNMDA) class each agonist activates a different type of channel protein. However, evidence has since accumulated that AMPA, kainate, and glutamate activate similar conductance levels (Cull-Candy and Usowicz, 1987) and bind to common receptors in cultured neurones (Kiskin et al., 1990; Patneau and Mayer, 1991). Moreover, several of the cloned recombinant channels are activated by AMPA as well as kainate (e.g., Keinänen et al., 1990).

New Methods

The method of patch clamping in the brain slice (Edwards et al., 1989) seems to be the best available technique to study excitatory synaptic transmission at the molecular level. It has the advantage that native GluR channels of visually identified mature neurones of the CNS within an intact synaptic environment can be studied, and that enzymes, which eventually might alter the functional properties of these channels, are not necessary for exposing the neuronal membrane. Through the use of fast application of agonists to isolated membrane patches (Franke et al., 1987; Jonas and Sakmann, 1992b; Colquhoun et al., 1992), synaptic events can be compared with currents activated by glutamate pulses of defined length and concentration. Thereby it might be possible to gain a better understanding of which factors determine the time-course and the amplitude of the EPSC. On the other hand, a comparison of the functional properties of the native GluR channels with the recombinant channels (e.g., Hollmann et al., 1989; Keinänen et al., 1990) and with the expression pattern for different GluR subunits as revealed by in situ hybridization may cast light on the molecular identity and the subunit composition of the native channels.

In the following, I will summarize our attempts to trace the mechanisms underlying excitatory synaptic transmission from the whole synapse to the GluR channel molecule. I will focus on the hippocampal trisynaptic circuit between dentate gyrus granule cells, CA3, and CA1 pyramidal cells, which is of considerable interest for many neurophysiologists because long-term changes at the levels of its glutamatergic synapses have been demonstrated which presumably are related to the formation of short-term memory.

Gating of Native GluR Channels: Comparison of EPSCs and Currents Evoked by Fast Application

EPSCs at the MF-CA3 Synapse

Fast excitatory synaptic transmission in the CNS is predominantly mediated by GluR channels of the nonNMDA type. At some synapses, the nonNMDA component dominates (e.g., the mossy fiber CA3 cell (MF-CA3) synapse in the hippocampus; Monaghan et al., 1983; P. Jonas, unpublished). At other synapses nonNMDA and NMDA receptors equally contribute to the EPSC and seem to be colocalized (e.g., at excitatory synapses at interneurones of the visual cortex; Stern et al., 1992). A measurement of EPSC amplitude and kinetics requires sufficiently accurate voltage clamp conditions. Many excitatory synapses (e.g., synapses on CA1 pyramidal cells), however, are localized far away from the cell soma at the dendritic tree and therefore do not fulfill this requirement. Only a few excitatory synapses are located electrotonically close to the cell soma, among them the MF-CA3 synapse (Johnston and Brown, 93). A specific extracellular stimulation of a single presynaptic granule was achieved using a fine-tipped pipette and current pulses of low intensity (1 to 5 μA, 100 μs). Using this method, it was possible to evoke unitary MF-CA3 EPSCs, i.e., EPSCs elicited by the activity of a single presynaptic neurone (Jonas and Sakmann, 1992a). This avoids complications arising when EPSCs evoked at differenet electrotonic locations (presumably with different kinetics) are averaged into one compound EPSC.

The unitary MF-CA3 EPSCs show a rise time as low as 0.4 ms. The EPSCs with the fastest rise times, presumably under the best voltage clamp conditions, show apparent decay time constants of 3 to 6 ms (22°C; Figure 1A). However, simulations using morphology-based compartmental models of CA3 pyramidal cells suggest that the real decay at the postsynaptic membrane might be still faster, by more than 10% (Jonas et al., 1993). Taking this into account, the fastest "real" decay time constants at the MF-CA3 synapse at the postsynaptic site would be about 2.5 ms. The average apparent conductance change associated with a unitary event is 1 nS, and there is evidence that it is composed of eight quantal units (Jonas and Sakmann, 1992a; Jonas et al., 1993). As reported for nonNMDA components in other synapses, the MF-CA3 EPSC shows a linear peak curent-voltage relation (I-V; -100 to $+60$ mV) and is blocked by the potent and selective nonNMDA receptor antagonist 6-Cyano-7-nitroquinoxaline-2,3-dione (CNQX), with a half-maximal inhibitory concentration of 350 nM and a Hill coefficient close to one.

The measured kinetics of MF-CA3 synapse EPSCs is considerably faster than the nonNMDA GluR mediated EPSC components in CA1

64

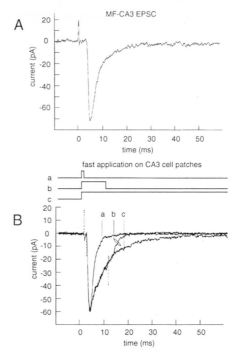

Figure 1. EPSCs in CA3 pyramidal cells as compared to macroscopic currents activated by glutamate pulses in outside-out membrane patches. A) Unitary MF-CA3 EPSC, stimulation of a single granule cell. B) Patch currents activated by a 1, 10, and 100 ms pulse of 1 mM glutamate. CA3 cell patch. Decay after removal of glutamate was considerably faster than desensitization. The decay time-course of the EPSC comes close to the decay kinetics of the patch current evoked by the 1 ms pulse. Membrane potential -70 mV in A, -50 mV in B. 5–10 single traces averaged. All recordings at room temperature. B) adapted from Colquhoun et al. (1992).

pyramidal cells (Hestrin et al., 1990a), presumably due to much worse space-clamp conditions in the latter cell type (Major, Jonas and Sakmann, unpublished). The fastest decay time constants at the MF-CA3 synapse are comparable to EPSCs in stellate interneurones of visual cortex (2.4 ms, 22°C; Stern et al., 1992). However, they are still slower than in spinal motoneurones (0.4 ms, 37°C; Finkel and Redman, 1983) and in cerebellar granule cells (1.3 ms, 23°C; Silver et al., 1992). It is unclear if this is due to differences in release properties of the presynaptic terminals, functional properties of the postsynaptic GluR channels, or voltage-clamp conditions.

In order to find out which parameters influence the kinetics and the amplitude of an EPSC, we attempted to mimic the processes during synaptic transmission using fast application of glutamate. Solution changes as fast as synaptic release are not possible on whole cells, but only on excised outside-out membrane patches (Franke et al., 1987;

Jonas and Sakmann, 1992b; Colquhoun et al., 1992). The critical, and still unproven, assumption for a comparison between EPSCs and currents activated in somatic patches is, however, that subsynaptic and extrasynaptic GluR channels have identical functional properties.

Long Pulses of Glutamate Applied to Isolated Membrane Patches

The method of fast application was originally developed by Dudel and coworkers (Franke et al., 1987). Several modifications are now available. We used a double-barrelled application pipette which was continuously perfused with control and test solution, respectively. The recording pipette with the outside-out patch was positioned in front of the application pipette. Using a piezo element, the application pipette was moved for a defined time interval. The 20–80% exchange time measured with an open recording pipette (without membrane patch) was about 100 μs, and the exchange time at an intact membrane patch, as measured by a switch between solutions of different K concentration during the opening of a Ca-activated K channel of large conductance, was about 200 μs (Colquhoun et al., 1992). With this system it was possible to apply glutamate pulses of arbitrary concentration and length (1 to 300 ms) and to thereby mimic the events at excitatory synapses.

Fast application of glutamate at a concentration of $>30 \mu$M activated macroscopic currents in almost every membrane patch isolated from the somata of hippocampal neurones in slices. The external solutions contained 1 mM Mg and no glycine; under these conditions, the glutamate-activated current seemed to be entirely mediated by nonNMDA receptors, since it was blocked by low concentrations of CNQX (see below). The size of the peak current with a 5 MΩ patch pipette varied between 10 and 500 pA at a membrane potential of -50 mV. The single-channel conductance estimated from nonstationary fluctuation analysis was 9 pS (P. Jonas, unpublished), suggesting that up to 1000 channels are present in one patch. The current evoked by long (100–300 ms) pulses of glutamate showed a fast rise and an exponential decay within tens of milliseconds (Jonas and Sakmann, 1992b; Colquhoun et al., 1992). The current decay in the maintained presence of the agonist, denoted as desensitization, is a property of all ligand-gated ion channels, e.g., also acetylcholine or GABA$_A$ receptor channels. Desensitization of the native nonNMDA GluR channels, however, develops faster than that of the other types of receptors. It seems fast enough to suspect that it might even determine the decay of an EPSC.

Glutamate concentrations of about 1 mM were necessary to produce near-maximum peak current responses; the dose-response curve for the peak current activated by long glutamate pulses in membrane patches from CA3 pyramidal cells showed a half-maximal activation with a concentration of 340 μM and a Hill coefficient of 1.3. This suggests that possibly two glutamate molecules bind to one channel (Jonas and Sakmann, 1992b). The rise time of the currents activated by long (100–300 ms) pulses of glutamate showed a pronounced concentration dependence, decreasing from values of about 3 ms at 100 μM glutamate to values as low as 200 μs with 3 mM glutamate. In contrast, the desensitization time constant showed a relatively weak concentration dependence, with time constants between 8 and 12 ms. Moreover, the current did not desensitize completely, the size of the nondesensitizing component being a few percent of the peak current (Jonas and Sakmann, 1992b). This nondesensitizing component appears to correspond to the current activated by slow bath application of glutamate to whole cells described in several previous studies.

The data obtained with long pulses of glutamate put constraints on the time-course of the glutamate concentration in the synaptic cleft. To mimic the rise time of the EPSC, the concentration of glutamate at its receptor during synaptic transmission must be relatively high (on the order of 1 mM). Even for the highest glutamate concentrations, the decay of the MF-CA3 EPSC was significantly faster than the desensitization. This suggests that desensitization alone cannot be responsible for the decay phase of the EPSC and that the presynaptic bouton might release a glutamate pulse of brief duration. In agreement with this finding, it has been observed that pharmacological inhibition of desensitization does not change the decay time-course of EPSCs (Mayer and Vyklicky, 1989).

Mimicking Synaptic Currents with Brief Glutamate Pulses

To experimentally address the question of how long the transmitter remains in the synaptic cleft after it has been released, we applied brief pulses of glutamate to isolated patches. Figure 1B shows current responses of an outside-out patch isolated from a hippocampal CA3 cell to 1 mM glutamate pulses of 1-, 10-, and 100-ms length. The average decay time constant (offset time constant) of the current activated by 1-ms pulses was 2.5 ms, and the desensitization time constant during the 100-ms pulse was 11.3 ms. The difference in the two time constants suggests that they were determined by different molecular steps: closing (or bursting) of channels after removal of the agonist and desensitization in the maintained presence of the agonist, respectively. The current evoked by 10-ms pulses initially declines slowly (with the desensitization time constant) and then more rapidly (with the offset time constant).

The apparent decay time constant of the MF-CA3 EPSC was between 3 and 6 ms, and the corresponding decay directly at the postsynaptic site may be even faster than this (see above). Therefore, the decay time constant of the EPSC comes close to the offset time constant of currents activated by brief pulses of glutamate in membrane patches. It is therefore likely that the glutamate pulse during excitatory synaptic transmission is about 1 ms long and that the decay phase of the EPSC reflects the distribution of GluR channel lifetimes after removal of transmitter from the synaptic cleft. Consistent with this hypothesis, several further similarities of EPSCs (Hestrin et al., 1990a; Keller et al., 1991; P. Jonas, unpublished) and patch currents evoked by brief glutamate pulses (Colquhoun et al., 1992) have been observed: 1) the I-V relation of the peak current EPSC was linear, similar to the I-V relation of the current activated by brief glutamate pulses in patches; 2) The decay time constant of the EPSC shows only a weak voltage dependence, similar to the offset time constant, and 3) The CNQX concentrations inhibiting EPSC and currents in membrane patches were almost identical. The half-maximal inhibition of currents activated by 1 ms pulses of 1 mM glutamate in CA3 cell patches was reached with a concentration of 117 nM, the Hill coefficient was 1.2.

Besides hippocampal neurones, few data of EPSCs and currents activated by fast application obtained from the same cell type are available. At stellate interneurones of visual cortex, brief glutamate pulses can mimic the time-course of nonNMDA as well as NMDA components of the EPSC (P. Stern and B. Sakmann, unpublished). As in hippocampus, desensitization does not seem to play a role in the time-course of the EPSC in these cells, implying that the synaptic glutamate pulse is brief. However, there is probably no unique answer to the question of which functional importance desensitization has for EPSC kinetics. In cultured chick spinal neurones (Trussell and Fischbach, 1989), motoneurones (Smith et al., 1991), and neurones from cochlear nucleus (Raman and Trussell, 1992), desensitization is faster than in hippocampus and may eventually curtail excitatory synaptic events. Differences in GluR subunit expression pattern may be responsible for the variability in desensitization kinetics between different cell types (see below).

If glutamate remains in the synaptic cleft only for a short time, as our data suggest, the question arises of which factors promote its elimination. No high-turnover enzyme analogous to the acetylcholine esterase at the neuromuscular junction exists at glutamatergic synapses. High-affinity glutamate uptake mechanisms are also unlikely to be involved because blockers of these mechanisms do not influence the decay kinetics of EPSCs in CA1 pyramidal neurones (Hestrin et al., 1990b). At the moment, the most likely explanation is that elimination of glutamate from the cleft occurs by diffusion. In fact, it has been shown that free diffusion can lower the transmitter concentration in the synaptic cleft extremely rapidly, within about 1 ms (Eccles and Jaeger, 1958).

Desensitization of GluR Channels – A Factor Setting the Amplitude of EPSCs?

Although it is unlikely that desensitization is a major determinant of the decay kinetics of the EPSC at hippocampal MF-CA3 synapses, the question arises of whether desensitization could influence the peak amplitude of the synaptic event, thus implying that desensitization could occur at high firing rates. Double-pulse experiments with two brief pulses separated by defined intervals (Figure 2A) show that, during the first pulse, about 50% of the GluR channels became desensitized. Recovery from desensitization was monoexponential in membrane patches from CA3 and CA1 pyramidal cells with time constants of 48 and 58 ms, respectively. Recovery from desensitization on membrane patches of granule cells of dentate gyrus was slower; it was biexponential with time constants of 33 and 450 ms (Figure 2B; Colquhoun et al., 1992). Interestingly, the frequency dependence of

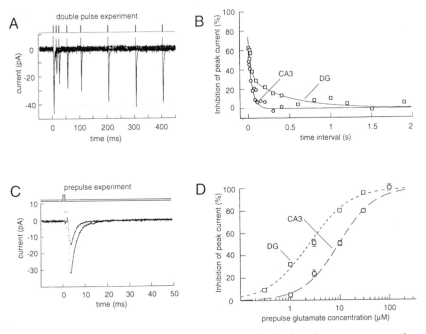

Figure 2. Recovery from desensitization and desensitization by low glutamate concentrations in the equilibrium. A) Current responses evoked with two 1 ms pulses of 1 mM glutamate, separated by intervals of different length. Current traces for different recovery intervals were superimposed. B) Recovery kinetics, data pooled from several experiments. C) Desensitization by low glutamate concentration in the equilibrium. Control trace and trace with a 15 s prepulse of 3 μM glutamate were superimposed. D) Dose-reponse curves for equilibrium desensitization. Squares: granule cells of dentate gyrus; circles: CA3 pyramidal cells. A) and C) from granule cell patches. Membrane potential -50 mV. Adapted from Colquhoun et al. (1992).

synaptic potentials shows a similar cell-specific difference: With increasing stimulus rate, depression occurs in granule cells, but potentiation is observed in CA3 and CA1 pyramidal cells (Alger and Teyler, 1976). Therefore, it appears likely that, at high firing frequencies of presynaptic neurones, depression can occur due to desensitization of postsynaptic receptors. Recovery from desensitization in neurones of the trisynaptic circuit was slower as compared to spinal neurones (Trussell and Fischbach, 1989), which provides further evidence for cell-specific differences of functional properties of GluR channels. A wide time range of recovery from desensitization has been reported for GluR channels at the neuromuscular synapse of the locust (Dudel et al., 1990), suggesting, at least for this preparation, a microheterogeneity of GluR channels with respect to recovery from desensitization.

The concentration of glutamate in the cerebrospinal fluid is about 1 μM (Lerma et al., 1986). Therefore, the question arises if low concentrations of this transmitter can cause significant desensitization. When a 15 s prepulse of low glutamate concentration, e.g., 3 μM, preceded the brief pulse of 1 mM glutamate, the current evoked by the latter pulse was considerably smaller as compared to control conditions (Figure 2C). A half-maximal inhibition could be observed with concentrations of 2.4, 9.6, and 4.2 μM for granule cells, CA3 pyramidal cells, and CA1 pyramidal cells, respectively (Figure 2D; Colquhoun et al., 1992). These data suggest that, under physiological conditions in vivo, a considerable fraction of the GluR channels would be in the desensitized state and that ambient glutamate might be a factor setting the efficacy of synaptic transmission. It has been shown that the glutamate concentration rises almost tenfold during ischemia (Benveniste et al., 1984), thus implying that desensitization may also play a (probably protective) role under pathophysiological conditions.

In conclusion, the data suggest that for hippocampal MF-CA3 synapses the EPSC decay is determined by channel gating after removal of transmitter from the synaptic cleft, implying that the glutamate pulse is short. Desensitization may play an important role in setting the peak amplitude of the EPSCs and thereby influence synaptic efficacy.

Pharmacology, Rectification, and Selectivity: Comparison of Cloned and Native Channels

Molecular Biology of NonNMDA-Type GluR Channels

Since the cloning and the functional expression of the first GluR channel subunit of the nonNMDA class (Hollmann et al., 1989) several new subunits have been identified. On the basis of structure homology, these can be divided into three subfamilies: 1) the GluR-A to -D (-1 to

-4) subfamily, also referred to as AMPA/kainate subfamily; 2) the GluR-5 to -7 subfamily, and 3) the KA-1, -2 subfamily (for review see, e.g., Sommer and Seeburg, 1992). The increasing knowledge about the set of cloned GluR subunits provides a chance to identify native channels on the molecular level, and to thereby elucidate the elementary events underlying synaptic transmission.

Channels assembled from subunits of different subfamilies differ in their agonist profile when expressed in mammalian cell lines or *Xenopus* oocytes. Channels assembled from the GluR-A to -D subfamily are preferentially activated by AMPA, but also by glutamate and kainate (Hollmann et al., 1989; Boulter et al., 1990; Keinänen et al., 1990). The responses to AMPA and glutamate show rapid desensitization, whereas the responses to kainate do not desensitize. Within the GluR-A to -D subfamily, recombinant homomeric channels differ in their rectification properties and their divalent permeability. Channels formed by GluR-A, -C, or -D subunits show a double rectification and a high Ca permeability. Homomeric channels assembled from GluR-B subunits, however, show outward rectification and low Ca permeability. These properties of GluR-B channels have been traced to the presence of an arginine residue in the second transmembrane segment (Verdoorn et al., 1991; Hume et al., 1991; Burnashev et al., 1992b). Quantitative differences in desensitization kinetics between GluR-A and -B homomeric channels have also been observed (Verdoorn et al., 1991).

Several lines of evidence suggest that, in expression systems, heterooligomeric channels are formed between different subunits of the GluR-A to -D subfamily (Nakanishi et al., 1990; Verdoorn et al., 1991). Most importantly, the expression rate in host cells is much higher when different subunits are coinjected. In heteromeric combinations the rectification and permeability of the steady-state current are dominated by the GluR-B subunit (Nakanishi et al., 1990; Hollmann et al., 1991; Burnashev et al., 1992b). The desensitization kinetics and the I-V of the peak current, however, derive properties from both subunit partners: The combination GluR-A/B shows monoexponential desensitization, with a time constant of 17 ms for 300 μM glutamate, just between the time constants for the respective homomeric channels (7 ms and 36 ms, respectively; Verdoorn et al., 1991), and shows an almost linear I-V for the peak current (which appears to be intermediate between outward and double rectification).

In contrast, channels from the GluR-5 to -7 subfamily are preferentially activated by kainate and show desensitizing responses upon application of this agonist (Egebjerg et al., 1991; Sommer et al., 1992). Subunits of the KA-1, -2 subfamily bind kainate with high affinity, but do not form any functional homomeric channels when expressed in host cells (Werner et al., 1991; Herb et al., 1992). However, there is evidence that the subunits of the KA-1, -2 subfamily form heteromeric channels with GluR-5 to -7 subunits, thereby modulating their functional properties (Herb et al., 1992).

Expression Pattern of GluR Channel Subunits in the CNS

In situ hybridization with subunit-specific oligonucleotides demonstrated that the GluR-A, -B, -C, and -D subunits are expressed almost ubiquitously in the CNS (Keinänen et al., 1990). In the hippocampus, CA3 pyramidal cells express GluR-A, -B, and -C, whereas CA1 cells express all four subunits. In the cerebellum, granule cells express GluR-B and GluR-D, Purkinje cells express GluR-A, GluR-B, GluR-C, and, rather uniquely, Bergmann glia cells express GluR-A and GluR-D, but not GluR-B.

The expression pattern of the subunits of the GluR-5 to -7 subfamily in the CNS is more restricted, and they are only expressed weakly in hippocampus: GluR-5 is not expressed, GluR-7 is weakly expressed in the dentate gyrus, and GluR-6 is expressed in CA3 region and denate gyrus (see references in Sommer and Seeburg, 1992). These GluR subunits appear to be predominantly expressed in peripheral nervous system, e.g., in dorsal root ganglia cells. KA-1 also shows restricted expression pattern, and KA-2 ubiquitous expression in the CNS (Herb et al., 1992).

In situ hybridization, although a very potent technique for qualitatively detecting the RNA encoding a particular subunit, cannot predict the functional properties of native GluR receptors for several reasons.

1) Subunits may be sorted to peripheral regions of the cell, e.g., to axons or dendrites, and may therefore not appear in the whole-cell current;
2) combinations with other still unknown subunits may cause unpredictable features of native channels;
3) posttranslational modifications may change the functional properties of the channels;
4) the half-lifetime of different subunits in the membrane may be different, and subunits with long half-lifetime may dominate.

The value of in situ hybridization is that it can 'guide the patch pipette' to cell types which show characteristic differences in their GluR subunit expression pattern. Expression pattern and characteristic functional properties together may then allow a molecular identification of native channels.

Pharmacology, Rectification, and Ca Permeability of Native Neuronal GluR Channels in Hippocampus

Rectification and permeability of nonNMDA GluR channels were characterized in detail using cultured neurones (Mayer and Westbrook, 1984; 1987). However, it turned out that, at least in some cases, there was a heterogeneity of functional properties between different cells. For instance, Iino and coworkers (1990) found outward rectification and low Ca permeability of kainate responses in the majority of cultured hippocampal neurones, but double rectification and high Ca permeability in a subset of cells. Although this was the first demonstration of Ca-permeable nonNMDA receptors, the interpretation of these data was not straightforward because the identity and the expression pattern of these cells were not known. We therefore studied the properties of GluR channels in visually identified neurones in slices from mature animals using patch-clamp and fast application techniques.

The native GluR channels in membrane patches obtained from the somata of the three principal neurones of the hippocampal circuit were preferentially activated by AMPA > glutamate > kainate. AMPA and glutamate responses were rapidly desensitizing, with little difference between AMPA and glutamate, whereas the kainate-activated current did not desensitize (Jonas and Sakmann, 1992b). With Na-rich solution on the outer side of the membrane, the I-V of the peak current activated by fast application of AMPA and glutamate was almost linear. In contrast, AMPA-, kainate-, and glutamate-activated steady-state currents show a pronounced outward rectification (as exemplified for a CA3 cell patch in Figure 3A). The reason for the different rectification of peak and steady-state current might be that one of the reaction steps leading to the desensitized state(s) of the channel is voltage dependent. The reversal potential with Na on the outer and Cs ions on the inner side of the membrane was close to 0, the relative Na permeability being $P_{Na}/P_{Cs} = 0.9$. When external Na ions were replaced by Ca ions, the reversal potential was shifted to negative values (Figure 3B), the relative Ca permeability being <0.05. Similar P_{Ca}/P_{Cs} values were obtained for the three principal neurones in the hippocampal trisynaptic circuit (Jonas and Sakmann, 1992b; Colquhoun et al., 1992).

The pharmacological features of native GluR channels in hippocampal neurones of the slice were similar to the cloned channels of the GluR-A to -D subfamily. The rectification and permeability were almost indistinguishable from a GluR-B homooligomer or a GluR-B containing heterooligomer (Verdoorn et al., 1991; Burnashev et al., 1992b). A current component with properties of the recombinant homomeric GluR-A, -C, or -D channels was not found, although the corresponding subunits are heavily expressed in these cells. Therefore, it is more likely that the native channels are heterooligomers than that

GluR-B cell type: CA3 cell (hippocampus)   non GluR-B cell type: fusiform glial cell (cerebellum)

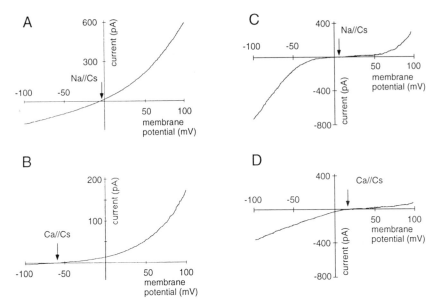

Figure 3. I-V of kainate-activated current in the equilibrium: Outward rectification and low calcium permeability in neurones of hippocampus (A, B), double rectification and high Ca permeability in fusiform glial cells of cerebellum (C, D). 300 μM kainate, outer solution either Ringer solution (135/140 mM NaCl) or high-Ca solution (100/110 mM CaCl$_2$). Response to a 2 s voltage ramp. Data in A, B were from CA3 pyramidal cell patches, data in C, D from a glial cell in the whole cell configuration. Arrows indicate reversal potentials. Adapted from Jonas and Sakmann (1992b) and Burnashev et al. (1992a).

they are a mosaic of different homooligomeric isoforms. Further evidence that the native channels are heterooligomers comes from their desensitization time constant, which is intermediate between GluR-A and GluR-B, and similar to a GluR-A/B combination. Therefore, it seems likely that native GluR channels in CA3 pyramidal cells are heterooligomeric channels containing a GluR-B subunit; possibly, they are GluR-A/B/C heteromers.

If all hippocampal native GluRs are similar to recombinant GluR channels of the GluR-A to -D subfamily, the question arises if whether the subunits of the GluR-5 to -7 class have any functional importance for the formation of native GluR channels. Desensitizing kainate responses have been described in the dorsal root ganglion cells (Huettner, 1990), and the half-maximal activating kainate concentration is much lower as compared to hippocampal neurones (15 as compared to 400 μM). It is likely that the functional importance of these subunits is mostly confined to the peripheral nervous system, and that their expression rate in hippocampus is too low to provide a significant contribution to the macroscopic current.

Cell-Specific Differences in Rectification and Ca Permeability: A Key for the Molecular Identification of the Native Channels

One of the key experiments elucidating the role of the GluR-B subunit for the functional properties of native GluR channels was performed using a cell type which, uniquely in the CNS, does **not** express GluR-B, but only GluR-A and -D: the Bergmann glia in the cerebellum (Keinänen et al., 1990; Burnashev et al., 1992a). A culture system for this glia subtype was developed (adapted from McCarthy and de Vellis, 1980), and it was verified by in situ hybridization experiments that, under culture conditions, the glial cells retain their GluR subunit expression pattern. With Na ions as predominant cations on the outer side of the membrane, the I-V relation showed a characteristic double rectification and a reversal potential close to 0 (Figure 3C). With Ca-rich solution on the outer side of the membrane the reversal potential was $+13$ mV (Figure 3D), the resulting permeability ratio P_{Ca}/P_{Cs} being 1.4 (Burnashev et al., 1992a). The functional properties of the glial GluR channels are therefore very similar to recombinant channels assembled from GluR-A, -C, or -D subunits. The doubly rectifying I-V relation and the influx of Ca ions through kainate-activated channels has also been reported in the slice preparation (Müller et al., 1992). Parallel fiber synapses on Purkinje cells are ensheathed by the Bergmann glial cells and it is therefore conceivable that, during activation of these synapses, glutamate spills over to the glial GluRs. By activation of these receptors and the influx of Ca ions, intracellular signal cascades in the glial cell could be triggered.

The striking difference between the functional properties of native GluRs in the hippocampal neurones and the cerebellar glial cells and the close correlation with the differential expression of the GluR-B subunit provides compelling evidence that the native GluR channels of neurones and glial cells are assembled from AMPA/kainate receptor subunits in a cell-specific way, and that the presence or absence of the GluR-B subunit determines their most important functional properties.

Conclusion

Combining the method of patch clamp in the brain slice, fast application, and molecular biological techniques, we can trace some properties of the EPSC to the molecular level. At the moment, however, neither the cascade of events mediating release in the presynaptic bouton nor the exact stoichiometry and the distribution of the native channels in the postsynaptic membrane are sufficiently understood. New technical developments will be necessary to address these points experimentally.

Acknowledgements
I would like to thank Drs. N. Burnashev, B. Sakmann, and N. Spruston for critically reading the manuscript, and Mrs. Dücker for typing the manuscript.

References

Alger BE, Teyler TJ (1976). Long-term and short-term plasticity in the CA1, CA3, and dentate regions of the rat hippocampal slice. Brain Res. 110:463–480.

Benveniste H, Drejer J, Schousboe A, Diemer NH (1984). Elevation of the extracellular concentrations of glutamate and aspartate in rat hippocampus during transient cerebral ischemia monitored by intracerebral microdialysis. J. Neurochemistry 43:1369–1374.

Boulter J, Hollmann M, O'Shea-Greenfield A, Hartley M, Deneris E, Maron C, Heinemann S (1990). Molecular cloning and functional expression of glutamate receptor subunit genes. Science 249:1033–1037.

Burnashev N, Khodorova A, Jonas P, Helm PJ, Wisden W, Monyer H, Seeburg PH, Sakmann B (1992a). Calcium-permeable AMPA-kainate receptors in fusiform cerebellar glial cells. Science 256:1566–1570.

Burnashev N, Monyer H, Seeburg PH, Sakmann B (1992b). Divalent ion permeability of AMPA receptor channels is dominated by the edited form of a single subunit. Neuron 8:189–198.

Colquhoun D, Jonas P, Sakmann B (1992). Action of brief pulses of glutamate on AMPA/ kainate receptors in patches from different neurones of rat hippocampal slices. J. Physiol. 458:261–287.

Cull-Candy SG, Usowicz MM (1987). Multiple-conductance channels activated by excitatory amino acids in cerebellar neurons. Nature 325:525–528.

Dudel J, Franke C, Hatt H, Ramsey RL, Usherwood PNR (1990). Glutamatergic channels in locust muscle show a wide time range of desensitization and resensitization characteristics. Neurosci. Lett 114:207–212.

Eccles JC, Jaeger JC (1958). The relationship between the mode of operation and the dimensions of the junctional regions at synapses and motor end-organs. Proc. Roy. Soc. B 148:38–56.

Edwards FA, Konnerth A, Sakmann B, Takahashi T (1989). A thin slice preparation for patch clamp recording from neurones of the mammalian central nervous system. Pflügers Arch. 414:600–612.

Egebjerg J, Bettler B, Hermans-Borgmeyer I, Heinemann S (1991). Cloning of a cDNA for a glutamate receptor subunit activated by kainate but not AMPA. Nature 351:745–748.

Finkel AS, Redman SJ (1983). The synaptic current evoked in cat spinal motoneurones by impulses in single group 1a axons. J. Physiol. 342:615–632.

Franke C, Hatt H, Dudel J (1987). Liquid filament switch for ultra-fast exchanges of solutions at excised patches of synaptic membrane of crayfish muscle. Neurosci. Lett. 77:199–204.

Herb A, Burnashev N, Werner P, Sakmann B, Wisden W, Seeburg PH (1992). The KA-2 subunit of excitatory amino acid receptors shows widespread expression in brain and forms ion channels with distantly related subunits. Neuron 8:775–785.

Hestrin S, Nicoll RA, Perkel DJ, Sah P (1990a). Analysis of excitatory synaptic action in pyramidal cells using whole-cell recording from rat hippocampal slices. J. Physiol. 422:203–225.

Hestrin S, Sah P, Nicoll RA (1990b). Mechanisms generating the time course of dual component excitatory synaptic currents recorded in hippocampal slices. Neuron 5:247–253.

Hollmann M, O'Shea-Greenfield A, Rogers SW, Heinemann S (1989). Cloning by functional expression of a member of the glutamate receptor family. Nature 342:643–648.

Hollmann M, Hartley M, Heinemann S (1991). Ca^{2+} permeability of KA-AMPA-gated glutamate receptor channels depends on subunit composition. Science 252:851–853.

Huettner JE (1990). Glutamate receptor channels in rat DRG neurons: activation by kainate and quisqualate and blockade of desensitization by Con A. Neuron 5:255–266.

Hume RI, Dingledine R, Heinemann SF (1991). Identification of a site in glutamate receptor subunits that controls calcium permeability. Science 253:1028–1031.

Iino M, Ozawa S, Tsuzuki K (1990). Permeation of calcium through excitatory amino acid receptor channels in cultured rat hippocampal neurones. J. Physiol. 424:151–165.

Johnston D, Brown TH (1983). Interpretation of voltage-clamp measurements in hippocampal neurons. J. Neurophysiol. 50:464-486.

Jonas P, Sakmann B (1992a). Unitary stimulus-evoked excitatory postsynaptic currents in CA3 pyramidal cells of rat hippocampal slices as resolved by patch clamp techniques. J. Physiol. 446:515P.

Jonas P, Sakmann B (1992b). Glutamate receptor channels in isolated patches from CA1 and CA3 pyramidal cells of rat hippocampal slices. J. Physiol. 455:143-171.

Jonas P, Major G, Sakmann B (1993). Quantal analysis of unitary EPSCs at the mosssy fibre synapse on CA3 pyramidal cells of rat hippocampus. J. Physiol. In press.

Keinänen K, Wisden W, Sommer B, Werner P, Herb A, Verdoorn TA, Sakmann B, Seeburg PH (1990). A family of AMPA-selective glutamate receptors. Science 249:556-560.

Keller BU, Konnerth A, Yaari Y (1991). Patch clamp analysis of excitatory synaptic currents in granule cells of rat hippocampus. J. Physiol. 435:275-293.

Kiskin NI, Krishtal OA, Tsyndrenko AY (1990). Cross-desensitization reveals pharmacological specificity of excitatory amino acid receptors in isolated hippocampal neurons. European J. Neurosci. 2:461-470.

Lerma J, Herranz AS, Herreras O, Abraira V, Martin Del Rio R (1986). In vivo determination of extracellular concentration of amino acids in the rat hippocampus. A method based on brain dialysis and computerized analysis. Brain Res. 384:145-155.

Mayer ML, Vyklicky L (1989). Concanavalin A selectively reduces desensitization of mammalian neuronal quisqualate receptors. PNAS 86:1411-1415.

Mayer ML, Westbrook GL (1984). Mixed-agonist action of excitatory amino acids on mouse spinal cord neurones under voltage clamp. J. Physiol. 354:29-53.

Mayer ML, Westbrook GL (1987). Permeation and block of N-methyl-D-aspartic acid receptor channels by divalent cations in mouse cultured central neurones. J. Physiol. 394:501-527.

McCarthy KD, De Vellis J (1980). Preparation of separate astroglial and oligodendroglial cell cultures from rat cerebral tissue. J. Cell Biol. 85:890-902.

Monaghan DT, Holets VR, Toy DW, Cotman CW (1983). Anatomical distributions of four pharmacologically distinct ^{3}H-L-glutamate binding sites. Nature 306:176-179.

Müller T, Möller T, Berger T, Schnitzer J, Kettenmann H (1992). Calcium entry through kainate receptors and resulting potassium-channel blockade in Bergmann glial cells. Science 256:1563-1566.

Nakanishi N, Shneider NA, Axel R (1990). A family of glutamate receptor genes: evidence for the formation of heteromultimeric receptors with distinct channel properties. Neuron 5:569-581.

Patneau DK, Mayer ML (1991). Kinetic analysis of interactions between kainate and AMPA: Evidence for activation of a single receptor in mouse hippocampal neurons. Neuron 6:785-798.

Raman IM, Trussell LO (1992). The kinetics of the response to glutamate and kainate in neurons of the avian cochlear nucleus. Neuron 9:173-186.

Silver RA, Traynelis SF, Cull-Candy SG (1992). Rapid-time-course miniature and evoked excitatory currents at cerebellar synapses in situ. Nature 355:163-166.

Smith DO, Franke C, Rosenheimer JL, Zufall F, Hatt H (1991). Glutamate-activated channels in adult rat ventral spinal cord cells. J. Neurophysiol. 66:369-378.

Sommer B, Burnashev N, Verdoorn TA, Keinänen K, Sakmann B, Seeburg PH (1992). A glutamate receptor channel with high affinity for domoate and kainate. EMBO J. 11:1651-1656.

Sommer B, Seeburg PH (1992). Glutamate receptor channels: novel properties and new clones. TIPS 13:291-296.

Stern P, Edwards FA, Sakmann, B (1992). Fast and slow components of unitary EPSCs on stellate cells elicited by focal stimulation in slices of rat visual cortex. J. Physiol. 449:247-278.

Trussell LO, Fischbach GD (1989). Glutamate receptor desensitization and its role in synaptic transmission. Neuron 3:209-218.

Verdoorn TA, Burnashev N, Monyer H, Seeburg PH, Sakmann B (1991). Structural determinants of ion flow through recombinant glutamate receptor channels. Science 252:1715-1718.

Werner P, Voigt M, Keinäne K, Wisden W, Seeburg PH (1991). Cloning of a putative high-affinity kainate receptor expressed predominantly in hippocampal CA3 cells. Nature 351:742-744.

Mechanically Sensitive
Cation Channels

Nonselective Cation Channels: Pharmacology, Physiology and Biophysics
ed. by D. Siemen & J. Hescheler
© 1993 Birkhäuser Verlag Basel/Switzerland

Mechanically Sensitive, Nonselective Cation Channels

Xian-Cheng Yang* and Frederick Sachs

Division of Biology 156-29, California Institute of Technology, Pasadena, CA 91125, USA; Department of Biophysical Sciences, 120 Cary Hall, State University of New York at Buffalo, Buffalo, NY 14214, USA

Summary
Mechanically sensitive channels (MSCs) are ubiquitous in plant and animal cells. They respond primarily to membrane tension, thus making them good transducers for forces derived from osmotic or hydraulic gradients and shear stress. They may also be modulated by membrane voltage and various ligands. MSCs are most commonly cation selective, passing calcium as well as monovalent ions, but some are K^+ selective, and a few are anion selective. MSCs occur at a density of about $0.2-5$ per μm^2. The universal distribution and biophysical properties of MSCs make them the ideal mechanotransducers in a wide variety of cellular processes.

Introduction

MSCs are found in almost every tissue (Table 1) of organisms that span the evolutionary tree (Morris, 1990). MSCs include stretch-activated channels (SACs) and stretch-inactivated channels (SICs). Both SACs and SICs can be further classified according to their ionic selectivities (Table 1). This brief, idiosyncratic review focuses on the biophysical properties of nonselective cation SACs, their potential roles in cells and some possible molecular biological approaches to establishing their structure. Interested readers can refer to Sachs (1988, 1992) and Morris (1990) for more comprehensive reviews on MSCs.

Biophysical Properties of Nonselective MSCs

Permeation

Single-channel studies of nonselective cation SACs (Cooper et al., 1986; Taglietti and Toselli, 1988; Yang and Sachs, 1990; Franco and Lansman, 1990a) have shown that: 1) The conductance is independent of the

*Present address: American Cyanamid Company, Medical Research Division, The CV/CNS Section, Pearl River, NY 10965, USA.

Table 1. Mechanically sensitive ion channels: stretch-activated (SA) and stretch-inactivated (SI)

Gating	Selectivity[a]	Blockers[b]	Tissues	Possible functions[c] (1–3)
SA	Cations: Ca: 7–20 pS Na: 22–50 pS K: 22–70 pS $V_{rev} \sim 0$ mV	Gd (3) Amiloride (3, 29) Streptomycin (3)	Muscle: cardiac (1, 4)	Hypertrophy, ANF secretion, ME feedback (30)
			skeletal (1, 5)	Hypertrophy, ME feedback
			smooth (1, 6)	Contraction and tone (31)
			Neurons: stretch receptors (7)	Sensory transduction
			hair cells (8, 9)	Sensory transduction
			Neuroblastoma (1)	Volume regulation
			Glia: astrocytes (10)	Taurine secretion (32), volume regulation, K transport
			Muller cells (11)	K transport and volume regulation
			Epithelia (1, 12–14)	Osmolarity and volume regulation
			Endothelia (1)	Vascular tone regulation
			Fibroblasts (1)	Endocytosis and motility
			Oocytes (1)	Cell division
			Bone: osteoblast-like cells (15)	Growth and metabolism
			Hepatocytes and hepatoma cells (16)	Volume regulation
			Fungi (17, 18)	Germling growth and differentiation
	K: 30–360 pS $V_{rev} \sim E_K$	Quinidine (3) Ba and Cs (10)	Cardiac muscle (1, 19)	Secretion, ME feedback
			Skeletal muscle (1)	Hypertrophy, ME feedback
			Neurons: soma and growth cones (1)	Neurite outgrowth and growth cone motility
			Glia: astrocytes (8)	Taurine secretion, volume regulation, K transport
			Epithelia (1)	Osmolarity and volume regulation
			Endothelia (20)	Vascular tone regulation
			Bone: osteoblast-like cells (15)	Growth and metabolism

	Location	Conductance / V_{rev}	Blockers	Function
	Blood: red cells (1) erythroleukemia cells (21) Early embryos (1)	Cl: up to 970 pS V_{rev} = N.D (other) or 0 mV (E. coli)		Osmolarity and shape regulation Regulation of protooncogene expression Embryogenesis
	Epithelia (22) Plant: roots (1) E. coli (1)		Cl blockers (Ep, 22) N.D (other)	Cl and drug metabolites transport, volume regulation Gravity sensing Osmoregulation
	Smooth muscle: mesangial cells (23) Epithelia (1, 24) Yeast (25) Gram-positive bacteria (26)	Ca, Na, K and Cl 10 pS – 8.5 nS (B) 20–60 pS (other) V_{rev} = −26–0 mV	Gd (yeast, 25) N.D (other)	Contraction and tone Osmolarity and volume regulation Osmoregulation Osmoregulation
SI	Skeletal muscle: normal (27) Skeletal muscle: mdx (27, 28)	Cations: Na, Ca (8 pS) V_{rev} = 32 mV (Ca)	Gd (28)	ME coupling Muscular dystrophy
	Neurons: Soma and growth cones (1)	K: 6 pS (outward) $V_{rev} \sim E_K$		Neurite outgrowth and growth cone motility

Notation: a) V_{rev}: reversal potential; E_K: K equilibrium potential; N.D.: not determined; B: bacteria. Note that single-channel conductances and reversal potentials depend on ionic conditions and the shape of I-V curves. For specific information, readers should consult original references. Conductances and reversal potentials for non-physiological ions are not included. In several cases, only whole-cell mechanically sensitive currents are recorded (Refs: Corey and Hudspeth (1983); Ohmori (1988); Chan and Nelson (1992); Olesen et al. (1988) and Valverde et al. (1992)).

b) Gd, Ba, and Cs are gadolinium, barium, and cesium ions, respectively; Ep: epithelia.

c) ANF: Atrial natriuretic factor; ME feedback: Mechanoelectric feedback that transduces mechanical stimuli into electric signals.

References: 1. Morris (1990); 2. Sachs (1988); 3. Sachs (1992); 4. Bustamante et al. (1991); 5. Franco and Lansman (1990a); 6. Davis et al. (1992); 7. Erxleben (1989); 8. Corey and Hudspeth (1983); 9. Ohmori (1988); 10. Bowman et al. (1992); 11. Puro (1991); 12. Hunter (1990); 13 Chang and Loretz (1991); 14. Chan and Nelson (1992); 15. Davidson et al. (1990); 16. Bear (1990); 17. Zhou et al. (1991); 18. Garrill et al. (1992); 19. Ruknudin et al. (1993); 20. Olesen et al. (1988); 21. Arcangeli et al. (1987); 22. Valverde et al. (1992); 23. Craelius et al. (1989); 24. Hurst and Hunter (1990); 25. Gustin et al. (1988); 26. Zoratti et al. (1990); 27. Franco and Lansman (1990b); 28. Franco et al. (1991); 29. Lane et al. (1991); 30. Hansen et al. (1990); 31. Bevan and Lager (1991); 32. Martin et al. (1990).

82

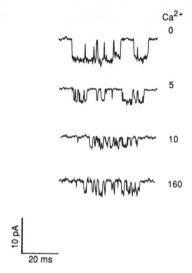

Figure 1. Ca^{2+} is permeable to SACs and affects the channel open time and conductance in *Xenopus* oocytes (Yang and Sachs, 1989). Cell-attached patches at a membrane potential of -100 mV with pipette solutions containing 150 mM NaCl and $CaCl_2$ (in mM) as indicated in the figure. The last trace has a vertical scale of 5 pA, recorded with 160 mM $CaCl_2$ and 0 mM Na in the pipette. (Copyright 1989 by AAAS, reproduced with permission.)

hydrostatic pressure used to activate the channel. 2) The channels are permeable to Ca^{2+} (Figure 1) with approximately linear current-vóltage (I-V) relationships. 3) I-V curves for monovalent cations are independent of anion species and show inward rectification with reversal potentials near zero. 4) The cation selectivity based on single-channel slope conductance follows the sequence: $K^+ > Cs^+ > Rb^+ > Na^+ > Li^+ > Ba^{2+} > Ca^{2+}$. 5) In mixtures of K^+ and Li^+ with a fixed total concentration (mole fraction tests), the single-channel currents vary monotonically with mole fractions. 6) Single-channel currents saturate with increasing external permeant ion concentrations. 7) Adding Ca^{2+} into monovalent cation solutions reduces single-channel conductances (Figure 1) and linearizes originally inwardly rectifying I-V curves.

Permeation Models

The fact that pressure across a patch has little effect on the single-channel conductance implies that the permeation pathway is either stiff or isolated from stress in the gating portion of the channel. Pressure itself constitutes an insignificant driving force for ion flow relative to the electrochemical potential (Sachs, 1988). Saturation of single-channel currents with increasing permeant ion concentration and the monotonic mole fraction behavior indicate that these SACs are one-ion pores, that is, only one ion is allowed at a time in the channel pore (Hille, 1992). The simplest permeation model for one-ion pores has two energy barriers separated by a central well (or site), the so-called 2B1S model. Where tested, the non-selective SACs are quantitatively consistent with this model (Taglietti and Toselli, 1988; Yang and Sachs, 1990). The ionic selectivity of the channel is determined by the way in which the barrier heights and the well depth changes with different ion species. The energy well is located within the membrane electric field and corresponds to an intra-channel ion binding site where each ion binds with a distinct affinity and residence time. For a given barrier height, a deeper well means stronger binding (or higher affinity) and longer residence time and, consequently, a smaller single-channel conductance and a weaker voltage dependence (linear I-V). For example, the intra-channel site has a higher affinity for Ca than the monovalent cations so that as Ca^{2+} is added to a solution of monvalent ions, the inwardly rectifying monovalent I-V is linearized towards that of Ca^{2+} and shows a smaller conductance. Using the energy peaks and wells for Na^+ and Ca^{2+}, determined from I-V curves, the 2B1S model predicts that in normal saline (150 mM Na^+ and 2 mM Ca^{2+}), the Ca^{2+} influx through a single SAC is ~ 20 fA (Yang and Sachs, 1989).

Kinetics

SACs in membrane patches can be activated by applying either suction or pressure to the patch pipette. Steady-state single-channel activities can be observed for tens of minutes with sustanied stimulus, showing that there is little inactivation or desensitization. The probability of a channel being open (P_0) increases sigmoidally with increasing membrane tension or depolarization (Sachs, 1988; Morris, 1990). The voltage-dependence of SACs is weak in most cases, ≥ 40 mV for an e-fold change in P_0 (Morris, 1990; Yang and Sachs, 1990; Franco and Lansman, 1990a). Voltage alone is usually not sufficient to activate SACs in the absence of mechanical stress.

Kinetic analysis of single-channel records indicates that SACs have one or two open times and at least two closed times. P_0 is mainly determined by the longest closed time, since it is most sensitive to tension and voltage, decreasing with increasing tension and depolarization. The open time is usually independent of both pressure and voltage, although ionic conditions can affect the open time. In *Xenopus* oocytes, the open time varied with external ionic species following the order (in ms) $Na^+(4.5) > Cs^+(3.8) > K^+(3.0) > Rb^+(2.3) > Li^+(1.5) > Ca^{2+}(0.48)$ (Yang and Sachs, 1990). In both oocytes and lens cells, increasing the Ca^{2+} concentration monotonically reduced the open time from a maximum with pure Na^+ to a minimum with pure Ca^{2+}

(Cooper et al., 1986; Moody and Bosma, 1989; Yang and Sachs, 1990). Mole fraction experiments with K^+ and Li^+ in oocytes showed that the open time decreased with increasing Li^+ concentration, leveling off to the minimum measured with pure Li^+ (Yang and Sachs, 1990). In mixtures of Na^+ and Gd^{3+}, Gd^{3+} at $< 10 \, \mu M$ reduced the open time in oocytes without showing a minimum and at $\geq 10 \, \mu M$ completely blocked the channel activity (Yang and Sachs, 1989). These effects of ions on the open time show very weak if any voltage-dependence (Yang and Sachs, 1989, 1990).

Models for Gating and Kinetics

For channels activated by mechanical forces, the dominant free energy to drive the conformational change is linear in the applied force, or tension (Sachs and Lecar, 1991; Sachs, 1992). However, which of the kinetic steps are sensitive to the applied force, and which are sensitive to the ion species can be difficult to determine. Since the effects of ions on open times are voltage-independent, there must be an allosteric ion binding site located outside the membrane electric field, perhaps in the outer vestibule of the channel. The allosteric site is different from the intra-channel ion binding site that resides within the membrane field and determines the permeation properties. Therefore, SACs have at least two distinct ion binding sites.

Analogous to allosteric enzyme kinetics, in the oocyte we have proposed a model to account for both SAC kinetics and interactions with ions (Yang and Sachs, 1990): C and O are the closed and open states, respectively, 1 and 2 indicate the corresponding ion species, and the states with asterisks are ion-bound. The channel is sensitive to tension and voltage at the transitions: C–C–O. Ions bind to the open channel at an allosteric site through $O–O_1^*$ and $O–O_2^*$. The ion-bound open channel undergoes allosteric closing transitions: $O_i^*–C_i^*$ ($i = 1, 2$).

$$
\begin{array}{c}
O_1^*\text{–}C_1^* \\
| \\
C\text{–}C\text{–}O \\
| \\
O_2^*\text{–}C_2^*
\end{array}
$$

(Model 1)

This model is quantitatively consistent with the kinetic data. The rate constants for ion association and dissociation vary with the ion species, accounting for the characteristic dependence of the open time on media composition. In mixtures of Na^+ and Ca^{2+}, Ca^{2+} (ion1) and Na (ion2) compete at the allosteric site, where Ca^{2+} has higher affinity over Na^+.

Therefore, the channel favors the path represented by the upper (or Ca^{2+}) branch of model 1. The Ca^{2+} bound channel also has a shorter open time so that the open time decreases from a maximum with Na^+ to a minimum with Ca^{2+} as increasing Ca^{2+} concentration. Similar explanations apply to the mole fraction and the Gd^{3+} block experiments. However, the complete block of SACs by Gd^{3+} at $\geq 10\ \mu M$ cannot be explained by binding to a single site, since the Hill coefficient for block is much larger than one (Yang and Sachs, 1989).

Adaptation

Examining a variety of preparations, several types of adaptation of SACs to mechanical stimuli can be observed. In response to a step pressure (Figure 2), channel activity may increase sharply and then decay to a steady-state level with lower P_0 (Gustin et al., 1988; Moody and Bosma, 1989; Hamill and McBride, 1992). SACs may enter long-lived closed states resembling desensitization in ligand-gated channels (Cooper et al., 1986). SACs in smooth muscle (Kirber et al., 1988) and chick cardiac muscle (Ruknudin et al., 1993) may increase their sensitivity to stretch with sustained stimulation. P_0's of SACs in fibroblasts show montonic or biphasic increases depending on stimulus paradigms (Stockbridge and French, 1988). Multiple-conductance SACs in a gram-positive bacterium switch from high conductance states to low conductance states with repeated membrane stretch (Zoratti et al., 1990). SACs in yeast also show adaptation to step changes in voltage (Gustin et al., 1988). These mechanisms of adaptation have not been worked out. It is possible that they reflect properties of the channels themselves, or alternatively, properties of the structures (cytoskeleton and membranes) that couple force to the channels. Adapta-

2 pA

200 ms

Figure 2. Adaptation of SACs in tunicate oocytes (Moody and Bosma, 1989). Arrow indicates the start of suction of 40 cm H_2O applied to the patch pipette. (Reproduced with permission.)

tion at the single-channel level may have physiological relevance in cellular mechanotransduction processes.

Pharmacology

None of the known blockers of MSCs (Table 1) can be considered specific. Among them, the lanthanide Gd^{3+} at $10-50\ \mu M$ completely blocks single nonselective cation SACs in several tissues and the mean current in a number of mechanotransduction processes (see Sachs, 1992). Gd^{3+} at $10\ \mu M$ blocks shrinkage-activated whole-cell currents (Chan and Nelson, 1992). However, Gd^{3+} blocks some Ca^{2+} channels with a dissociation constant, K_d, of 40 nM (Kasai and Neher, 1992). Amiloride produces voltage-dependent block of mechanosensitive currents in hair cells with $K_d = 50\ \mu M$ (at -50 mV) and a Hill coefficient of 1 (Jorgensen and Ohmori, 1988) and of nonselective cation SACs in *Xenopus* oocytes with a $K_d = 0.5$ mM (at -100 mV) and a Hill coefficient of 2 (Lane et al., 1991). Amiloride again is not specific for MSCs as it also blocks most of epithelium Na channels with $K_d = 0.1-0.5\ \mu M$ (Smith and Benos, 1991).

Physiological Functions of Nonselective MSCs

MSCs are ideal candidates for mechanical transduction in various cellular processes (Table 1). Figure 3 summarizes some ways in which the activation of nonselective cation SACs could transduce mechanical signals to cellular responses and emphasizes the interactions of SACs with voltage- and Ca^{2+}-dependent channels.

For the model in Figure 3 to be useful, the influxes of Na^+ and Ca^{2+} through activated SACs have to be large enough to depolarize membrane and to increase intracellular Na^+, $[Na^+]_i$, and Ca^{2+}, $[Ca^{2+}]_i$. Na influx through SACs in neuroblastoma cells depolarizes the membrane by ~ 20 mV that in turn activates delayed rectifier K^+ channels (Falke and Misler, 1988). In epithelial cells (Christensen, 1988) and retinal Müller cells (Puro, 1990) membrane stretch causes an increase in $[Ca^{2+}]_i$ as monitored by increasing activities of Ca^{2+}-activated K^+ channels, and correlates with the presence of SACs. Mechanically induced elevation of $[Ca^{2+}]_i$ (Figure 4) can arise from two sources: Ca^{2+} influx (Bear, 1990; Sigurdson et al., 1992) and Ca^{2+} release from internal stores (Snowdowne, 1986; Goligorsky, 1988; Wirtz and Dobbs, 1990). Sigurdson et al. (1992) provide strong evidence that SACs are transducers for the mechanically induced Ca^{2+} influx in chick heart: 1) Chick heart has Ca-permeable MSCs, which are blocked by 20 μM Gd^{3+}; 2) Mechani-

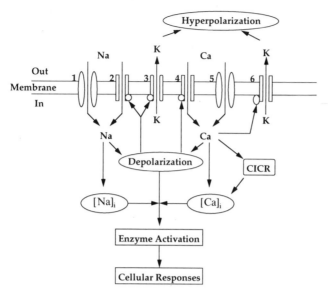

Figure 3. Nonselective cation SACs (1 and 5) transduce mechanical signals by inward Na^+ and Ca^{2+} fluxes that depolarize cell membrane and elevate intracellular Ca^{2+} and Na^+ concentrations ($[Ca^{2+}]_i$ and $[Na^+]_i$). Mechanically induced depolarization may activate voltage-gated Na^+, K^+ (or Cl^-) and Ca^{2+} channels (2–4). Mechanically induced Ca^{2+} influx increases $[Ca^{2+}]_i$ which may trigger Ca^{2+}-induced Ca^{2+} release (CICR) that increases $[Ca^{2+}]_i$ further and may activate Ca^{2+}-dependent K^+ (or Cl^-) channels (6). Inward Na^+ and Ca^{2+} currents through voltage-gated Na^+ and Ca^{2+} channels depolarize the membrane further (positive feedback). Outward K^+ (and/or Cl^-) currents offset mechanically induced depolarization and balance electrolytes and water contents (negative feedback). Note that a) nonselective cation SACs have reversal potentials near zero (Table 1) and thus pass inward currents carried by Na^+ and Ca^{2+} at the resting membrane potential; b) Depolarization also increases P_0, but decreases single-channel currents, i, of SACs with the overall effect depending on the product iP_0; c) For Cl^- channels, replace the label K^+ by Cl^- with arrows indicating outward Cl^- currents.

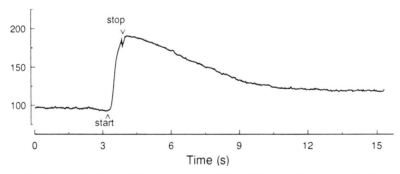

Figure 4. Mechanically induced changes in intracellular Ca^{2+} in a chick heart cell (Sigurdson et al., 1992). The cell is nudged at time "start" with a fire-polished pipette that is withdrawn at time "stop". Vertical scale is fluo-3 fluorescence intensity in arbitrary units. (Reproduced with permission.)

cally sensitive Ca^{2+} fluorescence depends on extracellular Ca^{2+} and is also blocked by $20\,\mu M$ Gd^{3+}; 3) Heart cells grown in the absence of chick embryo extract have neither MSCs nor mechanosensitive Ca^{2+} fluorescence, and 4) Most other ion channels are not mechanically sensitive. Based on a diffusion model and the ultrastructures of cardiac muscle, Sigurdson et al. (1992) also estimate that a Ca^{2+} current of 20 fA through SACs would increase $[Ca^{2+}]_i$ to a maximum concentration of $12.5\,\mu M$ that is sufficient to trigger Ca^{2+}-induced Ca^{2+} release.

Are MSCs an Artifact?

Morris and Horn (1991), working on cultured molluscan neurons, found that although they could readily record single-channel currents from MSCs, they were unable to record significant whole cell current with a variety of cell distortions including direct pushing with a pipette and osmotic swelling. They concluded that activation of MSCs was an artifact of patch formation and that they do not play a physiological role. The general conclusion was quickly shown to be false by Gustin (1991) and Zhou et al. (1991), who showed that in plant protoplasts the whole cell current could be well predicted from the single-channel current amplitude and channel density. In animal cells whole cell currents can be recorded from preparations that show MSCs (cf. Davis et al., 1992), although a saturating whole-cell current was not demonstrated. To demonstrate a saturating response requires that the stimulus be uniform over the whole cell. It is difficult to know the applied tension under patch conditions and virtually impossible under whole cell conditions. Morris and Horn never measured their stimulus (membrane tension) in whole-cell mode and assumed that they saturated the system. They may or may not have done so. Consider trying to predict the whole-cell current of a huge cell, like a cerebellar Purkinje cell, from single-channel data. With a pipette in the soma, the whole-cell current will be much less than predicted because the preparation is not space clamped; the voltage varies with position and time. With MSCs, the relevant stress is sensitive to geometry and to the compliance of the membrane and the bound cytoskeleton. Without knowledge of these factors, it is not possible to say whether or not enough force has been applied to activate a channel. In summary then, it is possible that in some preparations the channels are not available for activation without disrupting the cytoplasm by forming a patch, but in general that is not the case and there is every reason to believe that the MSCs do play a physiological role.

Future Directions: Molecular Biology of MSCs

What Have We Learned from Cloning of Shaker K Channels?

It is a challenge to clone MSCs. We are facing similar problems now for MSCs that workers had 5 years ago for voltage-gated K^+ channels: lack of specific blockers. One of the key elements in molecular cloning of *Drosophila shaker* K^+ channels was the *shaker* mutant. Similar genetic approaches may be taken to clone MSCs. The first choice of organism, in our opinion, is *C. elegans*. More than 95% of the worm genome is cloned and complete sequencing is underway (Roberts, 1990). There are only six touch receptor neurons among a total of 302 neurons in the worm. Eighteen genes have been identified to control touch receptor development and differentiation by screening touch-insensitive mutants (Chalfie and Au, 1989). Sequencing and expression of these genes may lead us to a MSC clone. Unfortunately, because of the extremely tough cuticle on the worms, electrophysiological studies of the worm itself have not yet succeeded. It may be possible, however, to express the clones in inert cell lines or even in oocytes, providing the ionic selectivity is different from the endogenous SACs. A second choice of organisms is yeast or *E. coli*, which both have MSCs and well worked out genetics.

Is P-Glycoprotein a MSC?

P-glycoprotein, encoded by the multiple drug resistance gene (MDR1), confers cancer cells resistance to chemotherapeutic drugs by pumping them out of the cell (Endicott and Ling, 1989). The primary structure of the protein suggests 12 membrane-spanning regions, similar to those of Cl^- channels activated by voltage (Jentsch et al., 1990) or cAMP (Anderson et al., 1991). Recently, Valverde et al. (1992) demonstrated that P-glycoprotein behaved as a volume-regulated Cl^- channel when transiently or stably expressed in NIH3T3 fibroblasts. However, this Cl^- channel activity has not been seen at the single-channel level and it is not known whether the channel itself is sensitive to membrane stress. A volume sensitive Cl^- conductance similar to P-glycoprotein has been found in chromaffin cells by Dorshenko and Neher (1992). They demonstrated that the Cl^- current was generated by activation of a volume-sensitive second messenger system involving G proteins, and not by a membrane-bound channel.

Is the Cytoskeleton Involved in Activation of MSCs?

Recent micromechanical experiments (Sokabe et al., 1991) support the idea (Guharay and Sachs, 1984) that MSCs may be coupled to the cytoskeleton. The suspected cytoskeletal component is of the spectrin/ fodrin family since reagents for actin filaments and microtubules do not abolish MSC activity (Guharay and Sachs, 1984; Yang and Sachs, unpublished). Because of the lack of specific reagents for spectrin/ fodrin, the connection between MSCs and the cytoskeleton has not been clearly identified. Molecular biology, in the form of antisense oligonucleotide hybrid arrest, may provide the tool to answer this question.

If the cDNA sequence encoding the protein one wants to block is known, antisense oligonucleotides can be synthesized and then introduced into cells by uptake or microinjection. Oligo-mRNA hybrids are selectively degraded by ribonuclease H, resulting in depletion of mRNA and loss of the protein. Depletion of the protein may be observed if the turnover time is shorter than the life span of the cell culture. This method has been successfully used to selectively inhibit a microtubule-associate protein (Caceres and Kosik, 1990), G-proteins (Kleuss et al., 1991), and neuronal acetylcholine receptors (Listerud et al., 1991). The specificity is obvious and inhibition of the protein is reversible after withdrawal of oligos in the culture media or after depletion of injected oligos. Since spectrin/fodrin have been cloned (Speicher and Marchesi, 1984; McMahon et al., 1987; Cioe et al., 1987), the antisense approach combined with patch clamp, optical imaging, and electron microscopy will allow us, in the near future, to clarify whether there is a spectrin/ fodrin link to MSCs.

Acknowledgements
This work was supported by a fellowship to X.-C. Yang from the American Heart Association, Greater Los Angeles Affiliate, and by USARO (LS-22560), NSF (BNS 90-09675) and a Fogarty grant to F.S.

References

Anderson MP, Rich DP, Gregory RJ, Smith AE, Welsh MJ (1991). Generation of cAMP-activated chloride currents by expression of CFTR. Science 251:679–682.
Arcangeli A, Wanke E, Olivotto M, Camagni S, Ferroni A (1987). Three types of ion channels are present on the plasma membrane of friend erythroleukemia cells. Biochem. Biophys. Res. Commun. 146:1450–1457.
Bear CE (1990). A nonselective cation channel in rat liver cells is activated by membrane stretch. Am. J. Physiol. 258:C421–C428.
Bevan JA, Laher I (1991). Pressure and flow-dependent vascular tone. FASEB J. 5:2267–2273.
Bowman CL, Ding J-P, Sachs F, Sokabe M (1992). Mechanotransducing ion channels in astrocytes. Brain Res. 584:272–286.
Bustamante JO, Ruknudin A, Sachs F (1991). Stretch-activated channels in heart cells: Relevance to cardiac hypertrophy, J. Cardiovasc. Pharmacol. 17 (Suppl.2):S110–S113.
Caceres A, Kosik KS (1990). Inhibition of neurite polarity by tau antisense oligonucleotides in primary cerebellar neurons. Nature 343:461–463.
Chalfie M, Au M (1989). Genetic control of differentiation of the *Caenorhabditis elegans* touch receptor neurons. Science 243:1027–1033.
Chan HC, Nelson DJ (1992). Chloride-dependent cation conductance activated during cellular shrinkage. Science 257:669–671.

Chang W, Loretz CA (1991). Identification of a stretch-activated monovalent cation channel from teleost urinary bladder cells. J. Exp. Zool. 259:304–315.

Christensen O (1987). Mediation of cell volume regulation by Ca^{2+} influx through stretch-activated channels. Nature 330:66–68.

Cioe L, Laurila P, Meo P, Krebs K, Goodman S, Curtis PJ (1987). Cloning and nucleotide sequencing of a mouse erythrocyte β-spectrin cDNA. Blood 70:915–920.

Cooper KE, Tang JM, Rae JL, Eisenberg RS (1986). A cation channel in frog lens epithelia responsive to pressure and calcium. J. Membrane Biol. 93:259–269.

Corey DP, Hudspeth AJ (1983). Kinetics of the receptor current in bullfrog saccular hair cells. J. Neurosci. 3:962–976.

Craelius W, El-Sherif N, Palant CE (1989). Stretch-activated ion channels in cultured mesangial cells. Biochem. Biophys. Res. Commun. 159:516–521.

Davidson RM, Tatakis DW, Auerbach AL (1990). Multiple forms of mechanosensitive ion channels in osteoblast-like cells. Pflügers Arch. 416:646–651.

Davis MJ, Donovitz JA, Hood JD (1992). Stretch-activated single-channel and whole-cell currents in vascular smooth cells. Am. J. Physiol. 262:C1083–C1088.

Dorshenko P, and Neher E (1992). Volume sensitive chloride conductance in bovine chromaffin cell membrane. J. Physiol. 449:197–218.

Endicott JA, Ling V (1989). The biochemistry of P-glycoprotein-mediated multidrug resistance. Ann. Rev. Biochem. 58:137–171.

Erxleben C (1989). Stretch-activated current through single ion channels in the abdominal stretch receptor organ of the crayfish. J. Gen. Physiol. 94:1071–1083.

Falke LC, Misler S (1989). Activity of ion channels during volume regulation by clonal N1E115 neuroblastoma cells. Proc. Natl. Acad. Sci. USA 86:3919–3923.

Franco A Jr, Lansman JB (1990a). Stretch-sensitive channels in developing muscle cells from a mouse cell line. J. Physiol. 427:361–380.

Franco A Jr, Lansman JB (1990b) Calcium entry through stretch-inactivated ion channels in *mdx* myotubes. Nature 344:670–673.

Franco A Jr, Winegar BD, Lansman JB (1991). Open channel block by gadolinium ion of the stretch-inactivated ion channel in *mdx* myotubes. Biophys. J. 59:1164–1170.

Garrill A, Lew RR, Heath IB (1992). Stretch-activated Ca^{2+} and Ca^{2+}-activated K^+ channels in the hyphal tip plasma membrane of the oomycete *Saprolegnia ferax*. J. Cell Sci. 101:721–730.

Goligorsky MS (1988). Mechanical stimulation induces Ca_i^{2+} transients and membrane depolarization in cultured endothelial cells. FEBS Lett. 240:59–64.

Guharay F, Sachs F (1984). Stretch-activated single ion channel currents in tissue-cultured embryonic chick skeletal muscle. J. Physiol. 352:685–701.

Gustin M (1991). Single channel mechanosensitive currents. Science 253:800.

Gustin MC, Zhou XL, Martinac B, Kung C (1988). A mechanosensitive ion channel in the yeast plasma membrane. Science 242:762–765.

Hamill OP, McBride DW Jr. (1992). Rapid adaptation of single mechanosensitive channels in *Xenopus* oocytes. Proc. Natl. Acad. Sci. USA 89:7462–7466.

Hansen DE, Craig CS, Hondeghem LM (1990). Stretch-induced arrhythmias in the isolated canine ventricle. Circulation 81:1094–1105.

Hille B (1992). Ionic Channels of Excitable Membranes. 2nd ed. Sinauer Associates, Sunderland.

Hunter M (1990). Stretch-activated channels in the basolateral membrane of single proximal cells of frog kidney. Pflügers Arch. 416:448–453.

Hurst AM, Hunter M (1990). Stretch-activated channels in single early distal tubule cells of the frog. J. Physiol. 430:13–24.

Jentsch TJ, Steinmeyer K, Schwarz G (1990). Primary structure of *Torpedo marmorata* chloride channel isolated by expression cloning in *Xenopus* oocytes. Nature 348:510–514.

Jorgensen F, Ohmori H (1988). Amiloride blocks the mechano-electrical transduction channel of hair cells of the chick. J. Physiol. 403:577–588.

Kasai H, Neher E (1992). Dihydropyridine-sensitive and ω-conotoxin-sensitive calcium channels in a mammalian neuroblastoma-glioma cell line. J. Physiol. 448:161–188.

Kirber MT, Walsh Jr JV, Singer JJ (1988). Stretch-activated ion channels in smooth muscle: a mechanism for the initiation of stretch-induced contraction. Pflügers Arch. 412:339–345.

Kleuss C, Hescheler J, Ewel C, Rosenthal W, Schultz G, Wittig B (1991). Assignment of G-protein subtypes to specific receptors inducing inhibition of calcium currents. Nature 353:43–48.

Lane JW, McBridge Jr DW, Hamill OP (1991). Amiloride block of the mechanosensitive cation channel in *Xenopus* oocytes. J. Physiol. 441:347–366.

Listerud M, Brussaard AB, Devay P, Colman DR, Role LW (1991). Functional contribution of neuronal AChR subunits revealed by antisense oligonucleotides. Science 254:1518–1520.

Martin DL, Madelian V, Seligmann B, Shain W (1990). The role of osmotic pressure and membrane potential in K^+-stimulated taurine release from cultured astrocytes and LRM55 cells. J. Neurosci. 10:571–577.

McMahon AP, Giebelhaus DH, Chappion JE, Bailes JA, Lacey S, Carritt B, Henchman SK, Moon RT (1987). cDNA cloning, sequencing and chromosome mapping of a non-erythroid spectrin, human α-fodrin. Differenciation 34:68–78.

Moody WJ, Bosma MM (1989). A nonselective cation channel activated by membrane deformation in oocytes of the Ascidian *Boltenia villosa*. J. Membrane Biol. 107:179–188.

Morris CE (1990). Mechanosensitive ion channels. J. Membrane Biol. 113:93–107.

Morris CE and Horn R (1991). Failure to elicit neuronal macroscopic mechanosensitive currents anticipated by single channel recording. Science 251:1246–1249.

Ohmori H (1984). Mechanoelectrical transducer has discrete conductances in the chick vestibular hair cell. Proc. Natl. Acad. Sci. 81:1888–1891.

Olesen S-P, Claphan DE, Davis PF (1988). Haemodynamic shear stress activates a K^+ current in vascular endothelial cells. Nature 331:168–170.

Puro DG (1991). Stretch-activated channels in human retinal Muller cells. Glia 4:456–460.

Roberts L (1990). The worm project. Science 248:1310–1313.

Ruknudin A, Sachs F, Bustamante JO (1993). Stretch-activated ion channels in tissue-cultured chick heart. Am. J. Physiol. 264:H960–H972.

Sachs F (1988). Mechanical transduction in biological systems. CRC Crit. Rev. Biomed. Engi. 16:141–169.

Sachs F and Lecar H (1991). Stochastic models for mechanical transduction. Biophys. J. 59:1143–1145.

Sachs F (1992). Stretch-sensitive ion channels: An update. In: Sensory Transduction. D. Corey, editor. NY: Rockefeller Univ. Press, pp 242–260.

Sigurdson W, Ruknudin A, Sachs F (1992). Calcium imaging of mechanically induced fluxes in tissue-cultured chick heart: role of stretch-activated ion channels. Am. J. Physiol. 262:H1110–H1115.

Smith PR, Benos DJ (1991). Epithelial Na^+ channels. Ann. Rev. Physiol. 53:509–530.

Snowdowne KW (1986). The effect of stretch on sarcoplasmic free calcium of frog skeletal muscle at rest. Biochim. Biophys. Acta 862:441–444.

Sokabe M, Sachs F, Jing Z (1991). Quantitative video microscopy of patch clamped membranes: stress, strain, capacitance and stretch channel activation. Biophys. J. 59:722–728.

Speicher DW, Marchesi VT (1984). Erythrocyte spectrin is comprised of many homologous triple helical segments. Nature 311:177–180.

Stockbridge LL, French AS (1988). Stretch-activated cation channels in human fibroblasts. Biophys. J. 54:187–190.

Taglietti V, Toselli M (1988). A study of stretch-activated channels in the membrane of frog oocytes: Interactions with Ca^{2+} ions. J. Physiol. 407:311–328.

Valverde MA, Diaz M, Sepulveda FV, Gill DR, Hyde SC, Higgins CF (1992). Volume-regulated chloride channels associated with the human multidrug-resistance P-glycoprotein. Nature 355:830–833.

Wirtz HRW, Dobbs LG (1990). Calcium mobilization and exocytosis after one mechanical stretch of lung epithelial cells. Science 250:1266–1269.

Yang X-C, Sachs F (1989). Block of stretch-activated ion channels in *Xenopus* oocytes by gadolinium and calcium ions. Science 243:1068–1071.

Yang X-C, Sachs F (1990). Characterization of stretch-activated ion channels in *Xenopus* oocytes. J. Physiol. 431:102–122.

Zhou XL, Stumpf MA, Hoch HC, Kung C (1991). A mechanosensitive channel in whole cells and in membrane patches of the fungus *Uromyces*. Science 253:1415–1417.

Zoratti M, Petronilli V, Szabo I (1990). Stretch-activated composite ion channels in *Bacilius subtilis*. Biochem. Biophys. Res. Commun. 168:443–450.

Nonselective Cation Channels: Pharmacology, Physiology and Biophysics
ed. by D. Siemen & J. Hescheler
© 1993 Birkhäuser Verlag Basel/Switzerland

Stretch-Activated Nonselective Cation Channels in Urinary Bladder Myocytes: Importance for Pacemaker Potentials and Myogenic Response

Marie-Cécile Wellner and Gerrit Isenberg

Department of Physiology, University of Cologne, D-50931 Köln, FRG

Summary

Filling of the bladder with urine stretches the myocytes in the wall. Stretch activates nonselective cation channels (SACs) thereby constituting a pacemaking mechanism. Once action potentials are triggered, Ca^{2+} influx through nifedipine-sensitive Ca^{2+} channels provides activator Ca^{2+} for the stretch-induced increase in wall tension (myogenic response). An additional component of myogenic response is independent of nifedipine and membrane potential; Ca^{2+} influx through SACs is large enough to induce Ca^{2+} release from intracellular stores.

Introduction

Miroelectrode studies on multicellular strips of the guinea-pig urinary bladder have shown spontaneous activity, i.e., pacemaker depolarization, action potentials and after-hyperpolarizations (approx. -40 mV, Creed, 1971). This electrical activity contrasts with the stable resting potential reported for myocytes isolated from the same tissue (-50 mV, Klöckner and Isenberg, 1985a). There are numerous differences between multicellular and single-cell preparations, e.g., the presence or absence of nerve endings and neurotransmitter. Here, we ask whether cell isolation abolished the spontaneous action potentials by removing the influence of mechanical stress exerted in vivo by the filling of the bladder or in studies with multicellular strips which are stretched to the length where force production becomes maximal. More specifically, we address the question of whether currents through stretch-activated nonselective cation channels (SACs; cf. Sachs, 1990) can constitute a pacemaking mechanism. As in most studies on SACs, we apply a negative pressure (i.e., suction) to the open end of the patch pipette; the deformation of the membrane-patch may stress the cytoskeleton that is interacting with the SAC channel protein (Sachs, 1990).

Occurrence and Function

Suction on the Patch Electrode Activates SACs

At potentials between -40 mV and -80 mV, the patches usually do not show any "spontaneous" channel activity. Appearance of single-channel inward currents, however, was induced when a negative pressure of approx. -4 kPa was applied. The effect of suction started promptly, was reversible, and could be repeated several times. The effect of suction should not be attributed to a nonspecific leakage because the closures between the individual channel openings show the current trace to have the same baseline as in the absence of suction. We think that the currents flow through SACs because the channel activity NP_0 increased with negative pressure and because the effect of suction was blocked by micromolar concentrations of Gd^{3+}. Our measurements suggest that SACs occur in the urinary bladder myocytes with density of approx. 0.5 μm^{-2}.

Single-Channel Conductance at Physiological Ionic Composition of the Bath

In the presence of continuous suction the patch potential was clamped to a variety of potentials between $+40$ and -90 mV (Figure 1). The amplitude of the unitary current was evaluated and plotted over the patch potential, resulting in linear i-v curves, the slope of which yielded the open-channel conductance. With 145 mM K^+ and 2 mM Mg^{2+} at the outer mouth of the channel, the conductance g was 40 ± 4 pS (mean $\pm$ S.D., n = 5). Substitution of 130 mM chloride by 130 mM less permeable aspartate did not change the conductance, as if chloride ions did

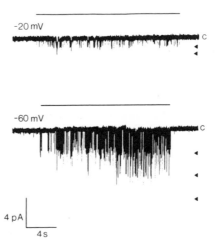

Figure 1. Stretch activates single-channel currents. Cell-attached recordings at -20 and -60 mV. Bath and electrode solution containing 140 mM KCl, 2 mM $MgCl_2$ and buffer. The experiments were performed at room temperature (22–25°C). Inward currents flowing from the pipette into the cell are shown as downward deflections. "C", closed state of the channel. Arrowheads mark amplitude of current through 1, 2 or 3 open channels. Horizontal lines mark time of negative pressure (-4 kPa). Note: more negative patch potentials (whole-cell conditions) increase the single-channel amplitude and the open probability.

not contribute to the single-channel current, i.e., that the channel is a cation channel. With physiological concentrations of Na^+, Ca^{2+}, and Mg^{2+}, the conductance was approx. 34 pS which is close to the 33 pS reported for SACs in smooth muscle cells from toad stomach (Kirber et al., 1988).

Hyperpolarization Increases the Open Probability of SACs

Suction or mechanical deformations such as the formation of the seal or the excision of the patch were the prerequisite for the appearance or currents through SACs. That is, hyperpolarization alone did not induce these currents in the guinea pig urinary bladder myocytes, which differs from the report on toad stomach myocytes where hyperpolarization did induce SAC-like activity in the absence of suction (Hishada et al., 1991). In our study the channel, once activated by suction, was voltage dependent since membrane hyperpolarization increased its activity (NP_0). In reference to NP_0 at -20 mV, NP_0 increased at -50 mV by a factor of 2 and at -80 mV by a factor of 3.3. Evaluation of the openings in histograms did not show a significant effect of suction on the open times or on the short shut times. However, the long intercluster closures (> 100 ms) became shorter and less frequent with membrane hyperpolarization.

In general, suction induces a nonselective conductance ($g \cdot NP_0$) which increases with hyperpolarization and operates as a pacemaker conductance, i.e., it depolarizes the membrane towards the threshold potential of -30 mV where the Ca^{2+} action potential originates (Klöckner and Isenberg, 1985b). As with other pacemaker systems, a higher degree of activation steepens the pacemaker-depolarization and shortens the interval between the action potentials. The feedback of the mechanical parameters through the SACs on the electrical activity seems to be crucial for the guinea pig urinary bladder myocytes. The isolated myocyte sinks without any external mechanical load to the coverslip. Obviously, it is the absence of the natural mechanical stress that abolishes the "spontaneous" appearance of pacemaker and action potentials.

SACs are Permeable for Ca^{2+} Ions

The myogenic response is defined as a contraction of the multicellular bladder strip in response to a change in length. The necessary increase in cytoplasmic activator Ca^{2+} could result from Ca^{2+} influx through voltage-gated L-type Ca^{2+} channels, activated by the action potentials. Here, we ask whether there is also Ca^{2+} influx through SACs and whether this influx could constitute another source of activator Ca^{2+}. Inward currents through SACs can be recorded when the outer channel mouth faces an electrode solution that contains only isotonic $CaCl_2$ and no other cations as charge carrier. With 110 mM Ca^{2+} the single-channel conductance g was 17 pS, which is 50% of the g measured with 140 mM Na^+. With 110 mM Ca^{2+}, the single-channel currents reversed polarity at approx. $+10$ mV. Provided the assumptions of the constant field equations can be applied (cf. Hille, 1992) the ratio between Ca^{2+} and K^+ permeability were approx. $pCa^{2+}:pK^+ \approx 1:1$. From this permeability ratio, one estimates for a bath solution containing 150 mM NaCl, 2 mM $CaCl_2$ etc. that about 5% of the total current is carried by Ca^{2+} (compare Yang and Sachs, 1989). Since the SACs appear at a density of approx. 0.5 channels per μm^2, the surface area of a bladder cell ($2231 \pm 179 \mu m^2$, Klöckner and Isenberg, 1985a) should bear approx. 1000 SACs. If stretch activates SACs with an open probability of 5%, one can expect at -50 mV a whole cell inward current of approx. -100 pA of which -5 pA may be carried by Ca^{2+} influx. Since this Ca^{2+} influx does not activate, it could increment the total calcium concentration of the cell (volume $6281 \mu m^3$) by $3 \mu M$. This increment due to Ca^{2+} influx through SACs is small in comparison with the one through L-type Ca^{2+} channels ($60 \mu M/0.1$ s, Klöckner and Isenberg, 1985b). Thus, it seems that activation of SACs and myogenic response are linked, not so much by the direct Ca^{2+} influx, but by pacemaker depolarization and activation of L-type Ca^{2+} channels.

Extracellular Ca^{2+} or Mg^{2+} Control Channel Conductance and Open Time

In the absence of bivalent cations (electrode solution without Ca^{2+} or Mg^{2+}) the conductance of SACs for inward currents was twice as high (82 pS) than in their presence. At positive patch potentials the high conductance fell to 45 pS, i.e., the SACs rectified. According to the literature (cf. Hille, 1992), we think that the outward current drives Mg^{2+} ions from the cytosol into the pore thereby blocking the channel. The results are compatible with the idea that the presence of divalent cations hinders the permeation of the monovalent K^+ cations, i.e., we suggest that intrachannel Ca^{2+} binding limits the permeation of the monovalent cations, as the rate of ion entry approaches the maximum rate of Ca^{2+} unbinding (see Hille, 1992).

The divalent cations also modulate the gating of the SACs. In general, the duration of the open state was shorter in the presence of 2 mM Mg^{2+} or 2 mM Ca^{2+} than in their absence. Reduction of the channel open time by Ca^{2+} ions (from 3 to 0.5 ms) has been described and modeled in the literature (Yang and Sachs, 1989). The histograms from SACs of urinary bladder myocytes demonstrated an additional open state with a duration longer than 12 ms; 110 mM Ca^{2+} as well as $5 \mu M$ Gd^{3+} specifically suppressed these long openings. Gating schemes including the Ca^{2+} effects on additional long-lasting open states have not been modeled up to now.

Gadolinium-Block of SACs

Gd^{3+} has been introduced as a specific blocker of SACs (Sachs, 1990). Gd^{3+} block of SACs was tested by using patch electrodes, the tips of which were filled with $GdCl_3$-free CsCl solution to a height of approximately 2 mm, whereas the $GdCl_3$ containing Cs solution was back-filled. $200 \mu M$ $GdCl_3$ completely blocked the SACs, i.e., the activity disappeared during the continuous suction and could not be restored. In addition, $20 \mu M$ $GdCl_3$ abolished SACs activity within approx. 1 min, though there was a transient 20-s period during which Gd^{3+} reduced the single-channel currents in regard to their amplitude and open time. In the presence of $5 \mu M$ $GdCl_3$, SAC activity could be recorded for a period up to 10 min; the long openings were missing, openings of < 4 ms were abbreviated, and the single-channel current has a reduced amplitude. The last two results confirm earlier reports of single-channel block by Gd^{3+}.

Extrapolation to the In Vivo Situation

Recently, SACs were recorded from myocytes isolated from porcine coronary arteries (Davis et al., 1992). Under whole-cell conditions the myocytes were stretched to 5–30% above their slack length while recording membrane potential or whole cell current. With the cells in physiological salt solution, stretch induced an inward current of -50 pA. More surprisingly, even in the presence of $5 \mu M$ nifedipine blocking the L-type Ca^{2+} channels, stretch-induced Ca^{2+} influx was sufficient to produce a significant increase in intracellular calcium and to trigger cell contraction (Davis et al., 1990).

We interpret this surprising result with the assumption that the SACs are part of an integrated cellular feedback system. A positive feedback could arise if the small Ca^{2+} influx through SACs triggered the release of Ca^{2+} from intracellular stores; the release could amplify the $[Ca^{2+}]_c$ signal and lead to contraction. An example of negative feedback is seen in the Ca^{2+} activation of BK-channels that re- or hyperpolarize the membrane. This negative feedback would terminate the Ca^{2+} influx through L-type Ca^{2+} channels, but not the Ca^{2+} influx through SACs

98

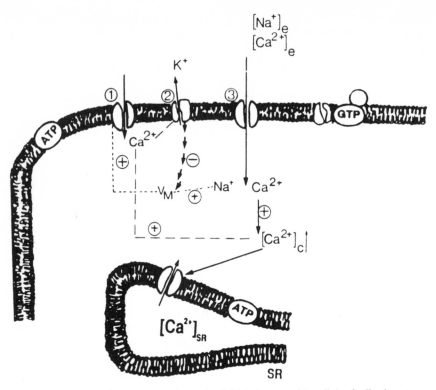

Figure 2. The myogenic response as the result of SACs integrated in cellular feedback systems. The membranes of the cell surface and of the sarcoplasmic reticulum (SR) are indicated together with a variety of channels as 1) L-type Ca^{2+} channel, 2) Ca^{2+}-activated BK channel, 3) stretch-activated channel (SAC). V_M is the membrane potential, $[Ca^{2+}]_c$ the cytosolic Ca^{2+} concentration and $[Ca^{2+}]_{SR}$ the Ca^{2+} concentration in the SR. Upon stretch, Ca^{2+} influx through SACs increases $[Ca^{2+}]_c$; this increase could trigger Ca^{2+} release from SR. The result is a positive feedback that could activate a contraction insensitive to Ca^{2+} channel antagonists like nifedipine. Nifedipine-sensitive components of the myogenic response may result from the membrane depolarization due to SACs (dotted lines, label $\oplus$). Depolarization opens voltage-gated L-type Ca^{2+} channels, induces Ca^{2+} influx, increases $[Ca^{2+}]_c$ (dashed lines) and activates contraction. Elevated $[Ca^{2+}]_c$ can Ca^{2+}-activate BK-channels with the result of a re- or hyperpolarization (short arrows, label $\ominus$), deactivation of L-type Ca^{2+} channels, termination of Ca^{2+} influx, decrease of $[Ca^{2+}]_c$ and relaxation (negative feedback).

(Figure 2). Forthcoming experiments that simultaneously measure $[Ca^{2+}]_c$ and whole-cell currents from stretched myocytes should test these predictions.

References

Creed KE (1971). Membrane properties of the smooth muscle membrane of the guinea-pig urinary bladder. Pflügers Arch. 326:115–126.
Davis MJ, Donovitz JA, Hood JD (1992). Stretch-activated single-channel and whole cell currents in vascular smooth muscle cells. Am. J. Physiol. 262:C1083–C1088.

Davis MJ, Hester FK, Donovitz JA, Montgomery CL, Meininger CJ (1990). Whole-cell current and intracellular calcium changes elicited by longitudinal stretch of single vascular smooth muscle cells (Abstract). FASEB J. 4:A844.

Hille B (1992). Ionic channels of excitable membranes. 2nd ed. Sinauer Associates, Sunderland, Mass.

Hishada T, Ordway RW, Kirber MT, Singer JJ, Walsh JW (1991). Hyperpolarization-activated cationic channels in smooth muscle cells are stretch sensitive. Pflügers Arch. 417:493–499.

Kirber MT, Walsh JW, Singer JJ (1998). Stretch-activated ion channels in smooth muscle: a mechanism for the initiation of stretch-induced contraction. Pflügers Arch. 421:339–345.

Klöckner U, Isenberg G (1985a). Action potential and net membrane currents of isolated smooth muscle cells (urinary bladder of the guinea-pig). Pflügers Arch. 405:329–339.

Klöckner U, Isenberg G (1985b). Calcium currents of cesium loaded isolated smooth muscle cells (urinary bladder of the guinea-pig). Pflügers Arch. 405:340–348.

Sachs F (1990). Stretch-sensitive ion channels. Neurosci. 2:49–57.

Yang XC, Sachs F (1989). Block of stretch-activated ion channels in Xenopus oocytes by gadolinium and calcium ions. Science 243:1068–1071.

Nonselective Cation Channels: Pharmacology, Physiology and Biophysics
ed. by D. Siemen & J. Hescheler
© 1993 Birkhäuser Verlag Basel/Switzerland

Mechanosensitive Nonselective Cation Channels in the Antiluminal Membrane of Cerebral Capillaries (Blood-Brain Barrier)

R. Popp, J. Hoyer† and H. Gögelein*

Max-Planck-Institut für Biophysik, Kennedyallee 70, D-60596 Frankfurt, FRG

Summary
Single stretch-activated (SA) cation channels have been investigated in the antiluminal membrane of freshly isolated brain capillaries. SA-channels did not distinguish between K^+ and Na^+ ions and were also permeable to Ca^{2+} and Ba^{2+} ions. With monovalent cations in the patch pipette the single-channel conductance was 37 pS and with the divalent cations Ba^{2+} and Ca^{2+} slope conductance was 16 and 19 pS, respectively. The open probability of the SA-channel increased with increasing negative pressure as well as with depolarization. Cell swelling induced by hypotonic shock activated the SA-channels in cell-attached experiments. The contribution of SA-channels to the regulation of cerebrospinal fluid in brain edema is discussed.

Introduction

The capillary endothelium of the brain (blood-brain barrier, BBB) plays an important role in homeostasis of the brain interstitial fluid. Complex tight junctions connecting the endothelial cells prevent uncontrolled paracellular exchange of water soluble substrates between blood and brain. Therefore, the passage of nutrients and ions through the BBB must be mediated by specific transport systems which are differently located in the luminal and antiluminal membrane of the endothelial cells. An active Na^+/K^+-ATPase in the antiluminal membrane extrudes Na^+ from cell inside into the cerebrospinal fluid in exchange for K^+ ions (Betz et al., 1980). On the luminal side, an amiloride-sensitive Na^+ influx pathway as well as a furosemide-inhibitable Na^+/Cl^- cotransport system was described (Betz, 1983). This could explain the observed netflux of NaCl from the blood into the cerebrospinal fluid. It is assumed that BBB is also involved in homeostasis of the cerebral K^+ concentration (Bradbury and Stulcova, 1970; Goldstein, 1979). K^+ transport mainly occurs from brain to blood (Hansen et al., 1977). Antiluminal K^+ uptake into the cell is mediated by the antiluminal

†Klinikum Steglitz, Hindenburgdamm 30, D-12203 Berlin, FRG.
*Hoechst AG, Department of Pharmacology, D-65926, Frankfurt, FRG.

N^+/K^+-ATPase and K^+ ions leave the cell through luminal K^+ channels.

Electrophysiological findings from other vascular endothelial cells show mechanosensitive currents, which could act as mechanotransducer for hemodynamic stress in this kind of cells (Lansman et al., 1987; Olesen et al., 1988).

Our aim was to investigate the possible involvement of mechanosensitive channels in capillary endothelial cells. In the following, we report about stretch-activated nonselective cation channels (SA-channels) in the antiluminal membrane of porcine brain capillaries by use of the patch-clamp technique.

Results and Discussion

Intact porcine cerebral capillaries were directly approached by the patch-pipette at the antiluminal side. Seal resistances in the range of $1-10$ GΩ could be achieved. In cell-attached membrane patches SA-channels were recorded. Suction applied through the patch-pipette to previously silent membrane patches could directly activate SA-channels. The mean open probability of this channel increased with increasing suction (Figure 1). Membrane depolarization also led to an increase in channel activity.

The SA-channels were found to be cation selective and permeable to Na^+ and K^+ ions as well as to the divalent cations Ba^{2+} and Ca^{2+}, but were found to be impermeable to anions.

In cell-attached patches with KCl-solution in the pipette the mean single-channel conductance was 37 ± 2 pS at negative and 24 ± 2 pS at positive holding potentials (n $= 13$). With a pipette solution containing 70 mM $CaCl_2$ or $BaCl_2$ in substitution for KCl the current-voltage relationship was linear, and the conductance was 16 ± 3 pS (n $= 4$) and 19 ± 4 pS (n $= 9$), respectively.

The SA-channel is clearly different from a calciumsensitive nonselective cation channel observed in the same preparation which is not mechanosensitive and permeable to divalent cations (Popp and Gögelein, 1992).

Cell swelling induced by hypotonic shock activated the SA-channels in cell-attached experiments. Therefore, this channel could be involved in cell volume regulation (Christensen, 1987).

However, its localization in the antiluminal membrane of brain capillaries supports the idea that SA-channels may contribute to the regulation of K^+ transport from brain to blood under edema. During brain edema accumulation of cations and water in the interstitial fluid was observed (Betz et al., 1989). A loss of potassium by the brain cells increases interstitial K^+ concentration up to 40 mM (Hossmann et al.,

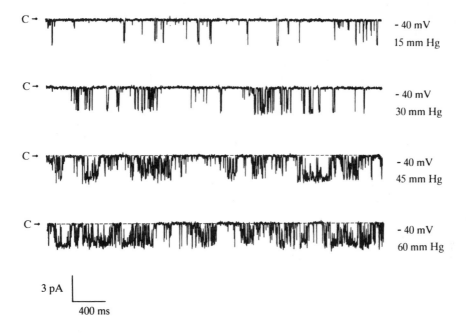

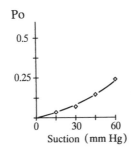

Figure 1. Dependence of channel activity in a cell-attached patch on pipette suction. a) SA-channel activity under various negative pressures, displayed next to the current trace. Channel openings are shown as downward deflections, C→ represents the closed state of the channel. The holding potential was −40 mV. Channel activity increased with increasing suction. b) Relationship between open probability and applied suction. (Reproduced with permission from Popp et al., 1992.)

1977; Hansen et al., 1980). At the same time hydrostatic pressure of the cerebrospinal fluid is enhanced. Based on these observations, the following model is proposed (Figure 2). The osmotic pressure could open the SA-channels and lead to an influx of sodium and calcium into the cells. Sodium influx will stimulate the Na^+/K^+ ATPase in the antiluminal membrane, resulting in an increased active uptake of K^+ from brain interstitial fluid into the cell. The simultaneous influx of Ca^{2+} ions through the SA-channels could activate Ca^{2+} sensitive K^+ channels

104

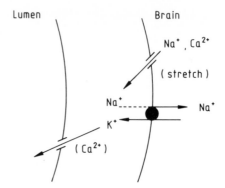

Figure 2. Schematic model for possible K$^+$ transport from brain to blood under edema.

located in the luminal membrane (Popp et al., 1992). Potassium ions could then leave the endothelial cell to enter the blood. Depolarization induced by the SA-channel would increase K$^+$ outflux across the luminal membrane, but would decrease the K$^+$ conductance at the antiluminal membrane, because of inward-rectifying properties of K$^+$ channels in this membrane (Hoyer et al., 1991).

References

Betz AL, Firth JA, Goldstein GW (1980). Polarity of the blood-brain barrier: Distribution of enzymes between the luminal and the antiluminal membranes of brain capillary endothelial cells. Brain Research 192:17–28.

Betz AL (1983). Sodium transport from blood to brain: Inhibition by furosemide and amyloride. J. of Neurochem. 41:1158–1164.

Betz AL, Iannotti F, Hoff JT (1989). Brain edema: A classification based on blood brain barrier integrity. Cerebrovascular and Brain Metabolism Reviews 1:133–154.

Bradbury MWB, Stulcova B (1970). Efflux mechanism contributing to the stability of the potassium concentration in cerebrospinal fluid. J. of Physiol. 208:415–430.

Christensen O (1987). Mediation of cell volume regulation by Ca^{2+} influx through stretch-activated channels. Nature 330:66–68.

Goldstein GW (1979). Relation of potassium transport to oxidative metabolism in isolated brain capillaries. J. of Physiol. 286:185–195.

Hansen AJ, Lund-Andersen H, Crone C (1977). K$^+$ permeability of the blood-brain barrier, investigated by aid of a K$^+$-sensitive microelectrode. Acta Physiol. Scand. 101:438–445.

Hansen AJ, Gjedde A, Siemkovicz F (1980). Extracellular potassium and blood flow in the post ischemic rat brain. Pflügers Arch. 389:1–7.

Hossmann KA, Sakaki S, Zimmermann V (1977). Cation activities in reversible ischemia of the cat brain. Stroke 8:77–81.

Hoyer J, Popp R, Meyer J, Galla H-J, Gögelein H (1991). Angiotensin II, vasopressin and GTP [Γ-S] inhibit inward rectifying K$^+$ channels in porcine cerebral capillary endothelial cells. J. Mem. Biol. 123:55–62.

Lansman JB, Hallam TJ, Rink TJ (1987). Single stretch-activated ion channels in vascular endothelial cells as mechanotransducers? Nature 325:811–813.

Olesen SP, Clapham DE, Davis PF (1988). Haemodynamic shear stress activates a K^+ current in vascular endothelial cells. Nature 331:168–170.

Popp R, Hoyer J, Meyer J, Galla H-J, Gögelein H (1992). Stretch-activated non-selective cation channels in the antiluminal membrane of porcine cerebral capillaries. J. of Physiol. 454:435–449.

Popp R, Gögelein H (1992). A calcium and ATP sensitive nonselective cation channel in the antiluminal membrane of rat cerebral capillary endothelial cells. Biochimica et Biophysica Acta 1108:59–66.

Gap Junction Channels

Nonselective Cation Channels: Pharmacology, Physiology and Biophysics
ed. by D. Siemen & J. Hescheler
© 1993 Birkhäuser Verlag Basel/Switzerland

The Gap Junction Channel

Rolf Dermietzel

Institute of Anatomy, University of Regensburg, Universitätsstrasse 31, D-93040 Regensburg

The Gap Junction Channel

One of the main avenues by which cells communicate with each other is through transmembranous channels which span the extracellular space, thereby allowing the direct transfer of small molecules and ions. The size range of admitted molecules allows the diffusion of current-carrying ions such as K^+, Na^+, Cl^-, and second messenger molecules such as cAMP, Ca^2, and inositol trisphosphate (IP3). Gap junction channels are clustered in the form of aggregates which can be best visualized by freeze-fracture electron microscopy. In conventional ultrathin electron micrographs these cell contacts are characterized by close appositions of the outer leaflets of the plasma membranes spaced by a gap of about 1.5 nm (Figure 1a, b). This characteristic ultrastructural feature has led to the somewhat inaccurate term "gap junction". In some tissues, such as neurons in the brain, gap junctions are small and may only contain a few of the intercellular channels in other tissues, such as liver or heart, gap junctions may be massive, containing as many as tens or hundreds of thousands of channels (Figure 1c).

Molecular Composition of the Gap Junction Channel

The recent extensive studies of gap junctions as important intercellular connections and the units that comprise them as an assembly of gated channels has led to the characterization of gap junctions in terms of a specific set of morphological, biochemical, and biophysical properties. I will first consider recent progress in the understanding of the molecular composition of gap junctions.

It has been known for some time that gap junction channels are composed of six subunits (connexin proteins) which form hemichannels or connexons contributed by each of the coupled cells (for recent reviews see Beyer et al., 1990; Dermietzel et al., 1990; Bennett et al., 1991) (Figure 2). These channels are permeant to molecules as large as

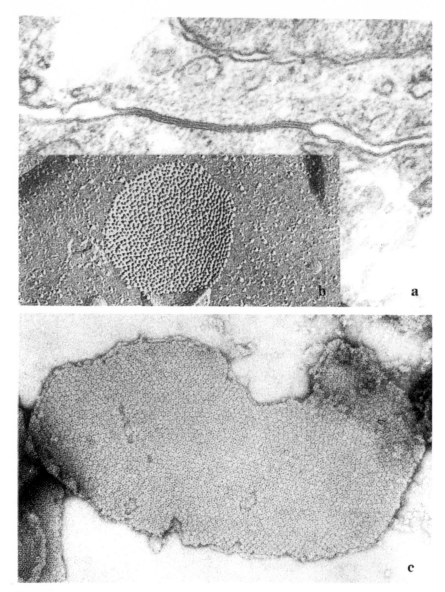

Figure 1. a) Thin-section electron micrograph of a gap junction between two hepatocytes. An electron lucent zone is evident between the outer layers of the apposed plasma membranes. b) Freeze-fracture electron micrograph of a liver gap junction. The intramembranous particles are assembled in a paracrystalline array. One particle represents a connexon which is formed by six subunits (the connexins). c) Negative stain image of an isolated gap junction plaque from liver. The stain (uranyl acetate) delineates the single channels. The electron dense deposits in the center of the channels mark the channel pore.

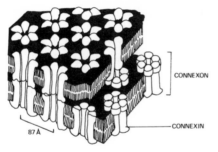

Figure 2. Diagram of the supramolecular composition of gap junctions as derived from x-ray diffraction studies (adopted from Makowsky et al., 1977).

1 kDa M_r, and in mammals are found in virtually every cell type except mature skeletal muscle, spermatozoa, and erythrocytes. Among the first tissues where gap junctions were documented structurally and electrophysiologically were crayfish nerve cord and Mauthner cells of goldfish brain (Furshpan and Potter, 1959; Robertson, 1963). However, because of the abundance of gap junctions in liver and heart, the search for the molecules forming these channels primarily utilized these tissues, leading ultimately to the elucidation of primary sequence and membrane topologies of three gap junction proteins (connexins 26 and connexin 32 from liver and connexin 43 from heart; Traub et al., 1982; Paul, 1986; Kumar et al., 1986; Beyer et al., 1987; Zhang and Nicholson, 1989). Most recently, low stringency hybridization and PCR cloning from genomic libraries have increased the number of putative mammalian connexins to about a dozen (Willecke et al., 1990; Willecke et al., 1991; Paul et al., 1991; Hoh et al., 1991; Hennemann et al., 1992; Haefliger et al., 1992; Gimlich et al., 1988) comprising a connexin supergene family with two distinct lineages termed class I (Bennett et al., 1991), or β group (Gimlich et al., 1988) in which connexins 32 and connexin 26 fall, and class II or α group, represented by connexin 43 (Gimlich et al., 1988). All family members possess similar gene structure, exhibit about 50% sequence homology at the amino acid level, and show diverse patterns of tissue distribution (Bennett et al., 1991; Dermietzel et al., 1991).

Functional studies in diverse cell types and in various expression systems have paralleled the progress obtained in the biochemical analyses. Gap-junction channels formed by different connexins have been shown to be regulated differently, both at the level of channel gating (such as different voltage sensitivities and responses to phosphorylating treatments (Bennett et al., 1991; Spray and Bennett, 1985; Spray et al., 1986; Spray and Sáez, 1987; Burt and Spray, 1988; Moreno et al., 1991; Swenson et al., 1989; Willecke et al., 1991) and at the level of channel expression (synthesis being altered, for example, by hormones, extracel-

112

lular matrix components, and cell cycle conditions (Spray et al., 1987; Sáez et al., 1986; Traub et al., 1987; Musil et al., 1990; Moreno et al., 1991; Meda et al., 1983; Dermietzel et al., 1987).

In addition to proof of the homology of various connexins, cloning experimentation has also provided insights into the orientation of the connexins within the plasma membrane. Hydropathicity plots show that connexins possess four hydrophobic domains which extend the length of the membrane. Each domain contains two loops on the cytoplasmic side and extracytoplasmic side, respectively (Figure 3). Topological studies, performed with the help of site-specific antibodies and proteolytic degradation (Zimmer et al., 1987; Hertzberg et al., 1988; Milks et al., 1988), have suggested that the N-terminus as well as the C-terminus of connexins are oriented toward the cytoplasmic side. The homology within these areas in the two major domains is approximately 40%; in contrast, the homology is 90% within the transmembranous domains and to about 100% in the extracytoplasmic loops. Thus, the domains responsible for the dogging mechanism of the hemichannels provided by each plasma membrane have remained highly conserved during the course of evolution, and variation in cytoplasmic residues may account for differential regulation of gap-junction channels in different tissues (see Spray and Burt, 1990). For example, connexin 32 is phosphorylated by cAMP-dependent kinase, while connexin 26 is not (Sáez et al., 1986; Traub et al., 1987, 1979).

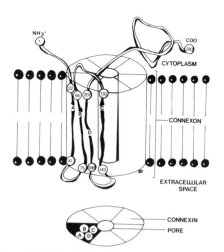

Figure 3. Topological model of the molecular arrangement of a single gap junction channel (connexon) which is composed of six subunits (connexins). The carboxyl- and amino-terminus of the connexins are predicted to be located on the cytoplasmic side. The four loops which span the membrane are designed as loops A-D. The amino acid transitions from the intramembranous position to the extramembranous part are numbered. The lower portion of the illustration depicts the hypothetical structure of á connexon in cross-section.

Future biomolecular investigation of connexins promises to provide additional insights into their manner of regulation and functioning. Dahl et al. (1987), for instance, were successful in injecting the mRNA of connexin 32 into *Xenopus* oocytes and observing the subsequent appearance of electrophysiological coupling between oocyte pairs. This successful attempt at transferring the messenger of a prominent gap-junction protein into an in vivo translation system in the *Xenopus* oocyte has made a number of other experiments possible which could be of extraordinary importance in characterizing the function of gap junction channels. For example, Swenson et al. (1989) recently introduced RNAs coding for different connexins (connexins 32 and 43) and were able to monitor the physiological properties of homologous as well as heterologous channel composites. Barrio et al. (1991) extended this kind of work and provided evidence that the biophysical properties of the gap-junction channels are largely influenced by the particular type of connexin. Other methods of studying gap junction properties using exogenous expression systems include the incorporation of gap junction proteins into bilayers (e.g., Spray et al., 1986; Young et al., 1987) and the stable transfection of communication deficient mammalian cells with connexin cDNA (Eghbali et al., 1990). The latter approach has allowed for the in situ monitoring of unitary conductances of truncated connexins (Spray et al., 1992).

Function of Gap Junction Channels

Gap junction channels permit the intercellular diffusion of ions and small molecules, including second messengers and tissue metabolites (Pitts, 1980; Flagg-Newton et al., 1981; Pitts and Finbow, 1986; Sáez et al., 1989a). Their channels exhibit little charge selectively; furthermore, the cutoff in terms of the size limit of permanent species is about 1 kDa (Simpson et al., 1977). Lucifer Yellow, a brilliantly fluorescent dye molecule with a molecular radius of about 0.7 nm, diffuses through gap junction channels, achieving a form of "dye coupling" which is commonly used to detect the presence of gap junctions between cells.

The electrical equivalent of permeance is conductance. The single-channel conductance for gap junctions in different tissues, for example, is about 50–150 pS (Neyton and Trautmann, 1985; Spray et al., 1986; Burt and Spray, 1988; Young et al., 1987).

Gap junction channels open and close in response to various treatments (for review, see Bennett and Goodenough, 1978; Spray and Bennett, 1985). Voltage dependence was first observed by Furshpan and Potter (1959) at the crayfish rectifying synapse. The observed symmetrical voltage dependence in which voltage across the junctional membrane closes the channels is now known to be a common property of embryonic vertebrate gap junctions, and gap junctions in certain tissues of adult mammals (Spray et al., 1979, 1981; Harris et al., 1981; Spray and Bennett, 1985). Inside-outside voltage sensitivity in which simultaneous hyperpolarization or depolarization of both cells changes junctional conductance is a common property of arthropod gap junctions, and is also seen in squid embryos (Obaid et al., 1983; Spray et al., 1984). The suggestion that voltage dependence may be due to redistribution of cytoplasmic ions such as Ca^{++} (Loewenstein, 1966) is erroneous; rather, the mechanism by which transjunctional voltage changes conductance of gap junctions is probably analogous to action in voltage-dependent, nonjunctional channels, although the sequences of the molecule comprising the sensor are presumably different.

In various tissues, intracellular pH gates gap junction channels, although the sensitivity varies (Turin and Warner, 1977; Spray et al., 1985). Gap junctions can also be closed by elevating Ca^{++} concentration to μM or even mM levels, as might occur during severe cell injury (Oliveira-Castro and Loewenstein, 1971; Spray et al., 1984). Other uncoupling treatments include lipophilic molecules such as heptanol, halothane, and octanol (Burt and Spray, 1989; Spray et al., 1985b). Phosphorylation of gap junction protein in liver cells and cell membranes has been temporarily correlated with increased junctional conductance in response to activation of kinases and cAMP (Sáez et al., 1986; Traub et al., 1987). In other tissues, phosphorylation and conductance studies indicate a variation in sensitivity. Second messenger molecules activating these kinases also affect expression of gap junction channels; furthermore, recent experiments indicate that effects of cAMP can be exerted at levels of transcription, mRNA stability, and turnover (Kessler et al., 1985; Spray et al., 1987; Sáez et al., 1989a, 1989b).

For each of these gating agents, the channel obeys the "all or nothing" principle in that the amplitude of single-channel currents and the relative permeability of the channels to molecules of various sizes are constant in response to these agents (Zimmermann and Rose, 1983; Verselis et al., 1986). Since channel substrates have not yet been resolved, gap junction channels appear simple in operation than nonjunctional channels gated by ligand binding.

Conclusion

Gap junctions are groupings of channels which, spanning the extracellular space between plasma membranes, couple neighboring cells. Thereby, a direct signal transfer from one cell to another is made possible. The orientation of the subunits, connexins, which surround hydrophilic channels in a hexameric arrangement, allows the channels to exist in both open and closed states. Biochemical as well as biomolecular studies have shown that connexins belong to a family of homologous but not identical proteins which are expressed in virtually every tissue. Despite the fact that the functions of the channels are still, to a certain extent, unknown, attempted blocking of the channels in differentiated tissue with specific antibodies has proven the significant role which these structures play during morphogenesis in multicellular organisms. The cooperative utilization of various investigatory methods, incorporating techniques from the areas of ultrastructural analysis as well as electrophysiology and molecular biology, provides promise in unveiling further structural and functional details within gap junctions.

References

Barrio LC, Suchyna T, Bargiello T, Xu LX, Roginski RS, Bennett MVL, Nicholson BJ (1991). Gap junctions formed by connexins 26 and 32 alone and in combination are differently affected by applied voltage. Proc. Natl. Acad. Sci. 88:8410–8414.

Bennett MVL, Goodenough DA (1978). Gap junctions, electrotonic coupling, and intercellular communication. Neurosciences Res. Prog. Bull. 16(3):373–485.

Bennett MVL, Barrio TA, Bargiello TA, Spray DC, Hertzberg E, Sáez JC (1991). Gap junctions: new tools, new answers, new questions. Neuron 6:305–320.

Beyer EC, Paul DL, Goodenough DA (1987). Connexin 43: A protein from rat heart homologous to a gap junction protein from liver. J. Cell Biol. 105:2621–2629.

Beyer EC, Paul DL, Goodenough DA (1990). Connexin family of gap junction proteins. J. Membrane Biol. 116:187–194.

Burt JM, Spray DC (1988). Ionotropic agents modulate gap junctional conductance between cardiac myocytes. Am. J. Physiol. 254:H1206–H1210.

Dahl G, Miller T, Paul D, Voellmy R, Werner R (1987). Expression of functional cell-cell channels from cloned rat liver gap junction complementary DNA. Science 236:1290–1293.

Dermietzel R, Yancey SB, Traub O, Willecke K, Revel J (1987). Major loss of the 28-Kd protein of gap junction in proliferating hepatocytes. J. Cell Biol. 105:1925–1934.

Dermietzel R, Hwang TK, Spray DS (1990). The gap junction family: structure, function and chemistry. Anat. Embryol. 182:517–528.

Eghbali B, Kessler JA, Spray DC (1990). Expression of gap junction channels in communication incompetent cells after stable transfection with complementary DNA encoding connexin 32. Proc. Natl. Acad. Sci. USA 87:1328–1331.

Flagg-Newton JL, Dahl G, Loewenstein WR (1981). Cell junction and cyclic AMP: I. Upregulation of junctional membrane permeability and junctional membrane particles by administration of cyclic nucleotide or phosphodiesterase inhibitor. J. Membrane Biol. 63:105–121.

Furshpan EJ, Potter DD (1959). Transmission at the giant motor synapses of the cray fish. J. Physiol. 145:289–325.

Gimlich RL, Kumar NM, Filula NB (1988). Sequence and developmental expression of mRNA coding for a gap junction protein in Xenopus. J. Cell Biol. 107:1065–1073.

Haefliger J-A, Bruzzone R, Jenkins NA, Gilbert DJ, Copeland NG, Paul DL (1992). Four novel members of the connexin family of gap junction proteins. [Molecular cloning, expression, and chromosome mapping]. J. Biol. Chem. 267:2057–2064.

Harris AI, Spray DC, Bennett MVL (1981). Kinetic properties of a voltage dependent junctional conductance. J. Gen. Physiol. 77:95–117.

Hennemann H, Kozjek G, Dahl E, Nicholson B, Willecke K (1992). Molecular cloning of mouse connexins 26 and 32: similar genomic organization but distinct promotor sequences of two gap junction genes. Europ. Cell Biol. 58:81–89.

Hertzberg EL, Diskert RM, Tiller AA, Zhou J, Cook RG (1988). Topology of the Mr 27000 liver gap junction protein: Cytoplasmic localization of amino- and carboxy-termini and a hydrophilic domain which is protease hyper-sensitive. J. Biol. Chem. 263:19105–19111.

Hoh JH, John SA, Revel J (1991). Molecular cloning and characterization of a new member of the gap junction gene family, connexin-31. J. Biol. Chem. 266:6524–6531.

Kessler JA, Spray DC, Sáez JC, Bennett MVL (1985). Development and regulation of electronic coupling between sympathetic neurons. In: Gap Junctions. Bennett and Spray editors. Cold Spring Harbor: Cold Spring Harbor Laboratories, pp. 231–240.

Kumar NM, Gilula NB (1986). Cloning and characterization of human and rat liver cDNAs coding for a gap junction protein. J. Cell Biol. 103:767–776.

Loewenstein WR (1966). Permeability of membrane junctions. Ann. N.Y. Acad. Sci. 137:441–469.

Makowsky L, Caspar DL, Phillips WC, Goodenough DA (1977). Gap junction structure. 2. Analysis of the X-ray diffraction data. J. Cell Biol. 74:629–645.

Meda P., Michaels RL, Halban PA, Orci L, Sheridan JD (1983). In vivo modulation of gap junctions and dye coupling between B-cells of the intact pancreatic islet. Diabetes 32:858–868.

Milks LC, Kumar NM, Houghten R, Unwin N, Gilula NB (1988). Topology of the 32-kd liver gap junction protein determined by site-directed antibody localizations. EMBO J. 7:2967–2975.

Moreno AP, Campos de Carvalho AC, Verselis V, Eghbali B, Spray DC (1991a). Voltage-dependent gap junction channels are formed by connexin 32, the major gap junction protein of rat liver. Biophys. J. 59(4):920–925.

Moreno AP, Eghbali B, Spray DC (1991b). Connexin 32 gap junction channels in stably transfected cells: Equilibrium and kinetic properties. Biophys. J. 60:1267–1277.

Musil LS, Cunningham BA, Edelmann GM, Goodenough DA (1990). Differential phosphorylation of the gap junction protein connexin 43 in junctional communication-competent and deficient cell lines. J Cell Biol. 111(5):2077–2088.

Neyton J, Trautmann A (1985). Single-channel currents of an intercellular junction. Nature 317:331–335.

Obaid AL, Socolar SJ, Rose B (1983). Cell-to-cell channels with two independently regulated gates in series: Analysis of junctional conductance modulation by membrane potential, calcium, and pH. J. Membrane Biol. 73:69–89.

Oliveira-Castro GM, Loewenstein WR (1971). Junctional membrane permeability: Effects of divalent cations. J. Membrane Biol. 5:51–77.

Paul DA (1986). Molecular cloning of cDNA for rat liver gap junction protein. J. Cell Biol. 103:123–134.

Paul DL, Ebihara L, Takemoto LJ, Swenson KI, Goodenough DA (1991). Connexin 46, a novel lens gap junction protein, induces voltage-gated currents in nonjunctional plasma membrane of Xenopus oocytes. J. Cell Biol. 115:1077–1089.

Pitts JD (1980). The role of junctional communication in animal tissues. In Vitro 16(12):1049–1056.

Pitts JD, Finbow ME (1986). The gap junction. J. Cell Sci. 4:239–266.

Robertson JD, Bodenheimer PS, Stage DE (1963). The ultrastructure of Mauthner cell synapses and nodes in goldfish brain. J. Cell Biol. 19:159–199.

Sáez C, Spray DC, Nairu A, Hertzberg EL, Grenngard P, Bennett MVL (1986). cAMP increases junctional conductance and stimulates phosphorylation of the 27 kDA principal gap junction polypeptide. Proc. Natl. Acad. Sci. (USA) 83:2473–2477.

Sáez JC, Connor JA, Spray DC, Bennett MVL (1989a). Hepatocyte gap junctions are permeable to the second messenger, inositol 1,4,5-triphosphate, and to calcium ions. Proc. Nat. Acad. Sci. USA 86:2708–2712.

Sáez JC, Gregory WA, Watanabe T, Dermietzel R, Hertzberg EL, Reid L, Bennett MVL, Spray DC (1989b). cAMP delays disappearance of gap junctions between pairs of rat hepatocytes in primary culture. Am. J. Physiol. 257:C1–C11.

Simpson IB, Rose RW, Loewenstein WR (1977). Size limit of molecules permeating the junctional channel. Science 195:294–296.

Spray DC, White RL, Campos de Cavalho AC, Harris AL, Bennett MVL (1984). Gating of gap junction channels. Biophys. J. 45:219–230.

Spray DC, Bennett MVL (1985). Physiology and pharmacology of gap junctions. Ann. Rev. Physiol. 47:281–303.

Spray D, White RL, Verselis V, Bennett MVL (1985). General and comparative physiology of gap junction channels. In: Gap Junctions. Bennett and Spray editors. New York: Cold Spring Harbor Laboratory, pp. 139–153.

Spray DC, Campos de Carvalho AC, Bennett MVL (1986). Sensitivity of gap junctional conductance to H ions in amphibian embryonic cells is independent of voltage sensitivity. Proc. Natl. Acad. Sci. USA 83:3533–3536.

Spray DC, Sáez JC (1987). Agents that regulate gap junctional conductance. Sites of action and specificities. Adv. Mod. Environ. Toxicol. 14:1–26.

Spray DC, Burt JM (1990). Structure-activity relations for the cardiac gap junction channel. Amer. J. Physiol. 258:C195–C205.

Swenson KI, Jordan JR, Beyer EC, Paul DL (1989). Formation of gap junctions by expression of connexins in Xenopus oocyte pairs. Cell 57:145–155.

Traub O, Janssen-Timmen U, Drüge PM, Dermietzel R, Willecke K (1982). Immunological properties of gap junction protein from mouse liver. J. Cell. Biochem. 19:27–44.

Traub O, Look J, Paul D, Willecke K (1987). Cyclic adenosine monophosphate stimulates biosynthesis and phosphorylation of the 26 kDa gap junction protein in cultured mouse hepatocytes. Europ. J. Cell Biol. 43:48–54.

Traub O, Look J, Dermietzel R, Brümmer F, Hülser D, Willecke K (1989). Comparative characterization of the 21 kDa and 26 kDa gap junction proteins in murine liver and cultured hepatocytes. J. Cell Biol. 108:1039–1051.

Turin L, Warner AE (1977). Carbon dioxide reversible abolishes ionic communication between cells of the early amphibian embryo. Nature 270:56–67.

Verselis V, White RL, Spray DC, Bennett MVL (1986). Gap junctional conductance and permeability are linearly related. Science 234:461–464.

Willecke K, Jungbluth S, Dahl E, Hennemann H, Heynkes R, Grzeschik K-H (1990). Six genes of the human connexin gene family coding for gap junctional proteins are assigned to four different human chromosomes. Europ. J. Cell Biol. 53:275–280.

Willecke K, Heynkes R, Dahl E, Stutenkemper R, Hennemann, H, Jungbluth S, Suchyna T, Nicholson BJ (1991). Mouse connexin 37: Cloning and functional expression of gap junction gene highly expressed in lung. J. Cell Biol. 114(5):1049–1057.

Young JDE, Cohn ZA, Gilula NB (1987). Functional assembly of gap junction conductance in lipid bilayers: Demonstration that the major 27 kd protein forms the junctional channel. Cell 48:733–743.

Zhang J, Nicholson BJ (1989). Sequence and tissue distribution of a second protein of hepatic gap junctions, Cx26, as deduced from its cDNA. J. Cell Biol. 109(6, Pt. 2):3391–3401.

Zimmer DB, Green CR, Evenas WH, Gilula NB (1987). Topological analysis of the major protein in isolate and intact rat liver gap junctions and gap junction derived from single membrane structures. J. Biol. Chem. 262:7751–7763.

Zimmerman AL, Rose L (1983). Analysis of cell-to-cell diffusion kinetics: Changes in junctional permeability without accompanying changes in junctional selectivity. Biophys. J. 41:216a.

Cyclic Nucleotide-Activated Nonselective Cation Channels

Nonselective Cation Channels: Pharmacology, Physiology and Biophysics
ed. by D. Siemen & J. Hescheler
© 1993 Birkhäuser Verlag Basel/Switzerland

Cyclic Nucleotide-Gated Nonselective Cation Channels: A Multifunctional Gene Family

Colin J. Barnstable

Department of Ophthalmology and Visual Science, Yale University School of Medicine, 330 Cedar Street, New Haven, CT 06510, USA

Introduction

The light-regulated flow of ions across the plasma membrane of vertebrate rod and cone photoreceptors passes through a nonselective cation channel that is directly gated by guanosine 3′,5′-cyclic monophosphate (cGMP). It is the purpose of this brief chapter to review the physiological and structural properties of this channel. As will be described below, these channels appear to have structural features in common with some other voltage-gated channels. Whether or not they also show structural relationships to other types of nonselective cation channels, such as those described in other chapters in this volume, remains to be determined.

Over the past few years it has become clear that cyclic nucleotide-gated cation channels represent a considerable gene family whose products are expressed in a wide variety of neuronal and nonneuronal cells. Thus, an understanding of the nonselective cation channels of a specialized sensory cell like the rod photoreceptor may help our understanding of physiological processes in many other cell types.

Physiological Properties of the Rod Photoreceptor Nonselective Cation Channel

Almost three decades ago it was shown that rod and cone photoreceptors exhibit a membrane hyperpolarization in response to light (Bortoff, 1964; Tomita, 1965). Subsequent work has shown that a light sensitive conductance is open in the dark and the cell is maintained in a depolarized state. Light initiates a series of reactions that lead to a reduction in the concentration of cGMP and a reduction or complete closing of the conductance.

The ion selectivity of the conductance was initially determined in intact photoreceptors. Using a rapid perfusion system it was found that apparent permeability ratios for the monovalent cations Li^+, K^+, Rb^+, and Cs^+ relative to Na^+ were 1.1, 0.7, 0.5 and 0.3 respectively (Yau and Nakatani, 1984; Hodgkin et al., 1985). Divalent cations were also found to carry current through the channel, but measurement of their permeability is difficult because they can also bind to a site within the channel and thus block current flow (Lamb and Matthews, 1988; Nakatani and Yau, 1988). Under physiological conditions with external concentrations of Na^+, Ca^{2+} and Mg^{2+} of approximately 110 mM, 1 mM and 1.0 mM apparent permeability ratios of $P_{Ca}/P_{Na} = 12.5$ and $P_{Mg}/P_{Na} = 2.5$ were obtained (Nakatani and Yau, 1988). Other divalent cations, including Ba^{2+}, Sr^{2+}, Mn^{2+}, Ni^{2+} and Co^{2+}, were also able to pass through the channel (Capovilla et al., 1983; Yau and Nakatani, 1984, 1985). Although the channel is more permeable to Ca^{2+}, at least 70% of the current is carried by Na^+ under physiological conditions. Similarly, the channel is also permeable to K^+, but under physiological conditions in the dark at a membrane potential of −40 mV the K^+, efflux is less than 20% of the Na^+ influx. The various components of the flux through the channel result in a reversal potential for the light sensitive current in isolated salamander rod photoreceptors that was found to be near zero millivolts (Bader et al., 1979).

Using a single electrode voltage clamp technique the current-voltage relationship in isolated salamander rods showed an outward rectification (Bader et al., 1979). The current-voltage relationship was essentially flat in the physiological range of membrane potentials, suggesting that the dark current is not influenced by changes in membrane potential.

Early measurements of the single-channel conductance of the rod photoreceptor channel suggested a value of about 0.1 pS (see Yau and Baylor, 1989, for a review). This value was almost certainly due to measurements being made when the channels were blocked by divalent cations. In the absence of divalents the channel exhibits a large conductance of about 25 pS and a number of subconductance states whose nature is not yet fully understood.

It is thought that the channels are present in the rod photoreceptor plasma membrane at a density of up to $1000\ \mu m^{-2}$ (Haynes et al., 1986). Normally, however, only 1–2% of channels are thought to be open even in complete darkness. The reasons for this are not clear but arguments have been made that this can increase the sensitivity and decrease the noise in photoreceptors (Yau and Baylor, 1988).

The most significant result from the first published account of recordings from excised patches from rod photoreceptors was that the channels were directly responsive to cGMP (Fesenko et al., 1985). The concentration of cGMP necessary to open half the channels was in the range of $10–50\ \mu M$. The reason for such a range is not clear but it is thought that binding of cGMP to the protein, and thus channel activation by cGMP, appears to be sensitive to depolarization. A sigmoid relationship between steady state macroscopic current and cGMP concentration has been found and from these curves a Hill coefficient of approximately 3 has been derived (see Yau and Baylor, 1989 for a summary of many studies).

Recordings from excised inside-out patches from cone photoreceptors have shown that they too have a nonselective cation channel gated by cGMP (Haynes and Yau, 1985). The current-voltage relationship for the cone channel has a different shape than for rods in the presence of divalent cations. This difference is thought primarily to be due to differences in the blocking interactions between the divalent cations and the channel protein.

Structural Characterization of the Rod Photoreceptor cGMP-Gated Cation Channel

When washed bovine rod outer segment membranes were dissolved in detergents and fractionated on ion exchange and affinity resins, a single protein of approximately 63 kD was found to exhibit channel activity when reconstituted into planar lipid membranes (Cook et al., 1987). The channels measured with purified protein had a single-channel conductance of 26 pS and showed cooperativity in response to cGMP with a Hill coefficient of between 2 and 3. The Km for cGMP in the reconstituted channels was 31 μM, the same order of magnitude as found for the channel in patches of rod photoreceptor membrane.

Partial amino acid sequences were obtained for the purified 63 kD protein and these were used to construct oligonucleotides for use as probes to screen a bovine retinal cDNA library (Kaupp et al., 1989). Clones were sequenced and shown to contain a single open frame that contained sequences corresponding to the independently derived peptide sequences. The open reading frame corresponded to a protein of 690 amino acids with a calculated molecular weight of 79,601 daltons. The protein purified from rods was clearly a proteolytic cleavage product with the N-terminal amino acid corresponding to serine 93 of the full length sequence. All the available evidence suggests that the 63 kD form is the major species found in rods. Where the cleavage takes place, and whether it has any biological purpose, remains to be shown.

Expression of the full length channel sequence in *Xenopus* oocytes confirmed that it was a cGMP-gated cation channel (Kaupp et al., 1989). The channel showed a Km for cGMP of 43–71 μm and a Hill coefficient of 1.5–2.1. The single-channel conductance and ion selectivities were similar to those found for the channel in membrane patches isolated from rod photoreceptors.

Using this sequence as a basis, other groups have now cloned the rod photoreceptor channels from human, mouse and rat. The deduced

124

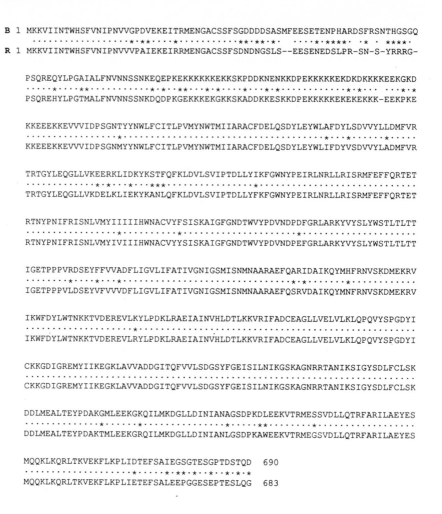

B 1 MKKVIINTWHSFVNIPNVVGPDVEKEITRMENGACSSFSGDDDDSASMFEESETENPHARDSFRSNTHGSGQ
· * · * * · · * · · · · · · * · * * * · · · · * · · · · · * · · · · · · · * · · · · · · · · · · · * * * *
R 1 MKKVIINTWHSFVNIPNVVVPAIEKEIRRMENGACSSFSDNDNGSLS--EESENEDSLPR-SN-S-YRRRG-

PSQREQYLPGAIALFNVNNSSNKEQEPKEKKKKKKEKKSKPDDKNENKKDPEKKKKKEKDKDKKKKEEKGKD
· · · · · · · · · · * · · · · · * * · · · · · · · · * · * · · * * · · · * · * · · · · · · · · · · * · · · · · · · · · · · · · · · · * * · · · · · · · * * · *
PSQREHYLPGTMALFNVNNSSNKDQDPKGEKKKEKGKKSKADDKKESKKDPEKKKKKEKEKEKKK-EEKPKE

KKEEEKKEVVVIDPSGNTYYNWLFCITLPVMYNWTMIIARACFDELQSDYLEYWLAFDYLSDVVYLLDMFVR
· * · * · · * · · · · · · · · · * · · · · ·
KKEEEKKEVVVIDPSGNMYYNWLFCITLPVMYNWTMIIARACFDELQSDYLEYWLIFDYVSDVVYLADMFVR

TRTGYLEQGLLVKEERKLIDKYKSTFQFKLDVLSVIPTDLLYIKFGWNYPEIRLNRLLRISRMFEFFQRTET
· · · · · · · · · · · · · * · · * · · · * * * · · · · · · · · · · · · · · · · · * ·
TRTGYLEQGLLVKDELKLIEKYKANLQFKLDVLSVIPTDLLYFKFGWNYPEIRLNRLLRISRMFEFFQRTET

RTNYPNIFRISNLVMYIIIIIHWNACVYFSISKAIGFGNDTWVYPDVNDPDFGRLARKYVYSLYWSTLTLTT
· · · · · · · · · · · · · · · * · · · · · · · · * · · · · · · · · · · · · · · * · · · · · · · · · · · ·
RTNYPNIFRISNLVMYIVIIIHWNACVYYSISKAIGFGNDTWVYPDVNDPEFGRLARKYVYSLYWSTLTLTT

IGETPPPVRDSEYFFVVADFLIGVLIFATIVGNIGSMISNMNAARAEFQARIDAIKQYMHFRNVSKDMEKRV
· · · · · · * · · · · * · * · * · · · · · · · · * · · · · ·
IGETPPPVLDSEYVFVVVDFLIGVLIFATIVGNIGSMISNMNAARAEFQSRVDAIKQYMNFRNVSKDMEKRV

IKWFDYLWTNKKTVDEREVLKYLPDKLRAEIAINVHLDTLKKVRIFADCEAGLLVELVLKLQPQVYSPGDYI
· · · · · · · · · · · · · · · · · · * ·
IKWFDYLWTNKKTVDEREVLRYLPDKLRAEIAINVHLDTLKKVRIFADCEAGLLVELVLKLQPQVYSPGDYI

CKKGDIGREMYIIKEGKLAVVADDGITQFVVLSDGSYFGEISILNIKGSKAGNRRTANIKSIGYSDLFCLSK
· ·
CKKGDIGREMYIIKEGKLAVVADDGITQFVVLSDGSYFGEISILNIKGSKAGNRRTANIKSIGYSDLFCLSK

DDLMEALTEYPDAKGMLEEKGKQILMKDGLLDINIANAGSDPKDLEEKVTRMESSVDLLQTRFARILAEYES
· · · · · · · · · · · * · · · · · · * · · · · · · · · · * · · · · · · * * · · · · · · · · * · · · · · · · · · ·
DDLMEALTEYPDAKTMLEEKGRQILMKDGLLDINIANLGSDPKAWEEKVTRMEGSVDLLQTRFARILAEYES

MQQKLKQRLTKVEKFLKPLIDTEFSAIEGSGTESGPTDSTQD   690
· * · · · · · * · * * * · * · · * · · * · * · *
MQQKLKQRLTKVEKFLKPLIETEFSALEEPGGESEPTESLQG   683

Figure 1. Comparison of the aminoacid sequences of the bovine (B) and rat (R) rod photoreceptor cGMP-gated cation channels. Identities are marked with a period, differences with an asterisk. Seven gaps have been introduced into the rat sequence so as to maintain the optimal alignment. 71 residues, or just over 10% of the sequence, differ between the two species.

amino acid sequences of the bovine and rat channels are shown in Figure 1. There is generally a high degree of homology throughout the molecule, with most of the variation clustered in the N- and C-terminal regions. In the N-terminal region there are a number of deletions in the rat (and mouse) sequence as compared with bovine. At present it is not known whether the channel proteins of these other species are cleaved in the same way as that of bovine rods.

A Structural Model of the Channel

Analysis of the hydrophobicity of the amino acids in the channel sequence led to the initial proposition of a secondary structure containing 4 or 6 transmembrane helices (Kaupp et al., 1989). It was then suggested that the channel had some homology with other voltage-gated channels including the *Shaker* K$^+$ channel from *Drosophila* (Jan and Jan, 1990). Subsequent cloning of other K$^+$ channel genes has strengthened this idea and has allowed a more precise alignment (Guy et al., 1992). In Figure 2 the *Shaker* and *eag* K$^+$ channels from *Drosophila* and the rat rod cGMP-gated channel are aligned. By analogy with these K$^+$ channels, the revised model for the rod cGMP-gated cation channel has six transmembrane segments. Since the cGMP binding site in the C-terminal region (see below) is cytoplasmic, both N- and C- termini should be on the cytoplasmic side of the membrane. Between transmembrane segment 5 and 6 is a stretch of amino acids that are believed to form a β-sheet that lines the aqueous pore. The importance of these residues has been tested tested experimentally for K$^+$ channels but not yet for the cGMP-gated cation channel (Yellen et al., 1991; Hartmann et al., 1991; Yool and Schwartz, 1991). The supposed pore sequence of the cGMP-gated channel contains a group of three prolines at its C terminus. Whether this distorts the pore structure is not yet known. K$^+$

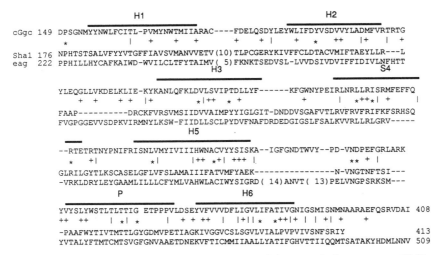

Figure 2. Alignment of the portions of the sequences of the rat rod photoreceptor cGMP-gated cation channel, the Drosophila Sha1 K+ and the Drosophila eag channel. The putative membrane spanning regions are marked as is the putative pore sequence. Identities are shown between all three molecules (∗), between the rat and the Sha1 sequences (|) and between the rat and eag sequences (+). Gaps introduced into the sequences to optimize alignment are shown as dashes or as the number of residues omitted from a particular sequence. Sequences are derived from our unpublished data (rat cGMP-gated channel), (Wei et al., 1990) (Sha1) and (eag).

channels are believed to associate as tetramers, giving a pore that is lined with 8 strands of β-sheet or a β-barrel. Examination of the pore-lining sequences does not provide any obvious clue as to why the cGMP-gated channel should act as a non-selective channel, nor does it show an obvious site for binding of divalent cations that can block the channel. What the model also does not show is which stretches of sequence are folded to lie close to the pore on either side of the membrane and with which divalent cations and other compounds could act to influence channel properties.

The *Shaker* family of K^+ channels is voltage-sensitive and this property is thought to be due to the series of five positively charged residues lying at every third residue in the S4 transmembrane segment. The cGMP-gated channel has four of these residues but in addition has several negatively charged residues. For example the net charge in this segment is $+5$ for the Shaker channel but only $+2$ for the rat rod photoreceptor channel. This difference may help explain why the cyclic nucleotide-gated channels do not show the same voltage-sensitivity as the voltage-gated K^+ channels.

Genomic Organization of the cGMP-Gated Channel

Genes for the rod cGMP-gated channels have been isolated from human and rat (Dhallan et al., 1992; our unpublished results). As indicated in Figure 3 they show near identity in exon/intron organization. Of interest is the observation that the carboxy terminal 2/3 of the molecule is contained in one large exon. This exon includes four of the six putative transmembrane segments, the pore-lining residues and the cGMP-binding site. The remaining amino terminal 1/3 of the protein is encoded by seven small exons. For the human gene, alternative splicing has been shown to remove one of the small exons (exon 8) to produce an in-frame deletion of 36 amino acids. It is not clear what function this alternatively spliced form of the channel carries out in the rod photoreceptor. When cDNA corresponding to this variant form of the channel was expressed in human embryonic kidney cells, no functional channels were detected. It is possible that the variant can form complexes with the full length protein and thus give rise to channels with modified properties.

Cyclic Nucleotide-Gated Cation Channels of Olfactory Sensory Neurons

Responses to odorant molecules take place on the specialized cilia of olfactory sensory neurons. Stimulation of olfactory receptors results in a cascade of reactions that leads to an increase in cAMP concentration. With the cloning of an olfactory system specific G-protein and olfactory receptors that have homology with the visual pigment protein opsin, it

Figure 3. Gene structure of the rod photoreceptor cGMP-gated channel. Exons 1–4 are not fully characterized in the rat, but the others are organized identically to the human. Coding regions are shown hatched and noncoding regions are shown as solid bars. It is not clear that species other than human have an *alu* element as exon 2. The exon sizes are given in base pairs. Data for the rat from our unpublished results, for the human from Dhallan et al. (1992).

is now clear that the basic mechanisms of olfaction and vision have many similarities (Buck and Axel, 1991; Jones and Reed, 1989). An important difference, however, is that stimulation of the olfactory system results in an increase in cyclic nucleotide concentration whereas stimulation of the visual system leads to a decrease in cyclic nucleotide concentration.

Even though the cilia of olfactory sensory cells are only about $0.25\,\mu$m in diameter, it has been possible to use patch methods to study the conductance underlying the olfactory response (Nakamura and Gold, 1987). Bath application of cAMP caused a reversible increase in membrane conductance. Since this occurred in the absence of added nucleotide triphosphates, it was assumed to result from the direct action of the cyclic nucleotide. The conductance showed a reversal potential near 0 mV and could be blocked by divalent cations. Interestingly, the conductance showed a higher affinity for cGMP than for cAMP. Although most patches showed affinities in the low micromolar range, some had affinities in the millimolar range. This suggests the existence of either a different receptor or, more likely, a mechanism that modulates the activity of the channel.

With the isolation of the rod photoreceptor cGMP-gated channel, the important question of whether the olfactory conductance was a related channel arose. Screening of olfactory epithelial cDNA libraries with probes derived from the bovine rod cGMP-gated channel led to the isolation of clones that were shown by expression to encode such channels (Dhallan et al., 1990; Ludwig et al., 1990). Comparison of the coding sequences of the rod photoreceptor and olfactory channels show strong homology through much of the molecule, although there is considerable variation at the N and C termini. The putative amino acids lining the pore show 18 identities out of 20 residues.

Cyclic Nucleotide Binding Sites

In Figure 4 the sequences of a series of channels in the putative region of cGMP binding are shown. This region also shows good homology with the two cGMP-binding domains of cGMP-dependent protein kinase, and with the cAMP-binding domain of catabolite activator protein (CAP) from *E. coli*. Since the crystal structure of CAP has been determined, it has been possible to derive a reasonable molecular model of the nucleotide-binding site in the rod channel (Kumau and Weber, 1992). The binding site forms a pocket and within this it is thought that Arg 552 (rod numbering) forms an ionic interaction with the phosphate of cGMP and Gly 536 and Glu 537 form hydrogen bonds with the ribose 2'-OH. Other hydrogen bonds may be formed by Thr 553, Ile 538 and Ser 539.

```
                                                      ****   *
cGgc 491 YSPGDYICKKGDIGREMYIIKEGKLAVV-ADDGITQFVVL---SDGSYFGEISILNIKGSK
cGKa 122 YGKDSCIIKEGDVGSLVYVMEDGKVEVT------KEGVKLCTMGPGKVFGELAIL------
cGKb 240 YENGEYIIRQGARGDTFFIISKGKVNVTREDSPNEDPVFLRTLGKGDWFGEKALQ------
CAP   23 YPSKSTLIHQGEKAETLYYIVKGSVAVLIKDEEGKE-MILSYLNQGDFIGELGLFEE----
eag  592 SAPGDLLYHTGESIDSLCFIVTGSLEVI-QDDEV--VAIL---GKGDVFGD-QFW--KDS--

         **
         AGNRRTANIKSIGYSDLFCLSKDDLMEALTEYPDAKTMLEEKGRQILMKDGLLDINIANLG
         YNCTRTATVKTLVNVKLWAIDRQCFQTIMMRTGLIKHTEYMEFLKSVPTFQSLPEE
         GEDVRTANVIAAEAVTCLVIDRDSFKHLIGGLDDVSNKAYEDAEAKAKYEAEAAFF
         -GQERSAWVRAKTACEVAEISYKKFRQLIQVNPDILMRLSAQMARRLQVTSEKVGN
         AVGQSAANVRALTYCDLHAIKRDKLLEVLDFYS
```

Figure 4. The putative cGMP binding region of the rat rod photoreceptor cGMP-gated cation channel (cGgc) as judged by homology with the two known cGMP-binding sites of cGMP-activated protein kinase (cGKa and cGKb), with the *E. coli* cGMP-binding domain of catabolite activator protein (CAP) and with the *Drosophila eag* K$^+$ channel (eag). Residues identical in at least four of the sequences, or showing only very conservative substitutions, are shown in bold. Residues specifically mentioned in the text are marked with an asterisk.

The designation of a region of the rod photoreceptor cGMP-gated channel as the cyclic nucleotide-binding site was originally made on the basis of sequence homology with a number of other cyclic nucleotide-binding proteins. A segment of 80 amino acids was similar to cAMP-binding proteins and to cGMP-dependent protein kinase. Of particular interest is a threonine residue at position 553 (rod numbering). This is present in cGMP-dependent protein kinases but in 23 of 24 cAMP-dependent protein kinases this residue is alanine. Both the rod and olfactory channels have threonine at this position. This, presumably helps to explain why the olfactory channel has a higher affinity for cGMP. Site-directed mutagenesis of full length cDNAs coding for the rod photoreceptor and olfactory channels has been used to substitute an alanine residue for this threonine (Altenhofen et al., 1991). The mutant clones were tested by expression in *Xenopus* oocytes. The sensitivity of the channels to cGMP was reduced by 30-fold but the sensitivity to cAMP was not affected. Substitution of the threonine residue by serine increased the sensitivity to cGMP by 2- to 5-fold. These results suggest that the hydroxyl group on the threonine residue interacts with cGMP but not cAMP, and may be important for ligand discrimination.

A More Extensive Gene Family of Cyclic Nucleotide-Gated Cation Channels

In addition to the well characterized photoreceptor and olfactory receptor channels, there is suggestive evidence that members of this gene family may also be expressed in other tissues. Nothern blot analysis carried out at high stringency using a rod photoreceptor channel probe has detected channel gene expression in heart, kidney and possibly other brain regions (Ahmad et al., 1990). The expression in kidney has been confirmed by cloning the sequences expressed in the M1 cell line derived from mouse kidney cortical collecting duct (Ahmad et al., 1992a; see also chapter by Korbmacher and Barnstable, this volume).

Within retina there is also physiological evidence for cGMP-gated channels in ON-bipolar cells (Nawy and Jahr, 1990; Shiells and Falk, 1992). Activation of a glutamate receptor is thought to lead to an increase in cGMP-phosphodiesterase activity which in turn hydrolyses cGMP and leads to channel closure. Immunocytochemical studies using an antibody prepared against bovine rod photoreceptor channel reacted with rods but not other retinal cells (Wässle et al., 1992). This result suggests that either the channel expressed ON-bipolar cells is not the same as that expressed in rod photoreceptors, or it is expressed at much lower levels.

To explore further the expression of this channel gene family in retina we have carried out both in situ hybridization on cryostat sections of adult rat retina and PCR-based cloning from small pools of identified retinal ganglion cells. In situ labelling obtained with antisense probes was localized over the cell bodies (outer nuclear layer) and inner segments of the rod photoreceptors as expected. More detailed examination at higher power also showed clusters of silver grains over cells in the inner muclear layer that from their size and location were almost certainly bipolar cells (Ahmad et al., 1992b). More surprising, however, was the detection of a clearly labelled subpopulation of cells in the ganglion cell layer. A PCR analysis on small groups of up to 15 dissociated cells, identified as ganglion cells by retrograde transport of a fluorescent tracer, has been carried out using primer sets corresponding to the most conserved region of the rod photoreceptor cGMP-gated ion channel (Ahmad et al., 1992b). Cloning and sequencing of the PCR products has confirmed that ganglion cells express a channel gene very similar or identical to that found in rod photoreceptors.

To determine whether the cGMP-gated ion channel sequences identified might be functional, we carried out preliminary whole cell patch clamp recording on cultured retinal ganglion cells. Application of cGMP or 8-Br-cGMP caused a small (approx. 5 mV) depolarization in 6 of 18 cells tested (unpublished results in collaboration with D. Cummins). Although confirmation that these effects are due to a direct effect of the cyclic nucleotide will require single-channel recording from excised patches, these results in combination with the in situ hybridization do suggest that a subpopulation of retinal ganglion cells express a channel that can be activated by cGMP.

It has recently been shown that many cells in the ganglion cell layer express a soluble guanylate cyclase (Ahmad and Barnstable, 1993), This enzyme can be activated by nitric oxide. In the rat retina a subpopulation of amacrine cells express an NADPH diaphorase enzyme, which is thought to be a form of nitric oxide synthase (Sandell, 1985; Sagar, 1990; Hope et al., 1991; Dawson et al., 1991). Activation of the enzyme within these cells could lead to release of nitric oxide which might lead to a depolarization of a subpopulation of ganglion cells. Although this effect may not be sufficient to elicit action potentials within these cells, it may well be sufficient to modulate the responsiveness of the cells to other inputs. The specificity of such an effect would reside in the short lifetime of nitric oxide and the restricted distribution of cGMP-gated channels in the ganglion cell layer. It will be interesting to determine whether the cGMP responsive ganglion cells form a homogeneous physiological subclass and whether these cells contain other molecules that could modulate or effect cGMP responses such as cGMP phosphodiesterases or cGMP-dependent protein kinases.

Conclusions

There is a growing family of cyclic nucleotide-gated nonselective cation channels. Most of our knowledge about their structure and function comes from the large amount of work on the channel in rod photoreceptors, although information about the channel in olfactory sensory neurons is growing rapidly. What we have is a large body of data on the detailed electrophysiological properties of the channel on the one hand, and an amino acid sequence on the other. In the future we may obtain true structural information that will allow us to explain features such as the voltage dependence of cGMP binding, ion selectivity, divalent cation block or cooperativity of cGMP binding. In addition, we may begin to understand how the same or very similar channels can serve so many different functions, from sensory transduction, to neuronal modulation, to ion transport across epithelia. As we determine how many genes constitute this channel family, we can also investigate the mechanisms leading to selective expression in certain tissues and cell types. Finally, we can begin to determine the relationships between the cyclic nucloetide-gated channels and other channel families. Earlier the possible relationship with some K^+ channels was mentioned, linking the cyclic nucleotide-gated channels to the large interrelated family of voltage-gated channels. Other channels, including other nonselective cation channels, may also show structural similarities. It will be particularly interesting to know whether calcium-dependent nonselective cation channels are at all related to the cyclic nucleotide-gated channel family. Finally, we hope that structural information can be of some help in elucidating the functions of nonselective cation channels in the wide array of tissues in which they have been described.

Acknowledgements
Work referred to in this chapter from my own laboratory has been supported by NIH grants NS 20483 and EY 07119 as well as by a Jules and Doris Stein Research to Prevent Blindness, Inc. Professorship and the Darien Lions.

132

References

Ahmad A, Barnstable, CJ (1993). Molecular cloning and differential laminar expression of particulate and soluble gunaylate cyclases in rat retina. Exp. Eye Res. 56:51–62.

Ahmad L, Cummins D, Barnstable, CJ (1992b). Rat retinal Ganglion cells express a cGMP-gated ion channel. Invest. Ophthalmol. Vis. Sci. Suppl. 33:940.

Ahmad I, Korbmacher C, Segal AS, Cheung P, Boulpaep EL, Barnstable, CJ (1992a). Mouse cortical collecting duct cells show nonselective cation channel activity and express a gene related to the cGMP-gated rod photoreceptor channel. Pros. Natl. Acad. Sci. 89:10262–10266.

Ahmad I, Redmond LJ, Barnstable, CJ (1990). Developmental and tissue-specific expression of the rod photoreceptor cGMP-gated ion channel gene. Biochem. Biophys. Res. Comm. 173:463–470.

Altenhofen W, Ludwig J, Eisman E, Kraus W, Bönigk W, Kaupp UB (1991). Control of ligand specificity in cyclic nucleotide-gated channels from rod photoreceptors and olfactory epithelium. Proc. Natl. Acad. Sci. 88:9868–9872.

Bader CR, MacLeish PR, Schwartz EA (1979). A voltage-clamp study of the light response in solitary rods of the tiger salamander. J. Physiol. 296:1–26.

Bortoff A (1964). Localization of slow potential responses in the Necturus retina. Vision Res. 4:627–635.

Buck L, Axel R (1991). A novel multigene family may encode odorant receptors: A molecular basis for odor recognition. Cell 65:175–187.

Capovilla M, Caretta A, Cervetto L, Torre V (1983). Ionic movements through light-sensitive channels of toad rods. J. Physiol. 343:295–310.

Cook NJ, Hanke W, Kaupp UB (1987). Identification, purification, and functional reconstitution of the cyclic GMP-dependent channel from rod photoreceptors. Proc. Natl. Acad. Sci. 84:585–589.

Dawson TD, Bredt DS, Fotuhi M, Hwang RM, Snyder S (1991). Nitric oxide synthase and neuronal NADPH diaphorase are identical in brain and peripheral tissues. Proc. Natl. Acad. Sci. 88:7797–7801.

Dhallan RS, Macke JP, Eddy RL, Shows TB, Reed RR, Yau K-W, Nathans J (1992). Human rod photoreceptor cGMP-gated channel: Amino acid sequence, gene structure, and functional expression. J. Neurosci. 12:3248–3256.

Dhallan RS, Yau K-W, Schrader KA, Reed RR (1990). Primary structure and functional expression of a cyclic nucleotide-activated channel from olfactory neurons. Nature 347:184–187.

Fesenko EE, Kolesnikov SS, Lyubarsky AL (1985). Induction by cyclic GMP of cationic conductance in plasma membrane of retinal rod outer segment. Nature 313:310–313.

Guy HR, Durell SR, Warmke J, Drysdale R, Ganetzky B (1992). Similarities in amino acid sequences of Drosophila eag and cyclic nucleotide-gated channels. Science 254:730.

Hartmann HA (1991). Exchange of conduction pathways between two related K^+ channels. Science 251:942–944.

Haynes LW, Yau K-W (1985). Cyclic GMP-sensitive conductance in outer segment membrane of catfish cones. Nature 317:61–64.

Haynes LW, Kay AR, Yau K-W (1986). Single cyclic GMP-activated channel activity in excised patches of rod outer segment membrane. Nature 321:66–70.

Hodgkin AL, McNaughton PA, Nunn BJ (1985). The ionic selectivity and calcium dependence of the light sensitive pathway in toad rods. J. Physiol. 358:447–468.

Hope BT, Michael GJ, Knigge KM, Vincent SR (1991). Neuronal NADPH diaphorase is a nitric oxide synthase. Proc. Natl. Acad. Sci. 88:2811–2814.

Jan LY, Jan YN (1990). A super family of ion channels. Nature 345:672.

Jones DT, Reed RR (1989). Golf: An olfactory specific G-protein involved in odorant signal transduction. Science 244:790–795.

Kaupp UB, Niidome T, Tanabe T, Terada D, Bönigk W, Stühmer W, Cook NJ, Kanagawa K, Matsuo H, Hirose T, Miyata T, Numa S (1989). Primary structure and functional expression from complementary DNA of the rod photoreceptor cyclic GMP-gated channel. Nature 342:762–766.

Kumar VD, Weber IT (1992). Molecular model of the cyclic GMP-binding domain of the cyclic GMP-gated ion channel. Biochemistry 31:4643–4649.

Lamb TD, Matthews HR (1988). External and internal actions on the response of salamander retinal rods to altered external calcium concentration. J. Physiol. 403:473–494.

Ludwig J, Margalit T, Eismann E, Lancet D, Kaupp UB (1990). Primary structure of cGMP-gated channel from bovine olfactory epithelium. FEBS Lett. 270:24–29.

Nakamura T, Gold GH (1987). A cyclic nucleotide-gated conductance in olfactory receptor cilia. Nature 325:442–444.

Nakatani K, Yau K-W (1988). Calcium and magnesium fluxes across the plasma membrane of the toad rod outer segment. J. Physiol. 395:695–729.

Nawy S, Jahr CE (1990). Suppression by glutamate of cGMP-activated conductance in retinal bipolar cells. Nature 346:442–444.

Sagar SM (1990). NADPH-diaphorase reactive neurons of the rabbit retina: Differential sensitivity to excitotoxins and unusual morphologic features. J. Comp. Neurol. 300:309–319.

Sandell JH (1985). NADPH-diaphorase cells in the mammalian inner retina. J. Comp. Neurol. 238:466–472.

Shiells RA, Falk G (1992). The glutamate-receptor linked cGMP cascade of retinal on-bipolar cells is pertussis and cholera toxin-sensitive. Proc. Roy. Soc. B 247:17–20.

Tomita T (1965). Electrophysiological study of the mechanisms subserving color coding in the fish retina. Cold Spring Harbor Symp. Quant. Biol. 30:559–566.

Wässle H, Grünert U, Cook NJ, Molday RS (1992). The cGMP-gated channel of rod outer segments is not localized in bipolar cells of the mammalain retina. Neurosci. Lett. 134:199–202.

Yau, K-W, Baylor DA (1989). Cyclic GMP-activated conductance of retinal photoreceptor cells. Ann. Rev. Neurosci. 12:289–327.

Yau K-W, Nakatani K (1984). Cation selectivity of light-sensitive conductance in retinal rods. Nature 309:352–354.

Yau K-W, Nakatani K (1985). Study of the ionic basis of visual transduction in vertebrate retinal rods. In: Contemporary Sensory Neurobiology. Correia MJ, Perachio AA, editors. New York: Alan Liss, pp. 245–256.

Yellen G, Jurman ME, Abramson T, MacKinnon R (1991). Mutations affecting internal TEA blockade identify the probable pore-forming region of a K^+ channel. Science 251:939–942.

Yool AJ, Schwartz T (1991). Alteration of ion selectivity of a K^+ channel by mutation of the H5 region. Nature 349:700–704.

Nonselective Cation Channels: Pharmacology, Physiology and Biophysics
ed. by D. Siemen & J. Hescheler

Cyclic AMP-Gated Cation Channels of Olfactory Receptor Neurons

Frank Zufall

Section of Neurobiology, Yale University, School of Medicine, New Haven, Connecticut 06510, USA

Summary
Odor-induced electrical activity in vertebrate olfactory receptor neurons is, at least in part, the result of the direct cyclic AMP-dependent activation of a nonselective cation channel. Single-channel recordings from extraciliary regions of isolated salamander olfactory receptor neurons have greatly improved our knowledge about distinctive properties of the cAMP-gated channel such as channel kinetics, modulation through divalent cations, and pharmacology. Because of the central role of these channels in the transduction cascade, these efforts have led to a better understanding of the physiology of olfactory transduction.

Introduction

There is increasing interest in sensory ion channels in olfactory receptor neurons gated by internal ligands such as cyclic nucleotide and phospholipid second messengers of both vertebrate (for review see: Firestein, 1992) and invertebrate olfactory systems (Zufall and Hatt, 1991; Fadool and Ache, 1992). Here, my focus will be on cyclic nucleotide-gated channels in vertebrate olfactory cells.

In response to odors, vertebrate olfactory receptor neurons produce a net inward ionic current which leads to depolarization of the cell and, finally, to the generation of action potentials. This current is at least in part the result of the stimulation of a receptor-mediated G-protein-coupled second messenger system (Pace et al., 1985; Jones and Reed, 1989; Breer et al., 1990; Buck and Axel, 1991) which utilizes cAMP to directly activate an ion channel (Nakamura and Gold, 1987; Firestein et al., 1991b; Zufall et al., 1991a). Although the odor-sensitive ion channel can be gated by both cAMP and cGMP, it is designated here *cAMP channel* because cAMP mediates the initial large and rapid response to natural odor ligands. In a more general sense, the term cyclic nucleotide-gated channel (cN channel) may also be used. Recent cloning of the channel from several species and functional expression in heterologous cell types has revealed a close relation of the olfactory cAMP channel to the cGMP-sensitive channel found in photoreceptors, as was evidenced by a high degree of homology in their amino acid sequences (see Barnstable, this volume). The following sections will review important properties of the olfactory cN channel with respect to their functional significance.

Odor Substances and Cyclic Nucleotides Activate a Common Ion Channel

Early studies on the whole-cell currents activated by odors gave strong evidence that the conductance generating the odor response and the cyclic nucleotide-gated conductance of olfactory receptor neurons (ORNs) were identical (Kurahashi 1990; Firestein et al., 1991a; Frings and Lindemann, 1991). However, it was necessary to provide a direct demonstration of this identity at the single-channel level. Previous work had established the existence of the cN conductance in the olfactory cilia (Nakamura and Gold, 1987; Kurahashi and Kaneko, 1991). The high channel density, however, together with the unfavorable recording conditions in the cilia prevented isolation of single-channel events. An alternative was offered by the fact that cN channels also occur at low density in the membrane of the dendrites and somata of salamander ORNs where recordings could be obtained more easily and reliably (Firestein et al., 1991b; Zufall et al., 1991a). The strategy used was therefore to record from membrane patches in which the channel density was sufficiently low to allow quantitative analysis of single-channel currents.

Figure 1A shows odor-activated single-channel activity in a cell-attached recording activated by a 150 ms duration pulse of an odor

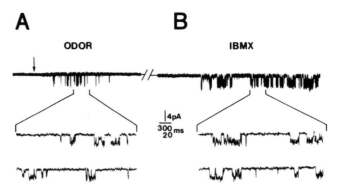

Figure 1. Odor- and cyclic nucleotide-induced activity in the same membrane patch (cell-attached). A) A 150 ms pulse of the odor stimulus (amyl acetate, acetophenone and cineole at 1 mM in Ringer) delivered at the *arrow* elicited some single-channel currents. B) Perfusion of the cell with 0.1 mM IBMX then elicited very similar activity in the same patch of membrane. Pipette potential, 40 mV. (From Firestein et al., 1991b.)

mixture. Membrane depolarization after odor receptor activation was prevented by bathing the cells in a choline-solution (olfactory cN channels are not permeable for choline; Kurahashi, 1989). However, the patch of membrane sealed within the electrode remained in high Na^+ and low Ca^{2+} solution, so that channels would still conduct current when opened by intracellular factors. Under these conditions, inwardly directed single-channel events were dependent on stimulus duration. A longer odor pulse elicited more single-channel activity and decreased the latency to the first opening (Firestein et al., 1991b).

If the odor-sensitive channels recorded here were the same as those underlying the previously identified cN conductance of olfactory cilia, it would be expected that treatments that increase the intracellular cAMP-concentration would also induce channel opening. Figure 1B shows that exogenously applied IBMX, a phosphodiesterase inhibitor, induced continuous single-channel opening that resembled the odor-induced ion channel. In fact, analysis of channel kinetic parameters gave no significant differences between those currents activated by odors or by IBMX. The same result was obtained by bath application of membrane permeable cyclic nucleotides such as 8-bromo-cGMP (Firestein et al., 1991b).

These results showed conclusively that the channels activated during the odor response are also gated by intracellular cyclic AMP, thus providing a direct link between the odor-induced production of cyclic AMP and the odor-sensitive current.

Characteristics of Single-Channel Currents

Subsequent characterization of odor-induced single-channel currents was performed in the inside-out recording mode. Single-channel characteristics described here were obtained in symmetrical high Na^+ solution under divalent cation free conditions. Patches mostly contained only one channel. One of the most striking features of all cN channels investigated so far including the olfactory channel is the absence of desensitization in the continued presence of saturating concentrations of agonists (Zufall et al., 1991a; Frings et al., 1992). This finding facilitated the interpretation of single-channel kinetics under steady-state conditions of stimulation.

An experimental advantage of the salamander cN channel is its high single-channel conductance of 40–45 pS, a value that is significantly higher than that from other preparations reported thus far, e.g., 28 pS in the newt (Kurahashi and Kaneko, 1991), 19 pS in the frog (Kolesnikow et al., 1990), 12–15 pS in the rat (Frings et al., 1992; Zufall, unpublished). Maximal channel open probability never exceeded 0.7 (Zufall et al., 1991a). Neither single-channel conductance nor open probability showed significant voltage-dependence in the salamander

(Zufall et al., 1991a). The current-voltage relation was linear under divalent cation free conditions. A slight voltage dependence of the open probability was reported for a cloned cN channel from catfish olfactory epithelium (Goulding et al., 1992). These results are particularly important since the S4 and H5 regions of voltage-gated channels, which are thought to be voltage sensor and pore lining, respectively, are unexpectedly conserved in the olfactory channel (Kaupp, 1991).

Kinetic analysis of single-channel events revealed a mean open time of 1–1.5 ms. The closed time distribution consisted of at least three components representing intraburst gaps ($200–300 \, \mu s$), interburst gaps (2–3 ms) and intercluster gaps (> 10 ms) (Zufall et al., 1991b). The mean duration of intercluster gaps strongly depended on the agonist concentration, thus it is the duration of individual clusters of single-channel currents that vary with the concentration (Zufall, in preparation). A very similar result has been obtained in ion channels mediating olfactory transduction in insects (Zufall and Hatt, 1991). This finding may be important for explaining gating mechanisms in olfactory channels.

In all different preparations studied so far, there has been general agreement that the olfactory channel can be gated both by cAMP and cGMP. In some preparations, there appears to be little difference between cAMP and cGMP, with $K_{1/2}$ values ranging from $2 \, \mu M$ to $40 \, \mu M$ (Nakamura and Gold, 1987; Kolesnikow et al., 1990; Goulding et al., 1992); in other preparations, the channel was more sensitive to cGMP than to cAMP, with sensitivity ratios varying from about 40 (Dhallan et al., 1990; Altenhofen et al., 1991) to 5 (Zufall et al., 1990a) and with $K_{1/2}$ values for cGMP ranging from $1 \, \mu M$ to $4 \, \mu M$. The high Hill coefficient of the dose-response relation of more than two indicated that channel activation seems to be a cooperative process (Zufall et al., 1991a; Altenhofen et al., 1991). The high sensitivity of the olfactory channel for cGMP gave rise to the idea that nitric oxide, the endogenous activator of the soluble form of guanylate cyclase, may play a crucial role as an intercellular messenger in the epithelium (Breer et al., 1992).

Block of the Ionic Pore by External Divalent Cations

As stated above single-channel recordings of the cN channel were made under conditions where the free Ca^{2+} and Mg^{2+}-concentrations were buffered below 10 nM. Under these conditions the channel displayed nearly ohmic behavior. Macroscopic odor-activated currents under physiological ionic conditions (i.e., with millimolar external Ca^{2+} and Mg^{2+}-concentrations), however, showed strong nonlinear characteristics as a function of membrane potential with outwardly rectifying properties and a negative conductance region at negative membrane potentials (Kurahashi, 1989). Since this discrepancy in the voltage dependence between single-channel and macroscopic currents could possibly be attributable to the effects of external divalent cations, the effect of these ions on single-channel cN currents was investigated and quantified (Zufall et al., 1992a; Zufall and Firestein, 1993).

The cAMP-evoked single-channel currents at different external Ca^{2+}-concentrations (Figure 2) illustrate that the apparent mean open-channel current was strongly reduced with increasing external Ca^{2+}-concentration. As a consequence, single cN currents under physiological ionic conditions and at the resting potential of the cell are below the resolution limit. A similar effect was obtained with external Mg^{2+}-ions although the effect was less pronounced (Table 1). Biophysical analysis showed that, in analogy with voltage-gated Ca^{2+} channels (Hess et al., 1986), the effects of external cations were broadly consistent with a model in which these ions bind to a site located within the ionic pore and thus, they lower the flux rate. From the voltage dependence of the K_D-value for the divalent cation-induced channel block, which is rather small with an e-fold increase per 128 mV of depolarization, the location of the blocking site can be derived; it is probably near the extracellular

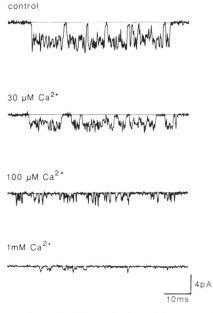

Figure 2. Block of the cyclic nucleotide-gated channel by external Ca^{2+}. Records were obtained from inside-out patches at a membrane potential of -100 mV exposed to $100\,\mu M$ cAMP. At increasing Ca^{2+}-concentrations the flickering behavior of the channel increases and the apparent open-channel amplitude decreases. The control solution contained less than 10 nM free Ca^{2+} and Mg^{2+}. (From Zufall and Firestein, 1993.)

Table 1. Comparison of the properties of the two different Ca^{2+}-binding sites at the olfactory cN channel. (Data are from 1) Zufall and Firestein, 1993 and 2) Zufall et al., 1991b.)

	Site A[1]	Site B[2]
Effect	reduction of channel open time reduction of channel conductance	stabilization of a closed state without reduction of channel conductance
Proposed mechanism	fast open-channel block	allosteric effect
Affinity	$Ca^{2+} > Mg^{2+}$ $K_D(Ca^{2+}) = 10\ \mu M$ $K_D(Mg^{2+}) = 300\ \mu M$	$K_D(Ca^{2+}) = 0.9\ \mu M$ Mg^{2+} ineffective
Location at the channel	sensing 10% of electric field from the outside	near the inside
Physiological significance	enhancement of signal-to-noise ratio	mediates fast desensitization

mouth and senses about 10% of the electric field (Zufall and Firestein, 1993). Analysis of current-voltage curves further indicated that Ca^{2+} is not only a blocker, but also a permeator of the channel.

In summary, channel block by divalent cations can account, together with Ca^{2+}-dependent inactivation (see below), for the nonlinearities of macroscopic odor-activated currents. Another physiological consequence of the channel block seems to be an increase of signal reliability by reducing response variability (Yau and Baylor, 1989). This is because the variance associated with the random behavior of a large population of small channels is much less than that generated by the activity of a few large channels, an argument that is especially important at the high input resistance of olfactory neurons.

Calcium-Dependent Inactivation as a Mechanism for Termination of the Odor Response

As pointed out above, cN currents in excised membrane patches do not desensitize. Macroscopic odor-activated currents in response to a maintained stimulus, however, display a transient time-course with a decay time constant of less than 5 s. Consequently, there seem to exist one or several specific mechanisms for termination (adaptation) of the odor response. The finding that the odor response could be transformed from transient to sustained by removing Ca^{2+} from the extracellular bathing solution (Zufall et al., 1991b) indicated that adaptation appeared to depend on a Ca^{2+} influx through the odor-sensitive conductance. Potentially, each component of a second messenger cascade could be subject to regulation. However, the finding that fast adaptation still occurred in the presence of IBMX, i.e., under conditions where the second messenger removal was inhibited, strongly suggested that the ion channel itself could be one target of a Ca^{2+}-mediated negative feedback loop.

To further test this hypothesis, the effect of intracellular Ca^{2+} on single-channel kinetics was investigated in inside-out patches. Interestingly, intracellular Ca^{2+} strongly reduced the open probability of the CNG channel without reducing channel conductance or channel mean open time (Zufall et al., 1991b). The $K_{1/2}$ value for this effect was 0.9 μM. Mg^{2+} ions were not effective. Thus, the effect of intracellular Ca^{2+} was clearly different from the channel block caused by extracellular Ca^{2+}. A subsequent kinetic analysis suggested that the Ca^{2+}-dependent inactivation could be due to an allosteric mechanism stabilizing a closed state. Hence, there must exist at least two different Ca^{2+}-binding sites with distinct properties at the olfactory channel (summarized in Table 1). The action of Ca^{2+} ions at an intracellular site would constitute a rapid and effective negative feedback loop which could account for the short-term adaptation of the odor response, an idea that has been confirmed very recently in catfish olfactory neurons (Kramer and Siegelbaum, 1992). However, adaptation at the level of the ion channel seems to be only one of several different mechanisms to terminate the odor response (Boekhoff and Breer, 1992).

Pharmacology of the Olfactory Channel

Pharmacological properties of the olfactory cN channel have not been well studied. Recent evidence suggested a surprising similarity between the ionic pore of Ca^{2+} channels and that of cN channels, both at the structural and at the functional level (Goulding et al., 1992; Zufall et al., 1992a; Heginbotham et al., 1992; Zufall and Firestein, 1993). This is reflected in the result that micromolar concentrations of the L-type Ca^{2+} channel antagonist nifedipine blocked the cN channel (Zufall and Firestein, 1993). Likewise, the blocking sequence of divalent cations exactly mirrored the picture from Ca^{2+} channels in that Cd^{2+} was a strong blocker compared to Ca^{2+}; Mg^{2+} was less effective than Ca^{2+}. Another similarity shared with voltage-gated Ca^{2+} channels is the effect that basic extracellular pH decreases some of the noise associated with the open channel, a phenomenon called proton block. Likewise, replacement of Na^+ as charge carrier with Li^+ significantly relieves the H^+ block (Zufall and Firestein, 1993).

Kolesnikov et al. (1990) and Frings et al. (1992) reported that derivatives of amiloride such as l-cis-diltiazem are effective inhibitors of olfactory cN currents. A description of more specific blockers for the cN channel could be very useful in future experiments for determining the portion of prospective parallel transduction pathways in the odor response.

Pulsed Application of Second Messengers to cN Channels

A long-standing problem in olfactory physiology is in the difference in time-course between the odor-induced second messenger accumulation and the electrical response of the cell. The cyclic AMP production is very fast and transient, having a peak after less than 50 ms after odor stimulation and thereafter decaying back to baseline level within a few hundred milliseconds (Breer et al., 1990). Since the time-course of the odor-induced current is much slower it was questioned how rapid second messenger kinetics could explain odor-response kinetics. We therefore initiated a series of experiments in which pulses of second messengers were applied to inside-out patches containing either the native salamander cN channel at low density or a cloned rat cN channel at high density (Zufall et al., 1992b). Agonists were applied using a piezo-switch device allowing solution exchange times in the range of a few hundred microseconds (Dudel et al., 1990). Surprisingly, these experiments showed that both onset and offset kinetics of the olfactory cN channel were rather slow. Especially, activation kinetics of the channel seemed to be rate-limited by the binding reaction of the second messenger, even at a concentration as high as 1 mM cGMP. Therefore, the traditional idea that cN channels behave like a static sensor of the second messenger concentration does not seem to hold true in olfactory cells. Instead, it is the slow intrinsic time-course of channel gating that determines the time-course of the electrical response, giving the ion channel the role of an integrator for the rapid second messenger pulses.

Conclusions

It is now well established that the cN channel, as described here, mediates olfactory transduction in vertebrate olfactory receptor neurons. Figure 3 summarizes what has been called a "consensus working model" of the main steps in the transduction pathway (Shepherd, 1991). One of the most exciting aspects of this scheme is the close relation to visual, hormone, and neurotransmitter signal transduction (see other chapters of this book). Knowledge about the final step in this cascade, the opening of a nonspecific cation channel, has been greatly facilitated by the use of isolated salamander olfactory receptor neurons and, especially, by the possibility of recording from these neurons in extraciliary regions with low channel density. The measured properties of the ion channel give a physiological basis for a number of observations such as short-term adaptation, Ca^{2+}-dependent block of the odor response

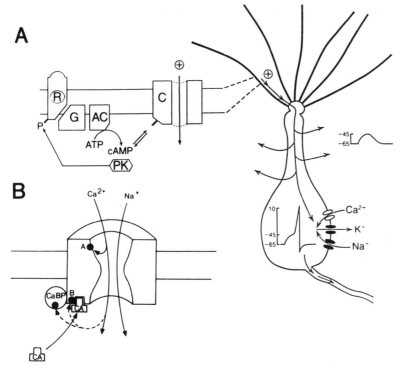

Figure 3. Scheme of biochemical and electrical events in an olfactory receptor neuron (on the right). The earliest steps in olfactory transduction occur in the fine cilia (as illustrated in the enlarged schematic (A) on the left) containing a receptor protein (R), a GTP-binding protein (G), adenylate cyclase (AC), and an ion channel (C) that is directly gated by cAMP. Protein kinase A (PK) has been shown to down regulate the cAMP production (Boekhoff and Breer, 1992). B) The ion channel itself is subject to sensitivity regulation by Ca^{2+} at two different binding sites. Site A represents a blocking site within the pore. Site B refers to allosteric regulation of cAMP (CA) sensitivity (see summarization in Table 1). It is not clear whether the allosteric effect of Ca^{2+} at site B is direct or due to activation of an intermediate Ca^{2+}-binding protein (CaBP). (Modified from Firestein, 1991 and Kramer and Siegelbaum, 1992.)

and kinetic properties of the odor current. Much interest now focuses on the ability to measure the kinetic behavior of cN channels under nonsteady state conditions to add the temporal aspect of odor transduction to our knowledge.

Acknowledgements
I wish to thank Dr. Stuart Firestein and Prof. Gordon M. Shepherd who contributed to the work described in this chapter.

References

Altenhofen W, Ludwig J, Eismann E, Kraus W, Bönigk W, Kaupp UB (1991). Control of ligand specificity in cyclic nucleotide-gated channels from rod photoreceptors and olfactory epithelium. Proc. Natl. Acad. Sci. USA 88:9868–9872.

144

Boekhoff I, Breer H (1992). Termination of second messenger signaling in olfaction. Proc. Natl. Acad. Sci. USA 89:471–474.

Breer H, Boekhoff I, Tareilus E (1990). Rapid kinetics of second messenger formation in olfactory transduction. Nature 345:65–68.

Breer H, Klemm T, Boekhoff I (1992). Nitric oxide mediated formation of cyclic GMP in the olfactory system. Neuroreport. 3:1030–1032.

Buck L, Axel R (1991). A novel multigene family may encode odorant receptors: A molecular basis for odor recognition. Cell 65:175–187.

Dhallan RS, Yau K-W, Schrader KA, Reed RR (1990). Primary structure and functional expression of a cyclic nucleotide-activated channel from olfactory neurons. Nature 347:184–187.

Dudel J, Franke C, Hatt H (1990). Rapid activation, desensitization, and resensitization of synaptic channels of crayfish muscle after glutamate pulses. Biophys. J. 57:533–545.

Fadool DA, Ache BW (1992). Plasma membrane inositol 1,4,5-trisphosphate activated channels mediate signal transduction in lobster olfactory receptor neurons. Neuron 9:907–918.

Firestein S (1991). A noseful of odor receptors. Trends Neurosci. 14:270–272.

Firestein S, Darrow B, Shepherd GM (1991a) Activation of the sensory current in salamander olfactory receptor neurons depends on a G protein-mediated cAMP second messenger system. Neuron 6:825–835.

Firestein S, Zufall F, Shepherd GM (1991b). Single odor-sensitive channels in olfactory receptor neurons are also gated by cyclic nucleotides. J. Neurosci. 11:3565–3572.

Firestein S (1992). Electrical signals in olfactory transduction. Current Opinion Neurobiol. 2:444–448.

Frings S, Lindemann B (1991). Current recording from sensory cilia of olfactory receptor cells in situ. I. The neuronal response to cyclic nucleotides. J. Gen. Physiol. 97:1–16

Frings S, Lynch JW, Lindemann B (1992). Properties of cyclic nucleotide-gated channels mediating olfactory transduction. Activation, selectivity, and blockage. J. Gen. Physiol. 100:45–67.

Goulding EH, Ngai J, Kramer RH, Colicos S, Axel R, Siegelbaum SA, Chess A (1992). Molecular cloning and single-channel properties of the cyclic nucleotide-gated channel from catfish olfactory neurons. Neuron 8:45–58.

Heginbotham L, Abramson T, MacKinnon R (1992). A functional connection between the pores of distantly related ion channels as revealed by mutant K^+ channels. Science 258:1152–1155.

Hess P, Lansman JB, Tsien RS (1986). Calcium channel selectivity for divalent and monovalent cations. J. Gen. Physiol. 88:293–319.

Jones DT, Reed RR (1989). G_{olf}. An olfactory neuron-specific G-protein involved in odorant signal transduction. Science 244:790–795.

Kaupp UB (1991). The cyclic nucleotide-gated channels of vertebrate photoreceptors and olfactory epithelium. Trends Neurosci. 14:150–157.

Kramer RH, Siegelbaum SA (1992). Intracellular Ca^{2+} regulates the sensitivity of cyclic nucleotide-gated channels in olfactory receptor neurons. Neuron 9:897–906.

Kolesnikov SS, Zhainazarov AB, Kosolapov AV (1990). Cyclic nucleotide-activated channels in the frog olfactory receptor plasma membrane. FEBS Lett. 266:96–98.

Kurahashi T (1989). Activation by odorants of cation-selective conductance in the olfactory receptor cell isolated from the newt. J. Physiol. 419:177–192.

Kurahashi T (1990). The response induced by intracellular cyclic AMP in isolated olfactory receptor cells of the newt. J. Physiol. 430:355–371.

Kurahashi T, Kaneko A (1991). High density cAMP-gated channels at the ciliary membrane in the olfactory receptor cell. Neuroreport 2:5–8.

Nakamura T, Gold GH (1987). A cyclic nucleotide-gated conductance in olfactory receptor cilia. Nature 325:442–444.

Pace U, Hanski E, Salomon Y, Lancet D (1985). Odorant-sensitive adenylate cyclase may mediate olfactory transduction. Nature 315:255–258.

Shepherd GM (1991). Sensory transduction: Entering the mainstream of membrane signalling. Cell 67:845–851.

Yau K-W, Baylor DA (1989). Cyclic GMP-activated conductance of retinal photoreceptor cells. Ann. Rev. Neurosci. 12:289–327.

Zufall F, Hatt H (1991). Dual activation of a sex-pheromone dependent ion channel from insect olfactory dendrites by protein kinase C activators and cyclic GMP. Proc. Natl. Acad. Sci. USA 88:8520–8524.

Zufall F, Firestein S, Shepherd GM (1991a). Analysis of single cyclic nucleotide-gated channels in olfactory receptor cells. J. Neurosci. 11:3573–3580.

Zufall F, Shepherd GM, Firestein S (1991b). Inhibition of the olfactory cyclic nucleotide gated ion channel by intracellular calcium. Proc. R. Soc. Lond. B246:225–230.

Zufall F, Shepherd GM, Firestein S (1992a). Block of the olfactory, cyclic nucleotide gated channel by extracellular calcium and magnesium ions. Biophys. J. 61:283a.

Zufall F, Hatt H, Firestein S (1992b). Activation and deactivation kinetics of cyclic nucleotide gated channels from olfactory receptor neurons studied with concentration jump techniques. Soc. Neurosci. Abstr. 18:249.

Zufall F, Firestein S (1993). Divalent cations block the cyclic nucleotide gated channel of olfactory receptor neurons. J. Neurophysiol. 69:1758–1768.

Nonselective Cation Channels: Pharmacology, Physiology and Biophysics
ed. by D. Siemen & J. Hescheler
© 1993 Birkhäuser Verlag Basel/Switzerland

Renal Epithelial Cells Show Nonselective Cation Channel Activity and Express a Gene Related to the cGMP-Gated Photoreceptor Channel

Christoph Korbmacher[+][*] and Colin J. Barnstable[#]

[+] Department of Cellular and Molecular Physiology and [#] Department of Ophthalmology and Visual Science, Yale University School of Medicine, 333 Cedar Street, New Haven, CT 06510, USA

Summary

Nonselective cation channels have been found in various parts of the nephron and represent a heterogeneous group of channels. We briefly review their putative physiological function. Renal epithelial nonselective cation channels may play a role in volume regulation, calcium entry, cell proliferation, and sodium reabsorption.

In some renal epithelia cGMP seems to be involved in the regulation of nonselective cation channels. Furthermore, there is evidence that a gene related to the cGMP-gated photoreceptor channel, a well-characterized, nonselective cation channel, is also expressed in whole rat kidney tissue. In the context of these observations, we review recent findings from our own work on a nonselective cation channel in the M-1 mouse cortical collecting duct cell line. We could demonstrate that M-1 cells show nonselective cation channel activity in inside-out patches and express a gene related to the cGMP-gated photoreceptor channel (Proc. Natl. Acad. Sci. USA 89:10262–10266, 1992). The possibility of a relation between the kidney channel and the photoreceptor channel is discussed.

Introduction

Nonselective cation channels have been described in a large variety of excitable and nonexcitable tissues (Partridge and Swandulla, 1988; Cook et al., 1990). Their common property is a high selectivity for cations over anions and a poor discrimination between sodium and potassium. They constitute a heterogeneous group of channels with vastly different biophysical properties including ligand-gated channels, such as the nicotinic acetylcholine receptor (reviewed by Changeux et al., 1984), cyclic nucleotide-gated channels, such as the cGMP-gated photoreceptor channel (reviewed by Kaupp et al., 1991), and various epithelial nonselective cation channels (reviewed by Cook et al., 1990).

So far, it is not clear how the various subtypes and families of nonselective cation channels should be appropriately classified. Single-

---

Correspondence to be sent to Dr. Christoph Korbmacher at his present address: Zentrum der Physiologie, Klinikum der Johann Wolfgang Goethe-Universität, Theodor Stern Kai 7, D-60596 Frankfurt am Main, FRG.

channel current recordings using the patch-clamp method allow a functional characterization and classification of the channels. However, knowledge of their molecular structure is necessary in order to determine whether and how the different nonselective cation channels are genetically related to each other. Whereas detailed information on the molecular structure of ligand-gated channels and cyclic nucleotide-gated channels is already available, comparable information on any of the epithelial nonselective cation channels is still lacking.

In the kidney, electrophysiological evidence for nonselective cation channels has been found in the proximal tubule (Gögelein and Greger, 1986; Merot et al., 1988; Marom et al., 1989; Filipovic and Sackin, 1991, 1992), in cortical thick ascending limb cells (Teulon et al., 1987; Paulais and Teulon, 1989; Merot et al., 1991), as well as in collecting duct cells (Light et al., 1988; Laskowski et al., 1990; Ling et al., 1991; Kizer et al., 1991; Korbmacher et al., 1992(a)). The physiological role of this heterogeneous group of channels is not yet well understood and may vary in different nephron segments. Renal epithelial nonselective cation channels may play a role in volume regulation, calcium entry, cell proliferation, and sodium reabsorption. In this article, we will briefly discuss these putative functions and the role of cGMP as a regulator of renal nonselective cation channels, in particular, in the collecting duct. Furthermore, we will review some recent observations (Ahmad et al., 1990; Ahmad et al., 1992) which raise the question whether a nonselective cation channel in the kidney is related to the cGMP-gated photoreceptor channel.

Possible Physiological Roles of Nonselective Cation Channels in the Kidney

Volume Regulation

Whenever the physiological role of a channel is unclear it has become popular to suggest that it may be involved in volume regulation, and such a function has also been proposed for nonselective cation channels in the kidney (Gögelein and Greger, 1986; Merot et al., 1988; Filipovic and Sackin, 1991, 1992). Indeed, Chan and Nelson recently reported a shrinking-induced, nonselective cation conductance in airway epithelial cells (Chan and Nelson, 1992). It is conceivable that in the kidney nonselective cation channels may also play a role in regulatory volume increase after hypertonic challenge. However, experimental evidence for such a function in the kidney is still missing.

Calcium Entry and Cell Proliferation

Nonselective cation channels may also function as calcium entry pathways. In *Necturus* kidney proximal tubule the apical nonselective cation channel, which is activated by stretch, has been shown to be calcium permeable (Filipovic and Sackin, 1991). Calcium-entry via nonselective cation channels may trigger other cellular responses, e.g., during volume regulation or cell proliferation. Interestingly, nonselective cation channel activation has been shown to play a role in the initiation of cell proliferation following stimulation with growth factors (Jung et al., 1992). As far as we know, a role of nonselective cation channels for cell proliferation has not yet been demonstrated in renal tissues. However, from our patch-clamp experiments we have the impression that, in cultured renal epithelial cells, nonselective cation channels occur more frequently and are more active in proliferating cells as compared to confluent cells (unpublished observations). This observation seems compatible with a potential role of the channel during cell proliferation.

Sodium Reabsorption

Nonselective cation channels may also be involved in sodium reabsorption. Amiloride sensitive sodium reabsorption is a characteristic transport function of the renal collecting duct epithelium (O'Neil and Boulpaep, 1979). Small conductance, amiloride-sensitive channels which are highly selective for sodium are believed to be the major route for sodium entry in sodium reabsorbing epithelia, including the cortical collecting duct (Garty and Benos, 1988). However, extracellular amiloride in a concentration of $0.5\,\mu M$ has also been shown to inhibit a nonselective cation channel in the rat inner medullary collecting duct. This channel is therefore believed to contribute to amiloride-sensitive sodium reabsorption (Light et al., 1988). A similar amiloride-sensitive nonselective cation channel has been found in endothelial cells of brain microvessels (Vigne et al., 1989). In contrast, the nonselective cation channel in cultured rabbit cortical collecting duct cells appears to be insensitive to extracellular amiloride, which argues against its role for Na-reabsorption (Laskowski et al., 1990; Ling et al., 1991). In rabbit urinary bladder epithelium the proteolytic enzymes urokinase and kallikrein have been reported to hydrolyze apical amiloride-sensitive sodium channels, possibly converting them into nonselective cation channels (Lewis and Hanrahan 1985; Lewis and Alles, 1986). However, as far as we know, direct patch-clamp evidence demonstrating a conversion of highly sodium-selective channels into nonselective cation channels is lacking. On the other hand, highly sodium-selective channels and nonselective cation channels have been shown to coexist in the same cell type (Laskowski et al., 1990; Kizer et al., 1991) and it is not yet clear whether or how these channels are interrelated. Thus, the contribution of the nonselective cation channel for amiloride-sensitive sodium reabsorption in the collecting duct is not yet completely resolved.

cGMP as a Regulator of Nonselective Cation Channels in the Kidney

Another interesting aspect is the possible involvement of cGMP as a regulator of nonselective cation channels in the kidney. cGMP has been shown to stimulate nonselective cation channels in toad urinary bladder (Das et al., 1991) and also in excised membrane patches of A6 cells (Ohara et al., 1991). The toad urinary bladder and the A6 cell line are widely used model systems to study distal tubule transport processes. On the other hand, in inner medullary collecting duct cells 10^{-4} M cGMP has been reported to reduce channel open probability by about 40% (Light et al., 1989, 1990). A complete channel shutdown has been reported in inner medullary collecting duct cells when cGMP was applied in the presence of cGMP-dependent protein kinase (Light et al., 1990). It has been suggested that, in the collecting duct, atrial natriuretic peptide (ANP) is the physiological stimulus which, *via* its second messenger cGMP, decreases the activity of the nonselective cation channel. If the nonselective cation channel indeed contributes to sodium reabsorption, its inhibition may in part be responsible for the natriuretic action of ANP (Light et al., 1989).

Kidney Tissue Expresses a Gene Related to the cGMP-Gated Photoreceptor Channel

A well characterized example of a cGMP-regulated nonselective cation channel is the cGMP-gated vertebrate photoreceptor channel which mediates the electrical response to light (Yau and Baylor, 1989). The channel is directly activated by cGMP without involving enzymatic reactions (for review see Kaupp, 1991). The amino-acid sequence of the cGMP-gated channel from bovine retinal photoreceptors has been deduced by cloning and sequencing its cDNA that, when expressed in *Xenopus* oocytes, encoded a cGMP-gated cation channel (Kaupp et al., 1989).

A similar protein has been isolated from olfactory epithelium (Dhallan et al., 1990; Goulding et al., 1992), suggesting that these nonselective cation channels constitute a gene family. Further evidence for such a gene family has been obtained by screening a number of rat tissues by Northern blot hybridization. A probe corresponding to an 788 base pair sequence of the rat rod photoreceptor channel detected a band of 3.2 kb in a Northern blot of rat retina, as expected (Ahmad et al., 1990). In the same experiment a band of the same size was detected in 20 μg of total RNA isolated from rat kidney and rat heart, but not in other tissues probed, namely, liver, muscle, thymus, testis and spleen. The hybridization signal with kidney and heart was present after stringent washing (0.1 $\times$ SSC, 68°C for 1 h). Thus, it suggested a strong homology between the photoreceptor sequence and the sequences of the transcripts detected in these tissues. Since the Northern blot analysis was carried out on whole kidney, it is not yet clear which site(s) within the kidney expresses the cGMP-gated channel gene. One potential site is the collecting duct.

M-1 Cells as a Model for Cortical Collecting Duct

The M-1 mouse cortical collecting duct cell line maintains epithelial transport properties characteristic of the collecting duct in vivo and provides a homogeneous cell population from a defined nephron segment. When grown on permeable collagen support, M-1 cells form a polarized epithelial monolayer with tight junctions and apical microvilli. They secrete potassium and show amiloride-sensitive electrogenic sodium reabsorption (Stoos et al., 1991; Fejes-Tóth and Náray-Fejes-Tóth, 1992; Korbmacher et al., 1991 and 1993(b)). Thus, the M-1 cells appear to be a good model for studying cortical collecting duct function in vitro.

In the following, we review data from our own laboratory which were obtained using M-1 mouse cortical collecting duct cells. We have carried out patch-clamp studies and have identified a nonselective cation channel in excised membrane patches of M-1 cells (Korbmacher et al., 1992a; Ahmad et al., 1992). Furthermore, we have used both Northern blot hybridization and PCR amplification to screen for expression of a cGMP-gated photoreceptor channel gene in M-1 cells. The PCR products which were obtained were subsequently cloned, sequenced, and compared to the sequence of the rod photoreceptor channel (Ahmad et al., 1992).

Nonselective Cation Channel Activity in M-1 Cells

In conventional whole-cell experiments, we have previously identified the following three conductance components in M-1 cells: 1) a sodium-selective, amiloride-sensitive conductance; 2) a barium- and glibenclamide-sensitive potassium conductance, and 3) a chloride conductance with a characteristic time-dependent decay upon depolarizing voltage steps (Korbmacher et al., 1991, 1992(a), 1992(b), and

1993(b)). Furthermore, transepithelial, ion-gradient measurements demonstrated the presence of potassium secretion and sodium reabsorption in M-1 cells grown on permeable support (Stoos et al., 1991; Korbmacher et al., 1993(b)). Thus, we were expecting to find single-channel activity of sodium-selective and potassium-selective channels in excised M-1 membrane patches. However, the only cation channel we have been able to clearly identify so far is a nonselective cation channel with an average conductance of 34 ± 2.3 pS (Korbmacher et al., 1992(a); Ahmad et al., 1992). The channel typically occurred in multi-channel patches and was similar to the nonselective cation channel described in rat inner medullary collecting duct cells (Light et al., 1988).

Ion Selectivity

Figure 1A shows two single-channel I/V plots obtained from an inside-out patch. The pipette contained Na_2SO_4/K_2SO_4 Ringer's solution and the batch initially contained NaCl-Ringer's solution and, subsequently, KCl-Ringer's solution. The two I/V plots are practically identical and reverse at 0 mV, indicating that the channel is cation selective and has a permeability ratio for sodium over potassium of about 1.

In the experiment shown in Figure 1B the pipette contained NaCl-Ringer's solution and the bath initially contained KCl-Ringer's solution. Under these conditions the I/V plot is linear and reverses at 0 mV. A subsequent change to sucrose-Ringer's solution containing only 20 mM cations resulted in a shift of the reversal potential in the positive direction. If one assumes a reversal potential of about $+40$ to $+50$ mV, the calculated permeability ratio for cations over chloride ranges from 11 to 48. Hence, this channel discriminates poorly between Na and K, but is highly selective for cations over anions.

Voltage-Dependence

An additional feature of the nonselective channel in M-1 cells is a slight voltage dependence of its open probability (NP_0), with the channel being more active at depolarizing voltages, as shown in Figure 1C. A similar voltage dependence of the single-channel open probability has been described, for example, in the nonselective cation channel in the rabbit proximal tubule (Gögelein and Greger, 1986), human colonic tumor cells (Champigny et al., 1991), human nasal epithelial cells (Jorissen et al., 1990) and also in the cloned bovine photoreceptor channel (Kaupp et al., 1989).

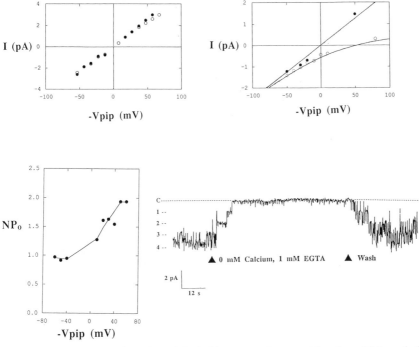

Figure 1. Nonselective cation channel in inside-out membrane patches from M-1 cortical collecting duct cells. Single-channel recordings were performed using conventional patch-clamp technique (Hamill et al., 1981). The trans-patch potential is the negative holding potential ($-V_{pip}$), or cytoplasmic potential referred to pipette potential. Single-channel I-V plots: A) Pipette contained Na_2SO_4/K_2SO_4-Ringer's solution. Bath solution was NaCl-Ringer's (open circles) or KCl-Ringer's (filled circles). NaCl-Ringer's contained (in mM): 140 NaCl, 5 KCl, 1 $CaCl_2$, 1 $MgCl_2$, 5 glucose, 10 Hepes adjusted to pH 7.5 with NaOH. KCl-Ringer's contained (in mM): 140 KCl, 5 NaCl, 1 $CaCl_2$, 1 $MgCl_2$, 5 glucose, 10 Hepes adjusted to pH 7.5 with KOH. Na_2SO_4/K_2SO_4-Ringer's solution contained (in mM): 12.5 NaCl, 30 Na_2SO_4, 12.5 KCl, 30 K_2SO_4, 60 sucrose, 10 Hepes adjusted to pH 7.5 with NaOH/KOH. B) Pipette contained NaCl-Ringer's solution. Bath solution was initially KCl-Ringer's (filled circles). The data were fitted by using the Goldman-Hodgkin-Katz equation for symmetrical 150 mM cations and 25-pS conductance. Subsequently, the bath solution was changed to sucrose Ringer's solution (open circles). The data were fitted by using the Goldman-Hodgkin-Katz equation for a cation ratio of 20 mM/150 mM ($E_{Rev} = 54$ mV) and a reference conductance of 25 pS. Composition of sucrose solution (in mM): 10 NaCl, 5 KCl, 1 $CaCl_2$, 1 $MgCl_2$, 260 sucrose, 5 glucose, 10 Hepes adjusted to pH 7.5 with NaOH. C) Voltage dependence of the open probability (NP_0). The pipette solution was KCl-Ringer's; the bath solution was NaCl-Ringer's. D) Effect of calcium removal from bath on nonselective cation channel activity in an inside-out patch at $-V_{pip} = -40$ mV. Downward deflections indicate inward current. The pipette solution was NaCl-Ringer's; the bath solution was KCl-Ringer's. Calcium removal was accomplished by superfusing with a nominally calcium-free KCl-Ringer's solution containing 1 mM EGTA. (Figure adapted from Ahmad et al., 1992.)

154

Blockers

Highly specific blockers for nonselective cation channels are not yet readily available. However, it has been reported that derivatives of diphenylaminecarboxylate (DPC), such as flufenamic acid, are potent blockers of nonselective cation channels, e.g., in rat exocrine pancreas (Gögelein and Pfannmüller, 1989; Gögelein et al., 1990), rat distal colon (Siemer and Gögelein, 1992), cerebral capillary endothelial cells (Popp and Gögelein, 1992), and mouse fibroblasts (Jung et al., 1992). In preliminary experiments we found that 100 μM flufenamic acid also reversibly inhibited open probability of the M-1 nonselective cation channel, usually by 50% or more.

We have not yet investigated the effect of extracellular amiloride on the nonselective cation channel in M-1 cells. However, in confluent M-1 cells we have demonstrated that the amiloride-sensitive, whole-cell conductance is sodium selective (Korbmacher et al., 1991, 1992(a), 1993(b)), which argues against a role of the nonselective cation channel in amiloride-sensitive sodium reabsorption in M-1 cells.

In order to clearly demonstrate a reversible amiloride effect on the single-channel level, one has to apply amiloride onto the extracellular surface of the patch membrane, since amiloride is believed to act on the channel pore from the outside (Garty and Benos, 1988). This can be achieved in cell-attached or inside-out patches via an internally perfused patch pitpette, or in outside-out patches directly via bath perfusion (Vigne et al., 1989). However, the internal pipette perfusion is a difficult technique. In inner medullary collecting duct cells the amiloride inhibition was detected via a voltage-dependent block of the channel in inside-out patch experiments with amiloride present in the pipette solution (Light et al., 1988). We do not consider this a feasible approach to demonstrate an amiloride effect on M-1 nonselective cation channels since they already show a similar voltage dependence of the open probability in the absence of amiloride.

Effect of Calcium Removal

Removal of calcium in the solution bathing the cytoplasmic surface of the channel abolished channel activity completely, as shown in Figure 1D. The effect was reversible upon readdition of cytoplasmic calcium. A similar calcium sensitivity has been reported in nonselective cation channels in various tissues, with threshold values for channel activation of about 10^{-6} M free calcium (Yellen, 1982; Thorn and Petersen, 1992; for review see Partridge and Swandulla, 1988).

Since intracellular calcium concentration is normally low, the observed calcium sensitivity raises the question of whether, under physiological conditions, these nonselective channels can be active. In M-1 cells intracellular calcium concentration has been measured using Fura-2 and it averaged 133 nM (Geibel et al., 1990, 1991; Korbmacher et al., 1993(a)). Thus, under resting conditions intracellular calcium concentration is probably too low to activate the nonselective cation channel in M-1 cells. This is compatible with our observation that in M-1 cells nonselective cation channel activity is rarely seen in the cell-attached configuration, but usually becomes apparent only after excision of the patch.

It is interesting to speculate whether hormonal stimuli may be required for channel activation. One possible candidate in the collecting duct is endothelin, which has recently been shown to increase intracellular calcium in collecting duct principal cells (Geibel et al., 1991, 1992; Naruse et al., 1991; Korbmacher et al., 1993(a)).

Effect of ATP

In a number of tissues, e.g., mouse pancreatic acinar cells (Thorn and Petersen, 1992), cerebral capillary endothelial cells (Popp and Gögelein, 1992), and thick ascending limb cells of Henle's loop (Paulais and Teulon, 1989), nonselective cation channel activity has been shown to decrease in inside-out patches in the presence of millimolar ATP on the cytoplasmic surface of the patch. In a few experiments, we have therefore tested the effect of ATP on the channel activity in M-1 cells. We found that ATP decreased channel open probability. However, the effect of ATP was quite variable and the decrease in NP_0 ranged from about 15% to more than 50% (unpublished observations).

Effect of cGMP

It has been reported that the nonselective cation channel in inner medullary collecting duct cells is inhibited by cGMP by a direct mechanism and by an indirect mechanism involving a cGMP dependent protein kinase (Light et al., 1990). Therefore, we have performed experiments in which we tested the effect of cGMP on the nonselective cation channel activity in inside-out patches of M-1 cells (Ahmad et al., 1992). In the recording shown in Figure 2, cGMP appears to reduce NP_0 with partial recovery after cGMP washout. NP_0 was calculated for 80 s during each period using amplitude histograms.

In 14 experiments similar to the one shown in Figure 2, NP_0 decreased in 10 out of 14 experiments in the presence of 10^{-4} M cGMP, as compared to the control period prior to application of cGMP (Ahmad et al., 1992). After washout of cGMP, NP_0 showed partial

156

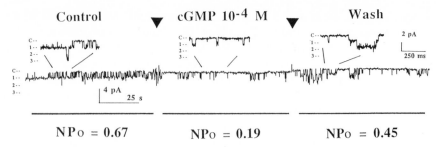

NPo = 0.67   NPo = 0.19   NPo = 0.45

Figure 2. Effect of 0.1 mM cGMP on a continuous recording of nonselective cation channel activity in an inside-out patch at $-V_{pip} = -50\,mV$. Downward deflections indicate inward current. cGMP (0.1 mM) was present between the arrow heads. Insets show current recordings on a faster time base. Horizontal bars indicate three 80-s intervals of recordings from which amplitude histograms were generated to calculate open probability (NP_0). (Figure adapted from Ahmad et al., 1992.)

recovery in five of these 10. In three of the 10 experiments channel activity ceased completely, either during the cGMP application or after washout, and did not recover thereafter. In four out of 14 experiments we did not see a decrease of channel activity, but instead a slight increase, although in two of these experiments open probability continued to increase even after cGMP washout.

After normalizing the data by dividing NP_0 by the apparent number of channels, the control P_0 averaged 0.40 ± 0.09, whereas P_0 in the presence of $10^{-4}\,M$ cGMP averaged 0.29 ± 0.07. Thus, in these 14 experiments M-1 nonselective cation channel activity was, on average, decreased by 27%, but with a p-value of just 0.05 (Ahmad et al., 1992). This cGMP effect in M-1 cells is of the same order as the effect observed in inner medullary collecting duct cells (Light et al., 1989, 1990).

The variability of the cGMP effect observed in our study (Ahmad et al., 1992) could be due to the presence or absence of factors such as membrane-bound protein kinases or certain substrates in an individual membrane patch micro-environment. Experiments are presently underway to investigate the role of some of these factors in the regulation of the activity of the M-1 nonselective cation channel.

Genes Related to the cGMP-Gated Photoreceptor Channel are Expressed in M-1 Cells

Having detected a nonselective cation channel in M-1 cells (Korbmacher et al., 1992(a); Ahmad et al., 1992) that can be affected by cGMP, and knowing that a molecule related to the photoreceptor channel is expressed in the kidney (Ahmad et al., 1990), we decided to see whether a similar transcript was present in M-1 cells. The same probe detected a band of 3.2 kb in a Northern blot of RNA from M-1 cells (Ahmad et al., 1992). To detect this band it was necessary to electrophorese 30 μg of poly A$^+$ RNA, suggesting that the RNA was present, but expressed at low levels in these cells.

To find out the exact relationship between the 3.2 kb band detected and the rod photoreceptor channel, we amplified the M-1 transcripts using PCR methods (Ahmad et al., 1992). Two sets of nested primers were used, as shown in Figure 3. Primers for PCR amplification were chosen, that corresponded to the sequence of conserved regions of the rat rod photoreceptor cGMP-gated ion channel. Primer set A was chosen to amplify a 788 bp sequence containing approximately 40% of the coding sequence and including the putative cGMP binding site. Primer set B was chosen to amplify a 458 bp sequence within the sequence amplified by primer set A.

Single-stranded cDNA was prepared from total RNA of the M-1 mouse cortical collecting duct cells using MuMLV-reverse transcriptase. Aliquots of this cDNA were then amplified in PCR reactions with either primer set A or β-actin primers as a control. Analysis of these reactions by gel electrophoresis and ethidium bromide staining showed clear bands using β-actin primers, but showed no bands using primer set A (Figure 4A).

However, when aliquots of these PCR reactions were further amplified using primer set B, strong bands consistent with the predicted size of 458 bp were observed (Figure 4B). A Southern blot analysis showed that these bands reacted with a probe corresponding to the rat

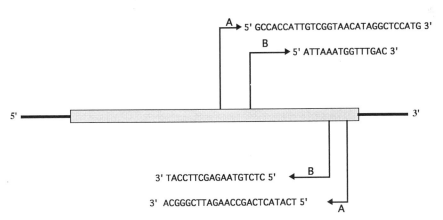

Figure 3. Two sets of nested primers used for polymerase chain reaction (PCR) to amplify transcripts expressed in M-1 cells. Primers for PCR amplification were chosen to correspond to the sequence of conserved regions of the rat rod photoreceptor cGMP-gated ion channel. Primer set A was chosen to amplify a 788 bp sequence containing approximately 40% of the coding sequence and including the putative cGMP binding site. Primer set B was chosen to amplify a 458 bp sequence within the sequence amplified by primer set A. (Figure adapted from Ahmad et al., 1992.)

158

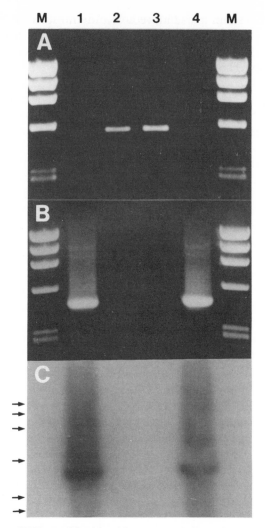

Figure 4. Analyses of PCR amplification with agarose gel electrophoresis and Southern blot. Single-stranded cDNA was prepared from total RNA of the M-1 mouse cortical collecting duct cells using MuMLV-reverse transcriptase. Aliquots of this cDNA were then amplified in PCR reactions with either primer set A or β-actin primers as a control. A) Gel electrophoresis and ethidium bromide staining of PCR products obtained with primer set A or β-actin primers. B) Aliquots of these PCR reactions were further amplified using primer set B. Resulting PCR products were electrophoresed and stained with ethidium bromide. C) Southern blot analysis of the gel shown in B with a probe corresponding to the rat rod photoreceptor channel. (Figure adapted from Ahmad et al., 1992.)

rod photoreceptor channel (Figure 4C). This experiment confirmed that in M-1 cells there are binding sites for the four primers separated by the same distance as predicted from the rat rod photoreceptor sequence.

Amplification by PCR made it possible to clone and sequence these M-1 cell transcripts (Ahmad et al., 1992). Amplified cDNAs were ligated into Bluescript plasmid vector and transformed into bacterial hosts. Putative positive clones were analyzed by restriction enzyme digestion and gel electrophoresis to confirm the presence of inserts of the correct sizes. The inserts of selected clones were sequenced on both strands of DNA.

Two sequences were obtained from separate M-1 clones (Ahmad et al., 1992). A comparison of these sequences with those of the rat rod channel and the rat olfactory channel is shown in Figure 5. One of the sequences (M-1 clone 1), obtained with primer set A, differed from the rat rod sequence by two bases out of 726. The other sequence (M-1 clone 2), obtained with primer set B, differed by 20 out of 426 bases, corresponding to about 5%. All differences were in redundant positions and thus gave a deduced amino acid sequence identical to that of the rod channel. This is in contrast to the rat olfactory channel which differs by 122 bases over the same 526 base region. This corresponds to a 29% variation at the DNA level and also results in an 11% variation at the amino acid level.

Based on the portion cloned and sequenced, one of these M-1 transcripts appears identical to that of the cGMP-gated cation channel of rod photoreceptors. However, differences might be present in other regions, particularly those closer to the N-terminus which are known to show significant variation between the rod and olfactory channels. Since two distinct sequences were detected in M-1 cells, it suggests that this cell line and, possibly, collecting duct cells in vivo express more than one gene coding for nonselective cation channels or channel subunits.

Based on the homology of the photoreceptor channel with *Shaker*-type potassium channels (Guy et al., 1991; Heginbotham et al., 1992), it has been suggested that the photoreceptor channel may exist as a tetramer. Alternatively, the photoreceptor channels may also form a pentameric structure similar to other ligand-gated channels. Indeed, analysis of the cGMP concentration dependence of channel activation suggests that complete channel activation requires the cooperative binding of at least four molecules of cGMP. Because only one cGMP molecule can be found per polypeptide monomer, the channel must be composed of four or more monomers (Kaupp, 1991). In either case, a multimeric structure raises the question of whether M-1 cells contain two types of homomultimeric or one type of heteromultimeric channels. Different subunit composition might help explain the different biophysical properties observed in the large family of nonselective cation channels.

At present, we do not know whether the detected genes are in fact responsible for the nonselective cation channel activity seen in our patch-clamp experiments. We have to consider the possibility that the detected genes encode a type of nonselective cation channel which, under the chosen experimental conditions, is not easily seen. For exam-

Rat retina   ATTTCCAATATGAATGCAGCCCGGGCAGAATTTCAATCAAGAGTTGATGCTATCAAACAGTACATGAATTTTCGAAATGTGAGCAAAGACATGGAAAAGAGAGTTATTAAATGGTTTGACT
Rat olf   --C---A-C---------CA-A--A--------G--C--GG-C-AGA----------C--------C-G--C----------------------GCC-AG--C--C--|||||||||||||
M1 clone 1 |||||||||||||||||||||||||||||||||||||||||||||||||||||||||||||||||||||||||||||||||||||||||||||||||||||||||||||||||||||||||

Rat retina   ACCTGTGGACCAACAAAAAGACAGTCGATGAGAGAAGTTCTGAGATACCTCCCTGACAAACTCAGGGCAGAGATTGCCATCAATGTTCACTGACACGTTAAAAAAGGTTCGTATCTT
Rat olf   --T-----T--G--------A----AC-------C--C-AGA----G--A-CA-G---------A----T------------T-GTC--TC-G---A--G--C--A--
M1 clone 1   --------T-------------------A----G-----------C----------------------A-------------C----T--------G-----------------
M1 clone 2   --A---------A---------------------C----------C------

Rat retina   TGCTGACTGTGAGGCTGGTCTGTTGGTGGAGTTGGTGTTGAAATTACACACCCAGGTGACAGTCCTGGAGATTACATATGCAAGAAAGGGACATTGGGCGGGAGATGTCATCATCAAG
Rat olf   CCAGGAT---A----AC---AC---GC-T-GT--T---C-TT------------C-TT--------CGT--G-------------------CAA---A------
M1 clone 1   ----------A---T------T--G-----------------------------C------
M1 clone 2   ------------------------------C---------------------------------------

Rat retina   GAAGGCAAACTTGCTGTGTGGCAGACGACGGAATCACACAGTTTGTGGTTGAGTCAGCGGCAGCTACTTTGGCGAGATCAGCATTCTTAACATCAAAGGCACAGGCTGCAACCGAA
Rat olf   --G-----GT-G--A-----A-T--T--T--CG-G--T----A--CCT--C-CTCA-CT---G------T---T---G------T--T--G--T-------AATG----T--C
M1 clone 1   ---T------------T--G--T-------------------------T----
M1 clone 2   --T--------------T-----------------T--------

Rat retina   GAACAGCCAATATTAAGGACGATTGGCTACTGGGACCTGTTCTGCCTCTCAAAGGATGACCTCATGGAAGCTCTTACAGAGTACCCCAGATGCCAAAACTATGTTGGAGGAGAAAGGGAGGCA
Rat olf   -T--T--T------CCGT--C-G--------A--T--C----T-G--C-------C-T--T--------G-A--T------T-------ACGG--T--G-
M1 clone 1   ---------------------C---------C-----------
M1 clone 2

Rat retina   GATCTTAATGAAAGACGGTCTACTGGATATAAACATTGCGAATTTGGGCAGTGAGTGACCCTAAAGCCTGGGAAGAGAAGGTCACTCGAATGGAGGGTCAGTGGACCTCCTGCAAACACGATT
Rat olf   --C-G-----G--A---------------GAG--TGAA-T-GCAGCTA-T-TG--GGTAG-T-TTCA--GA--CT--AACAGTTGGA-ACAAACATG-ATAC-T-GTACACTCGCTTTGC
M1 clone 1   ----------------------C---------C---------

Figure 5. Sequences of cDNAs cloned from M-1 cells. Amplified cDNAs were ligated into Bluescript plasmid vector and transformed into bacterial hosts. Putative positive clones were analyzed by restriction enzyme digestion and gel electrophoresis to confirm the presence of inserts of the correct sizes. The inserts of selected clones were sequenced on both strands of DNA. Two sequences were obtained from separate M-1 clones and sequences are compared with the sequence of the cyclic nucleotide-gated ion channel from rat rod photoreceptor (rat retina) and rat olfactory epithelium (rat olf). Dashes show identities with the rod photoreceptor sequence. M-1 clone 2 was derived from primer set B and is thus shorter than the other sequences. The sequences of oligonucleotides used in the PCR reactions have not been included in this figure. (Adapted from Ahmad et al., 1992.)

ple, the presence or absence of divalent cations may play a key role for the detection of certain subsets of nonselective cation channels. Whereas the nonselective cation channel in M-1 cells is calcium activated, the single-channel conductance of the photoreceptor channel is decreased from 20–25 pS in the absence of Ca and Mg to 0.1 pS in the presence of 1 mM Ca and Mg (Kaupp, 1991; Yau and Baylor, 1989). Thus, it will be interesting to see whether experiments performed in the absence of calcium and magnesium could reveal the presence of an additional nonselective cation channel in M-1 cells.

Conclusion

Obviously, there are remarkable differences in physiological properties between the rod photoreceptor (for review see Kaupp, 1991) and kidney nonselective cation channels.

The rod channel responds to cGMP by increasing its open probability, whereas the nonselective cation channel in inner medullary collecting duct cells and also in M-1 cells responds by decresing its open probability upon application of cGMP.

In addition, the channels are affected differently by divalent cations. In the photoreceptor channel the single-channel conductance in the absence of Ca or Mg is 20–25 pS and decreases to about 0.1 pS in the presence of 1 mM Ca or Mg. In contrast, the channel in M-1 cells is active in the presence of 1 mM Ca and 1 mM Mg, whereas calcium removal abolishes channel activity.

There is, however, a number of similarities between the photoreceptor channel and the nonselective cation channel in M-1 cells. These include a similar single-channel conductance, a similar selectivity for monovalent cations with a P_{Na} to P_K ratio close to one, a linear single-channel I-V relationship, and an increase in open probability at more depolarized voltages.

Future work will have to show whether the two observations in M-1 cells are purely coincidental or whether, indeed, the gene found is related to the nonselective cation channel activity seen.

Acknowledgements
This work was supported by grants DK-13844, DK-17433, EY-05206, and NS-20483 from the National Institutes of Health and a Forschungsstipendium of the Deutsche Forschungsgemeinschaft to C.K. The electrophysiological studies were performed by C.K. during a postdoctoral fellowship in the laboratory of E. L. Boulpaep at Yale and are presently being continued in the laboratory of E. Frömter at Frankfurt University. Computer programs for single-channel data acquisition and analysis were written by A. S. Segal who also participated in some of the patch-clamp experiments. I. Ahmad and P. Cheung participated in many of the molecular biology experiments carried out in the laboratory of C.B. We thank E. L. Boulpaep and E. Frömter for their helpful suggestions of improvements to this manuscript.

162

References

Ahmad I, Redmond LJ, Barnstable C (1990). Developmental and tissue-specific expression of the rod photoreceptor cGMP-gated ion channel gene. Biochem. Biophys. Res. Comm. 173:463–470.

Ahmad I, Korbmacher C, Segal AS, Cheung P, Boulpaep EL, Barnstable CJ (1992). Mouse cortical collecting duct cells show nonselective cation channel activity and express a gene related to the cGMP-gated photoreceptor channel. Proc. Natl. Acad. Sci. USA 89:10262–10266.

Champigny G, Verrier B, Lazdunski M (1991). A voltage, calcium, and ATP sensitive nonselective cation channel in human colonic tumor cells. Biochem. Biophys. Research Communications 176:1196–1203.

Chan HC, Nelson DJ (1992). Chloride-dependent cation conductance activated during cellular shrinkage. Science 257:669–671.

Changeux, J-P, Devillers-Thiéry A, Chemouilli P (1984). Acetylcholine receptor: An allosteric protein. Science 225:1335–1345.

Cook DI, Poronnik P, Young JA (1990). Characterization of a 25-pS nonselective cation channel in a cultured secretory epithelial cell line. J. Membrane Biol. 114:37–52.

Dhallan RS, Yau K, Schrader K, Reed R (1990). Primary structure and functional expression of a cyclic nucleotide-activated channel from olfactory neurons. Nature 347:184–187.

Das S, Garepapaghi M, Palmer LG (1991). Stimulation by cGMP of apical Na channels and cation channels in toad urinary bladder. Am. J. Physiol. 260:C234–C241.

Fejes-Tóth G, Náray-Fejes-Tóth A (1992). Differentiation of renal β-intercalated cells to α-intercalated and principal cells in culture. Natl. Acad. Sci. USA 89:5487–5491.

Filipovic D, Sackin H (1991). A calcium-permeable stretch-activated cation channel in renal proximal tubule. Am. J. Physiol. 260:F119–F129.

Filipovic D, Sackin H (1992). Stretch- and volume-activated channels in isolated proximal tubule cells. Am. J. Physiol. 262:F857–F870.

Garty H, Benos DJ (1988). Characteristics and regulatory mechanisms of the amiloride-blockable Na^+ channel. Physiological Reviews 68:309–373.

Geibel J, Korbmacher C, Fejes-Toth G, Boulpaep EL, Giebisch G (1990). AVP and endothelin increase intracellular Ca^{2+} in a mouse cortical collecting duct cell line. JASN 1(4):469.

Geibel J, Korbmacher C, Giebisch G, Boulpaep EL (1991). Endothelin (ET-1) stimulates a nifedipine sensitive Ca^{2+} entry mechanism in M-1 mouse cortical collecting duct cells. JASN 2(3):401.

Gögelein H, Greger R (1986). A voltage-dependent ionic channel in the basolateral membrane of late proximal tubules of the rabbit kidney. Pflügers Arch. 407(Suppl2):S142–S148.

Gögelein H, Pfannmüller B (1989). The nonselective cation channel in the basolateral membrane of rat exocrine pancreas. Pflügers Arch. 413:287–298.

Gögelein H, Dahlem D, Englert HC, Lang HJ (1990). Flufenamic acid, mefenamic acid and niflumic acid inhibit single nonselective cation channels in the rat exocrine pancreas. FEBS Letters 268:79–82.

Goulding EH, Ngai J, Kramer RH, Colicos S, Axel R, Siegelbaum SA, Chess A (1992). Molecular cloning and single-channel properties of the cyclic nucleotide-gated channel from catfish olfactory neurons. Neuron 8:45–58.

Guy HR, Durell SR, Warmke J, Drysdale R, Ganetzky B (1991). Similarities in amino acid sequences of Drosophila eag and cyclic nucleotide-gated channels. Science 254:730.

Hamill OP, Marty A, Neher E, Sakmann B, Sigworth FJ (1981). Improved patch-clamp techniques for high-resolution current recording from cells and cell-free membrane patches. Pflügers Arch. 391:85–100.

Heginbotham L, Abramson T, MacKinnon R (1992). A functional connection between the pores of distantly related ion channels as revealed by mutant K^+ channels. Science 258:1152–1155.

Jorissen M, Vereecke J, Carmeliet E, Van den Berghe H, Cassiman J-J (1990). Identification of a voltage- and calcium-dependent nonselective cation channel in cultured adult and fetal human nasal epithelial cells. Pflügers Arch. 415:617–623.

Jung F, Selvaraj S, Gargus JJ (1992). Blockers of platelet-derived growth factor-activated nonselective cation channel inhibit cell proliferation. Am. J. Physiol. 262:C1464–C1470.

Kaupp UB, Niidome T, Tanabe T, Terada S, Bönigk W, Stühmer W, Cook NJ, Kangawa K, Matsuo H, Hirose T, Miyata T, Numa S (1989). Primary structure and functional expression from complementary DNA of the rod photoreceptor cyclic GMP-gated channel. Nature 342:762–766.

Kaupp UB (1991). The cyclic nucleotide-gated channels of vertebrate photoreceptors and olfactory epithelium. TINS 14(4):150–157.

Kizer N, Hu J-M, Fejes-Tóth G, Stanton BA (1991). Whole cell conductances and single channel analysis of a cortical collecting duct (CCD) cell line isolated from a transgenic mouse. FASEB Journal 5(4):A689.

Korbmacher C, Segal AS, Fejes-Toth G, Giebisch G, Boulpaep EL (1991). Different expression of whole cell conductances in single and confluent M-1 mouse cortical collecting duct cells: effects of amiloride, barium and glibenclamide. JASN 2(3):743.

Korbmacher C, Segal AS, Boulpaep EL (1992(a)). Ion channels in a mouse cortical collecting duct cell line. Renal Physiology and Biochemistry 15:180.

Korbmacher C, Segal AS, Boulpaep EL, Schröder UH, Frömter E (1992(b)). Forskolin stimulates a chloride conductance in M-1 mouse cortical collecting duct cells. JASN Vol. 3. JASN 3(3):812.

Korbmacher C, Boulpaep EL, Giebisch G, Geibel J. (1993(a)). Endothelin increases intracellular calcium concentration in M-1 mouse cortical collecting duct cells by a dual mechanism. Am. J. Physiol. (in press).

Korbmacher C, Segal AS, Fejes-Tóth G, Giebisch G, Boulpaep EL. (1993(b)). Whole cell currents in single and confluent M-1 mouse cortical collecting duct cells. J. Gen. Physiol. (in press).

Laskowski FH, Christine CW, Gitter AH, Beyenbach KW, Gross P and Frömter E (1990). Cation channels in the apical membrane of collecting duct principal cell epithelium in culture. Renal Physiol. Biochem. 13:70–81.

Light DB, McCann FV, Keller TM, Stanton BA (1988). Amiloride sensitive cation channel in apical membrane of inner medullary collecting duct. Am. J. Physiol. 255:F278–F286.

Light DB, Schwiebert EM, Karlson KH, Stanton BA (1989). Atrial natriuretic peptide inhibits a cation channel in renal inner medullary collecting duct cells. Science 243:383–385.

Light DB, Corbin JD, Stanton BA (1990). Dual ion-channel regulation by cyclic GMP-dependent protein kinase. Nature 344:336–339.

Ling BN, Hinton CF, Eaton DC (1991). Potassium permeable channels in primary cultures of rabbit cortical collecting tubule. Kidney Internnational 40:441–452.

Marom S, Dagan D, Winaver J, Palti Y (1989). Brush-border membrane cation conducting channels from rat kidney proximal tubules. Am J. Physiol. 257:F328–F335.

Merot J, Bidet M, Gachot B, Le Maout S, Tauc M, Poujeoul P (1988). Patch clamp study on primary culture of isolated proximal convoluted tubules. Pflügers Arch. 413:51–61.

Merot J, Poncet V, Bidet M, Tauc M, Poujeol P (1991). Apical membrane ionic channels in the rabbit cortical thick ascending limb in primary culture. Biochim. Biophys. Acta 1070:387–400.

Naruse M, Uchida S, Ogata E, Kurokawa K (1991). Endothelin 1 increases cell calcium in mouse collecting tubule cells. Am. J. Physiol. 261:F720–F725.

Ohara A, Matsumoto P, Eaton DC, Marunaka Y (1991). A nonselective cation channel induced by cyclic GMP and nitroprusside in a distal nephron cell line (A6). FASEB Journal 5(4):A689.

O'Neil, RG, Boulpaep EL (1982). Ionic conductive properties and electrophysiology of the rabbit cortical collecting tubule. Am. J. Physiol. 243:F81–F95.

Paulais M, Teulon J (1989). A cation channel in the thick ascending limb of Henle's loop of the mouse kidney: Inhibition by adenine nucleotides. J. Physiol. 413:315–327.

Partridge LD, Swandulla D (1988). Calcium-activated nonspecific cation channels. Trends Neurosci 11:69–72.

Popp R, Gögelein H (1992). A calcium and ATP sensitive nonselective cation channel in the antiluminal membrane of rat cerebral capillary endothelial cells. Biochimica et Biophysica Acta 1108:59–66.

Siemer C, Gögelein H (1992). Activation of nonselective cation channels in the basolateral membrane of rat distal colon crypt cells by prostaglandin E_2. Pflügers Arch. 420:319–328.

Stoos BA, Náray-Fejes-Tóth A, Carretero OA, Ito S, Fejes-Tóth G (1991). Characterization of a mouse cortical collecting duct cell line. Kidney Int. 39:1168–1175.

164

Teulon J, Paulais M, Bouthier M (1987). A Ca^{2+} activated cation-selective channel in the basolateral membrane of the cortical thick ascending limb of Henle's loop of the mouse. Biochim. Biophys. Acta 905:125–132.

Thorn P, Peterson OH (1992). Activation of nonselective cation channels by physiological cholecystokinin concentrations in mouse pancreatic acinar cells. J. Gen. Physiol. 100:11–25.

Vigne P, Champigny G, Marsault R, Barbry P, Frelin C, Lazdunski M (1989). A new type of amiloride-sensitive cationic channel in endothelial cells of brain microvessels. J. Biol. Chem. 264:7663–7668.

Yau KW, Baylor DA (1989). Cyclic GMP-activated conductance of retinal photoreceptor cells. Ann. Rev. Neurosci. 12:289–327.

Yellen G (1982). Single Ca^{2+}-activated nonselective cation channels in neuroblastoma. Nature 296:357–359.

Nonselective Cation Channels: Pharmacology, Physiology and Biophysics
ed. by D. Siemen & J. Hescheler
© 1993 Birkhäuser Verlag Basel/Switzerland

Ciliary Cation Conductances in Olfactory Receptor Cells of the Clawed Toad *Xenopus laevis*

Detlev Schild

Physiologisches Institut der Universität, Humboldtallee 23, D-37073 Göttingen, FRG

Summary
One transduction pathway in olfactory receptor neurons is a cascade of receptors, a G-protein, adenylate cyclase, cAMP, and a cyclic nucleotide-activated cation conductance. Here, we show that this conductance is also present in olfactory cells of *Xenopus laevis*. With optical recordings from the cell's dendritic knob, we show that this conductance, when activated by odors, leads to an increase of intracellular calcium. It is further shown that there is a second cation conductance on the cilia of these cells which is modulated by calcium and can be activated by the application of odorants.

Introduction

Olfactory receptor cells are primary sensory cells. They have the bio-chemical and biophysical machinery for transducing chemical stimuli to receptor potentials as well as the conductances for generating action potentials. They are thus typical nerve cells but have the peculiarity that their input signals are odorous molecules rather than neurotransmitters or neurmodulators. It is therefore not astonishing that the voltage-gated conductances found in olfactory receptor neurones (Trotier, 1986; Firestein and Werblin, 1987; Schild, 1989; Lynch and Barry, 1991; Lucero et al., 1992) are well-known from other neurones. The first transduction pathway described is a second messenger cascade involving receptors, G-proteins, adenylate cyclase, cAMP, and a cyclic nucleotide sensitive cation conductance g_{cn} which is directly activated by cAMP (Nakamura and Gold, 1987). This pathway is best described in the tiger salamander (Firestein and Werblin, 1989) and in the newt (Kurahashi, 1989), but is also found in other species. However, it has also become clear that olfactory transduction mechanisms are not limited to this pathway: evidence for other transduction pathways has been reported, e.g., in silkmoth (Zufall and Hatt, 1991), catfish (Restrepo et al., 1990), and man (Restrepo et al., 1992). Other conductances which are modulated by odorants have been described recently (McClintock and Ache, 1989; Dionne, 1992), and it has been shown that odors of the same "odor class" can activate different second messenger pathways (Breer and Boekhoff, 1991).

Our model consists of olfactory receptor cells from the clawed toad *Xenopus laevis* and we herein delineate two cation conductances which are located primarily on the cilia and seem to be involved in olfactory transduction.

We obtained isolated receptor neurones by enzyme-free dissociation of olfactory mucosae in divalent ion-free Ringer's solution. The cells were studied in the whole-cell mode of the patch-clamp technique using standard procedures (Schild, 1989). As odorants we used two different cocktails, first, a mixture of citralva and amylacetate, and second, a mixture of triethylamine, phenylethylamine, and isovaleric acid. The nominal concentrations of all odorants used were 10 or 100 μM. Citralva and amylacetate have been shown to increase the concentration of cAMP in a cilia preparation (Sklar et al., 1986; Boekhoff et al., 1990; Boekhoff et al., 1990), while the odorants of the second cocktail given above are known to increase the concentration of IP_3 (Breer et al., 1990; Boekhoff et al., 1990), which is also true in *Xenopus laevis* (Schild et al., 1992). The odorants were applied through a pipette (as used for patch clamping) onto the cells' cilia.

Responses to Odorants That Increase cAMP

When the olfactory neurones were voltage clamped, the odorants citralva and amylacetate evoked inward currents at negative potentials (Figure 1A) and outward currents at positive potentials. The maximum response current as a function of holding potential is shown in Figure 1B. These recordings were made in a toad Ringer's solution with nominally no Ca^{2+} (actually 20 μM Ca^{2+}), and K^+ as the predominant cation in the pipette solution. Substituting Cs^+ for K^+ did not change the current-voltage relationship. However, when Ca_o^{2+} was increased to 5 mM, virtually no current responses were observable (Figure 1A). The conspicuous nonlinearity of the I-V curve suggested an influence of Mg^{2+} in the bath. It was therefore interesting to test the responses to odorants with and without Mg^{2+} in the bath. 5 mM Mg^{2+} induced a "shoulder" in the deactivation of the odor response which was absent with 0 Mg^{2+} in the bath.

The cAMP-gated generator conductance has been reported to be permeable for Ca^{2+} ions (Kurahashi and Shibuya, 1990). An increase in intracellular Ca^{2+} during stimulation with odorants could therefore be expected, at least in the distal compartments of the cells. We tried to measure this Ca^{2+} increase using a laser scanning microscope in its line-scanning mode as well as the fluorescent Ca^{2+}-indicator dye CaGreen/AM. The line to be scanned was determined in a full-frame image of the cell and located on the olfactory knob. The line was then scanned with a frequency of about 100 Hz (T = 11.4 ms) and the

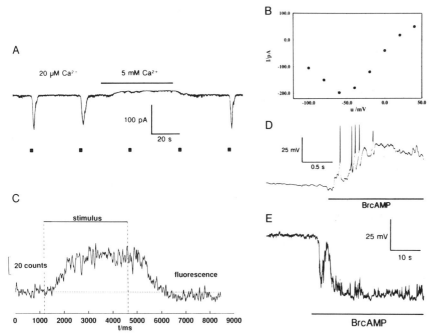

Figure 1. Effects of odorants and bromo-cAMP on *Xenopus* olfactory cells. A) Current trace recorded under voltage-clamp conditions at a holding potential of -80 mV. The filled squares indicate odorant applications. With $20 \mu M$ Ca^{2+} in the bath, every application of the odorants (citralva and amylacetate) is followed by an inward current pulse. This response is abolished when the calcium concentration in the bath is 5 mM. B) I-V curve taken from the recording, a part of which is shown in part A of the figure. Note the decrease of current amplitudes at voltages lower than -60 mV. C) Fluorescence signal indicating a calcium increase in the olfactory knob during stimulation. D) Current clamp recording of an olfactory cell. Upon addition of bromo-cAMP to the bath the cell depolarizes and generates spikes. E) Current clamp recording from a different cell which hyperpolarizes upon bath application of $100 \mu M$ BrcAMP.

fluorescence signal was recorded before, during, and after odorant application. Figure 1C gives a typical example of a fluorescence signal that increases during stimulus application with a delay of about 325 ms. Obviously, the intracellular Ca^{2+} concentration increased within the dendritic knob during stimulation, and this increase is high enough to be detected. In the near future, this might open the possibility to record responses to odors from more than one cell at the same time and without penetration of the cells' plasma membranes.

According to the notion that citralva and amylacetate increase cAMP and thereby gate a cAMP-sensitive generator conductance, one must expect that the extracellular application of bromo-cAMP would lead to an inward current or, when measured in the current clamp mode, to a depolarization. This, in fact, occurred as shown in Figure 1D. However, it has to be noted that in some cells bromo-cAMP led to a hyperpolarization as examplified in Figure 1E. This appears to indicate the possible existence of different second messenger pathways with different actions of the same second messenger upon membrane conductances.

Ca^{2+}-Modulated Cation Conductance and Responses to Odorants That Increase IP$_3$

The existence of different transduction pathways is further supported by the finding of a Ca^{2+}-modulated cation conductance in olfactory cilia of *Xenopus* (Schild and Bischofberger, 1991). This conductance g_c could be activated in the whole-cell configuration either by Ca^{2+}-influx through high-voltage-gated Ca^{2+}-channels or by Ca^{2+}-influx through the Ca/Na-transporter when driven in the reversed mode. In the first case, the conductance is activated at about its reversal potential, which is about 0 mV (Schild and Bischofberger, 1991), so that virtually no current through it results. Its deactivation is reflected by a characteristically long-lasting tail current after stepping back to the holding potential of -80 mV (Figure 2A). In the second case, the Ca^{2+}-influx elicits an outward current at positive membrane potentials (Figure 2B) and the deactivation tails are again seen after stepping back to the holding potential to -80 mV. This long-lasting kind of tail current was the only inward tail current present at -80 mV; it was sensitive to amiloride and to a decrease in Na-gradient across the membrane, as well as to Ba^{2+} (Schild and Bischofberger, 1991).

Remarkably, the same kind of tail current was also observed when the cells were stimulated with odors which increase the concentration of IP$_3$ in the cilia of *Xenopus* olfactory neurones. Using the combined voltage clamp and stimulus protocol given in Figure 2C, the membrane potential was stepped from a holding potential of -80 mV to depolarized voltages, and immediately after each command step the odorant mixture containing triethylamine, phenylethylamine, and isovaleric acid was injected onto the cilia. The resulting current responses are shown in Figure 2C. The currents are inward for negative potentials and outward for potentials higher than 10 mV. These experiments were carried out with toad Ringer's solution in the bath and with a pipette solution containing 10 mM TEACl and 40 mM Cs$_2$SO$_4$. The Nernst potential for chloride was -30 mV. Inward and outward currents were thus presumably carried by Na$^+$ and by Cs$^+$ (and possibly TEA$^+$), respectively. Note that for all command steps to voltages equal to or higher than -20 mV, the current responses to the odorants were not thoroughly deactivated at the end of the command step. Accordingly, deactivating tail currents appeared after stepping back to the holding potential of -80mV. The time constants of these tail currents were in the same range as those of the Ca^{2+}-modulated cation conductance.

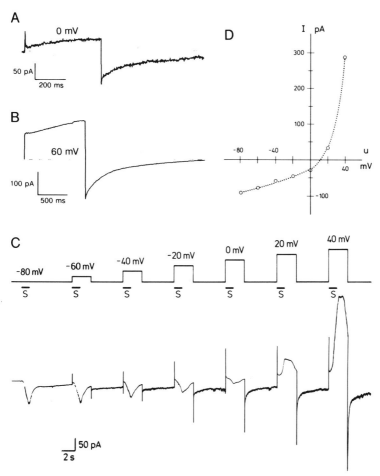

Figure 2. Ca^{2+}-modulated and odorant-activated cation conductance. A) Voltage-clamp recording showing a calcium inward current (plus leak) activated at -10 mV from a holding potential of -80 mV. Note the long-lasting tail current at $u = -80$ mV following the command step. B) Current response to a voltage-clamp step to 60 mV and subsequent tail current at -80 mV. See text for explanation. C) Upper trace: voltage clamp and odorant application protocol. The voltage is increased by 20 mV every step, and the odorant is applied at the onset of every command pulse. Lower trace: resulting current responses, which are inward for $u < 10$ mV. D) I-V curve of the odorant-activated current amplitudes.

The I-V-curve of the maximum current amplitudes of Figure 2C are shown in part D of this figure. It can be characterized by a marked outward rectification, and it does not show decreasing current amplitudes at potentials lower than -60 mV. By these features, it can be easily distinguished from the I-V curve of the cAMP-gated generator conductance.

This preliminary evidence shows that a class of odorants which increases the concentration or IP$_3$ activate current responses which seem

170

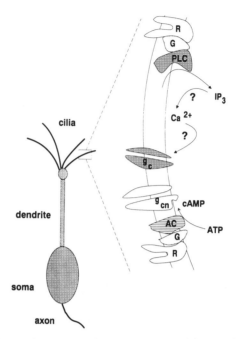

Figure 3. Sketch of an olfactory cell of *Xenopus laevis* and its transduction pathways. R: receptor proteins, G: G-proteins, AC: adenylate cyclase, g_{cn}: cyclic nucleotide-activated conductance, PLC: phospholipase C. The pathway shown in the upper half of the ciliary membrane is only partly understood.

to differ from the cAMP-gated currents and which share characteristics such as the reversal potential and deactivation time constants with the previously reported Ca^{2+}-modulated cation conductance.

Clearly, the above results open numerous questions and requirements: i) the effects of odorants upon intracellular calcium have to be shown directly and in detail, ii) do both pathways coexist in (at least some) olfactory neurones, and if so, iii) what are the interactions among them? One possible interaction has recently been reported: an increase in Ca^{2+} appears to shift the sensitivity of the cyclic nucleotide-gated conductance on cAMP to higher concentrations, i.e., the cAMP-gated conductance becomes less sensitive to cAMP with increasing Ca^{2+} (Lynch and Lindemann, 1992). A hint to another possible interaction is the hyperpolarizing effect of cAMP in some olfactory neurons (Figure 1E). Figure 3 is a sketch showing both pathways, including hypothetical steps (dashed lines) in the ciliary membrane of an olfactory neurone of *Xenopus laevis*. The question marks in the figure are to indicate that it is presently unknown how IP_3 leads to an increase of calcium and, second, how calcium activates the cation conductance g_c.

References

Boekhoff I, Tareilus E, Strotmann J, Breer H (1990). Rapid activation of alternative second messenger pathways in olfactory cilia from rats by different odorants. EMBO Journal 9:2453–2458.

Breer H, Boekhoff I (1991). Odorants of the same odor class activate different second messenger pathways. Chem. Senses 16:19–29.

Breer H, Boekhoff I, Tareilus E (1990). Rapid kinetics of second messenger formation in olfactory transduction. Nature 345:65–68.

Dionne VE (1992). Chemosensory responses in isolsated olfactory receptor neurons from Necturus maculosus. J. Gen. Physiol. 99:415–433.

Firestein S, Werblin F (1989). Odor-induced membrane currents in vertebrate-olfactory receptor neurons. Science 244:79–82.

Firestein S, Werblin FS (1987). Gated currents in isolated olfactory receptor neurons of the larval tiger salamander. Proc. Natl. Acad. Sci. USA 84:6292–6296.

Kurahashi T (1989). Activation by odorants of cation-selective conductance in the olfactory receptor cell isolated from the newt. J. Physiol. (London) 419:177–192.

Kurahashi T (1990). The response induced by intracellular cyclic AMP in isolated olfactory receptor cells of the newt. J. Physiol. (London) 430:355–371.

Kurahashi T, Shibuya T (1990). Ca^{2+}- dependent adaptive properties in the solitary olfactory receptor cell of the newt. Brain Res. 515:261–268.

Lucero MT, Horrigan FT, Gilly WF (1992). Electrical responses to chemical stimulation of squid olfactory receptor cells. J. Exp. Biol. 162:321–249.

Lynch JW, Lindemann B (1992). Divalent cations decrease sensitivity of cAMP-gated channels to cAMP in rat olfactory receptor cells. *ECRO Xth Congress*. Munich, p 96.

McClintock TS, Ache BW (1989). Histamine directly gates a cloride channel in lobster olfactory receptor neurons. Proc. Natl. Acad. Sci. USA 86:8137–8141.

Nakamura T, Gold GH (1987). A cyclic nucleotide-gated conductance in olfactory receptor cilia. Nature 325:442–444.

Restrepo D, Miyamoto T, Bryant BC, Teeter JH (1990). Odor stimuli trigger influx of calcium into olfactory neurons of the channel catfish. Science 249:1166–1168.

Schild D (1989). Whole-cell currents in olfactory receptor cells of *Xenopus laevis*. Exp. Brain Res. 78:223–232.

Schild D, Bischofberger J (1991). Ca^{2+} modulates an unspecific cation conductance in olfactory cilia of *Xenopus laevis*. Exp. Brain Res. 84:187–194.

Sklar PB, Anholt RRH, Snyder SH (1986). The odorant-sensitive adenylate cyclase of olfactory receptor cells. J. Biol. Chem. 261:15538–15543.

Trotier D (1986). A patch-clamp analysis of membrane currents in salamander olfactory receptor cells. Pflügers Arch. 407:598–595.

Zufall F, Hatt H (1991). A calcium-activated nonspecific cation channel from olfactory receptor neurones of the silkmoth Antheraea polyphemus. J. Exp. Biol. 161:455–469.

Channels Activated by Intracellular Calcium

Nonselective Cation Channels: Pharmacology, Physiology and Biophysics
ed. by D. Siemen & J. Hescheler
© 1993 Birkhäuser Verlag Basel/Switzerland

Control of Cell Function by Neuronal Calcium-Activated Nonselective (CAN) Cation Channels

L. Donald Partridge[+] and Dieter Swandulla[*]

[+]The University of New Mexico School of Medicine, Department of Physiology, Albuquerque, NM 87131, USA; *Institut für Experimentelle und Klinische Pharmakologie und Toxikologie, Molekulare Pharmakologie, Universität Erlangen-Nürnberg, D-91054 Erlangen, FRG

Much of the action of the nervous system can be ultimately associated with changes in the potential of individual neurons. Inhibition plays an essential role in neuronal interactions but it is the excitation provided by postsynaptic potentials and action potentials that generally determines neuronal output. Since excitation is usually associated with depolarization, maintained excitatory states of neurons require mechanisms of maintained depolarization. Sodium channels are one common channel type that select for ions whose electrochemical gradient produces an inward depolarizing current. These channels generally have strongly voltage-dependent inactivation mechanisms that prevent continued channel opening during depolarization. While this action prevents prolonged positive feedback action through the Hodgkin cycle, it makes these channels unavailable for maintaining a depolarized state of the channel. Calcium channels, the other large class of channels producing depolarization, exhibit both voltage-dependent and Ca^{2+}-dependent inactivation. One type of calcium channel, the HVA class, is capable of remaining open for hundreds of milliseconds but calcium channels in general are not suited to providing current for maintained depolarization.

CAN Channels – A Ubiquitous Class of Membrane Channels

Intracellular Ca^{2+} is an important second messenger that couples membrane excitation with cellular functions. Excitation-contraction coupling, and excitation-secretion coupling are important examples of this second messenger role in muscle cells, and secretory cells or presynaptic terminals. Excitation-gating coupling uses cytoplasmic Ca^{2+} in coupling channel activities that ultimately determine membrane potential. Calcium-activated nonselective (CAN) cation channels are a ubiquitous

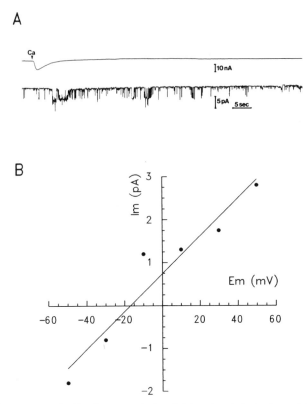

Figure 1. CAN currents. A) Simultaneous whole-cell (above) and single-channel (below) CAN currents recorded form a molluscan neuron. Pressure injection of Ca^{2+} into the neuron initiates an inward CAN current and increases the probability of opening of the individual CAN channel. B) Current-voltage relationship for a single CAN channel in a patch isolated from a vertebrate sensory neuron. This current reverses at the potential expected for a channel that is permeable to both Na^+ and K^+.

class of channels that are activated by sub-micromolar cytoplasmic Ca^{2+} and, when open, carry a depolarizing current (Fig. 1). While CAN channels are found in a number of different tissues (Partridge and Swandulla, 1988), they have an especially important role in neurons. In the case of CAN channels, excitation-gating coupling is a means of opening a non-inactivating, depolarizing channel in response to either voltage-dependent Ca^{2+} influx or the release of Ca^{2+} from cellular stores. This maintained depolarization can provide the depolarizing drive underlying bursts of action potentials (Hofmeier and Lux, 1981; Swandulla and Lux, 1985; Canavier et al., 1991), the current responsible for depolarizing afterpotentials (Adams, 1985; Schwindt et al., 1988; Hasuo et al., 1990), a source of Ca^{2+} in excitotoxicity (Choi, 1990; Currie and Scott, 1992); and the drive for ischemic depolarization (Schaeffer and Lazdunski, 1991; Rod and Auer, 1992).

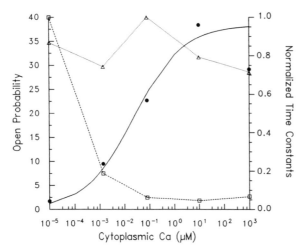

Figure 2. Effect of cytoplasmic Ca^{2+} on the opening of a CAN channel. The solid line is a best fit to points (10^{-11} M to 10^{-5} Ca^{2+}) for the opening of a single channel in a patch from a vertebrate sensory neuron. Open probabilities at millimolar cytoplasmic Ca^{2+} were typically depressed below the maximum value. The channel open and closed histograms at each Ca^{2+} value were fitted with a single exponential and the resultant time constants are plotted (($\triangle$) open time histograms, $\tau_{max} = 1.6$ ms: ($\square$) closed time histograms, $\tau_{max} = 56.4$ ms). Time constants fitted to open time histograms are rather insensitive to cytoplasmic Ca^{2+} while time constants fitted to closed time histograms decrease with increasing cytoplasmic Ca^{2+}.

Intracellular Ca^{2+} is heavily buffered so that steep gradients exist away from local concentration sources such as voltage-dependent calcium channels (Zucker and Fogelson, 1986; Smith and Augustine, 1988; Bers and Peskoff, 1991). Modelling studies demonstrate that CAN channel opening is determined by Ca^{2+} in the local micro-environment ($< 1 \mu$m) from a CAN channel (Partridge et al., 1991). Because the soma of neurons is reasonably isopotential, current through open CAN channels can have a much more generalized effect on the membrane potential of the whole soma. In this way, locally-modulated cytoplasmic Ca^{2+} can have a distributed effect over the soma of a neuron.

Ion channels are characterized by their selectivity, gating, and sensing (Hille, 1992). Although the molecular structure of neuronal CAN channels has not been determined, many characteristics of these channels have been determined. In the following we will briefly describe the biophysical properties of the class of neuronal CAN channels.

Biophysical Properties and Modulation of Neuronal CAN Channels

Neuronal CAN channels select strongly for cations over anions but pass a number of different cations. In addition to Na^+ and K^+, these permeant cations include: Cs^+ (Marcus and Eaton, 1990; Kim and Woodruff, 1991; Takeuchi et al., 1992), Li^+ (Yellen, 1982; Kramer and Zucker, 1985), Rb^+ (Marcus and Eaton, 1990; Takeuchi et al., 1992), NH_4^+ (Marcus and Eaton, 1990), TEA^+ (Swandulla and Lux, 1985; Simmonneau et al., 1987), choline (Swandulla and Lux, 1985; Zucker, 1988), Tris (Swandulla and Lux, 1985; Zucker, 1988), TMA (Zucker, 1988), and glucosamine (Swandulla and Lux, 1985; Zucker, 1988). Sodium and potassium are physiologically the two most important monovalent cations to which the CAN channel is permeant. In chick embryonic sensory neurons P_{Na}/P_K ratios are less than 1.0 indicating some preference for K^+ in these channels (Ranzani-Boroujerdi and Partridge, 1992). The P_{Na}/P_K ratio has not been thoroughly tested in other neuronal CAN channels.

The Ca^{2+} permeability of CAN channels is especially important because of the positive feedback of a channel permeant to its activating ion. Furthermore, mechanisms of producing a maintained source for cytoplasmic Ca^{2+} are coming increasingly under scrutiny because of the important role of Ca^{2+} in excitotoxicity. Helix CAN channels can probably pass Ca^{2+} (Swandulla and Lux, 1985) and Lymnaea neuronal CAN channels can at least pass Ba^{2+} (Yazejian and Byerly, 1989). Calcium permeability of other neuronal CAN channels may have escaped identification because of the difficulty in studying Ca^{2+} permeability in a channel that requires Ca^{2+} for its activation.

CAN channels in neurons exhibit single-channel conductances that are equivalent to those of CAN channels of other tissues (Partridge and Swandulla, 1988) ranging from 16 pS (Zufall et al., 1991) to 34 pS (Razani-Boroujerdi and Partridge, 1992). Two "maxi" CAN channels have been found in crustaceans, one with $\gamma = 213$ pS in sinus gland nerve terminals (Lemos et al., 1986) and another with $\gamma = 320$ pS in olfactory receptor cells (McClintock and Ache, 1990). These, however, represent a rather different channel from the typical 25 pS CAN channel of other neurons.

CAN channels are activated by Ca^{2+} ions on the cytoplasmic face of the membrane. Typical neuronal CAN channels have K_Ds that are somewhat less than $1 \mu M$ (Lipton, 1987; Partridge and Swandulla, 1987; Lando and Zucker, 1989; Zufall et al., 1991). In embryonic chick sensory neurons, we have found a $K_D = 0.5 \mu M$ and a Hill coefficient for Ca^{2+} of about 0.5. The implication of these values is that there is more than one Ca^{2+} binding site on the channel and that there is negative cooperativity in Ca^{2+} binding wherein the binding of one Ca^{2+} ion decreases the affinity for Ca^{2+} binding at the second binding site.

Channel open and closed probability histograms from inside-out patches of chick sensory neurons can be fitted with single exponential functions to determine the rate constants of channel opening and closing. Cytoplasmic Ca^{2+} has little effect on the closing rate constant but it increases the opening rate constant (Razani-Boroujerdi and Partridge, 1992; see also Figure 2). Thus the increased CAN current in the presence of higher Ca^{2+} concentrations results from the fact that the channels remain closed for shorter times although, once open, the channel stays open for about the same time period. In other words, Ca^{2+} destabilizes the closed state of the channel.

At cytoplasmic Ca^{2+} concentrations in embryonic chick sensory neurons in excess of 0.1 mM, the channel open probability falls to about one half of that at $100\ \mu M$ (Razani-Boroujerdi and Partridge, 1992). Similar high Ca^{2+} inhibition of the channel was reported by Yazejian and Byerly (1989). While such high cytoplasmic Ca^{2+} concentrations are rather unphysiological, such an effect is similar to the biochemical phenomenon of substrate inhibition (see Cornish-Bowden, 1979) and may indicate the nature of the Ca^{2+} binding site on the CAN channel.

With these K_Ds and Hill coefficients, the micro-environment around Ca and CAN channels is just on the border of supplying enough Ca^{2+} for effective CAN channel opening (Partridge et al., 1991). As is hypothesized for other channels (Roberts et al., 1990; Wang and Thompson, 1992), CAN channels are probably found in local concentrations with calcium channels. In this local domain, Ca^{2+} concentrations could rise to the necessary levels to activate the CAN channel while the Ca^{2+} in the remainder of the cytoplasm might not reflect this concentration change.

CAN currents are reduced by 5-HT, IBMX, 8Br-cAMP, dibutyryl-cAMP, and injection of cAMP, or the catalytic subunit of protein kinase A (PKA) (Partridge et al., 1990; see Figure 3). These results show that CAN current can be reduced by a transmitter-dependent channel phosphorylation mechanism. A complex role of G-proteins in this process is shown by the increase in CAN current following GTP-γ-S injection but depression of the current following application of cholera toxin (Partridge, 1989; Partridge et al., 1990). CAN channels in patches excised from embryonic chick sensory neurons decrease their open probability following application of the catalytic subunit of PKA. Thus the site of phosphorylation is the channel itself or another molecule that is closely enough allied to be isolated with the channel in the membrane patch (Razani-Boroujerdi and Partridge, 1992). In preliminary experiments, we have been able to reverse the phosphorylation-induced decrease in channel opening with the neutral phosphatase calcineurin.

Function of CAN Channels in Neuronal Tissue

The first descriptions of neuronal CAN channels were in bursting pacemaker neurons (Hofmeier and Lux, 1981; Lewis, 1984; Kramer and

180

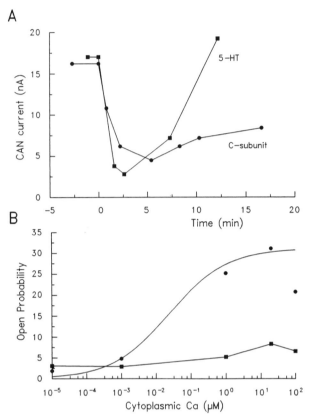

Figure 3. CAN channel phosphorylation. A) Externally applied serotonin (5-HT) or intracellularly injected catalytic subunit of protein kinase A (C-subunit) depress CAN current in molluscan neurons with a time course of several minutes. B) Phosphorylation of isolated vertebrate sensory neuron CAN channels by application of protein kinase A to the cytoplasmic side of an isolated membrane patch. Solid circles (●) show open probability at various Ca^{2+} concentrations in control conditions while filled squares (■) show open probabilities at these same Ca^{2+} concentrations 5 minutes or more after protein kinase A and ATP were added to the cytoplasmic side of the patch.

Zucker, 1985; Swandulla and Lux, 1985; Hasuo et al., 1990). The basis for bursting behavior in these neurons has been described in terms of the action of the CAN channel. Calcium initially enters through voltage-gated calcium channels and accumulates to a sufficient concentration to activate CAN channels. The inward CAN current then maintains the depolarization that causes the burst of action potentials. The burst is eventually terminated by activation of Ca^{2+}-activated potassium channels. Current through these channels predominates throughout the interburst period until another cycle again activates the CAN current.

Sensory receptors are another class of neurons whose normal function may rely on the activation of CAN channels. While cyclic nucleotide-activated nonselective channels are responsible for transduction in photoreceptors and olfactory receptors, CAN channels may play an important additional role in some sensory neurons. CAN currents are a possible component in the electrical response of taste receptor cells (Kinnaman and Roper, 1987) and photoreceptors (Payne et al., 1986) and CAN channels have been identified in certain olfactory receptor neurons (Zufall et al., 1991). Although CAN currents do not specifically generate the receptor response of these cells, they may function to shape the receptor potential. Another action of CAN currents in retinal ganglion cells may be a physiological role in neurite outgrowth.

While a maintained depolarization is essential to the normal function of bursting neurons, such depolarization, especially if associated with prolonged Ca^{2+} influx, can yield pathological conditions in other neurons. One such pathological state is excitotoxicity whereby neurons are killed by prolonged depolarization. Several authors have hypothesized a role of CAN channels in excitotoxicity (Choi, 1990; Currie and Scott, 1992). The initial Ca^{2+} influx through such pathways as the NMDA receptor channel or Ca^{2+} released from intracellular stores (Sawada et al., 1990; Currie and Scott, 1992) could trigger a prolonged Ca^{2+} influx through CAN channels. The resultant maintained high cytoplasmic Ca^{2+} concentration has been demonstrated to have numerous toxic effects on neurons. Maintained depolarization during ischemia that is resistant to calcium channel blockers has been further attributed to the inward current through CAN channels (Rod and Auer, 1992).

Spreading depression is a well known neurophysiological phenomenon that may underlie certain pathological conditions of the cortex. Recent evidence suggests that CAN channel activity contributes to this process (Siesjo and Bengtsson, 1989). An additional pathological role of CAN channels is in copper neurotoxicity (Kiss et al., 1991). Calcium released from intracellular stores by Cu^{2+} activates CAN currents that produce a toxic effect through maintained depolarization.

Neuronal CAN channels are a member of a class of channels that use Ca^{2+} as a second messenger in excitation-gating coupling. CAN channels are the only Ca^{2+}-activated channels that consistently produce depolarization of the neuron. This depolarizing action is important to both the normal and the pathological function of neurons. On the one hand, CAN channels couple local changes in Ca^{2+} concentration to the electrical state of the cell. On the other hand, by providing a source of maintained depolarization and Ca^{2+} influx, CAN channels can lead to neuronal cell damage.

182

References

Adams WB (1985). Slow depolarizing and hyperpolarizing currents which mediate bursting in *Aplysia* neurone R15. J. Physiol. 360:51–68.

Bers DM, Peskoff A (1991). Diffusion around a cardiac calcium channel and the role of surface bound calcium. Biophys. J. 59:703–721.

Canavier CC, Clark JW, Byrne JH (1991). Simulation of the bursting activity of neuron R15 in *Aplysia*: Role of ionic currents, calcium balance, and modulatory transmitters. J. Neurophysiol. 66:2107–2124.

Choi DW (1990). Cerebral hypoxia – some new approaches and unanswered questions. J. Neurosci. 10:2493–2501.

Cornish-Bowden A (1979). Fundamental of Enzyme Kinetics. London: Butterworths, pp. 93–94.

Currie KPM, Scott RH (1992). Calcium-activated currents in cultured neurones from rat dorsal root ganglia. Br. J. Pharmacol. 106:593–602.

Hasuo H, Phelan KD, Twery MJ, Gallagher JP (1990). A calcium-dependent slow afterdepolarization recorded in rat dorsolateral septal nucleus neurons in vitro. J. Neurophysiol. 64:1838–1846.

Hille B (1992). Ionic Channels of Excitable Menbranes. Sunderland, Mass., USA: Sinauer Assoc.

Hofmeier G, Lux HD (1981). The time courses of intracellular calcium and related electrical effects after injection of $CaCl_2$ into neurons of the snail *Helix pomatia*. Pflügers Arch. 391:242–251.

Kim YK, Woodruff ML (1991). Cobalt-dependent potentiation of net inward current density in *Helix aspersa* neurons. J. Neurosci. Res. 28:549–555.

Kinnaman SC, Roper SD (1987). Passive and active membrane properties of mudpuppy taste receptor cells. J. Physiol. 383:601–614.

Kiss T, Györi J, Osipenko ON, Maginyan SB (1991). Copper-induced non-selective permeability changes in intracellularly perfused snail neurons. J. Appl. Toxicol. 11:349–354.

Kramer RH, Zucker RS (1985). Calcium-dependent inward current in *Aplysia* bursting pacemaker neurones. J. Physiol. 362:107–130.

Lando L, Zucker RS (1989). "Caged calcium" in *Aplysia* pacemaker neurons characterization of calcium-activated potassium and nonspecific cation currents. J. Gen. Physiol. 93:1017–1060.

Lemos JP, Nordmann JJ, Cooke IM, Stoenkel, EL (1986). Single channel and ionic currents in peptidergic nerve terminals. Nature 319:410–412.

Lewis DV (1984). Spike afterpotentials in R15 of *Aplysia*: Their relationship to slow inward current and calcium influx. J. Neurophysiol. 51:387–403.

Lipton SA (1987). Bursting of calcium-activated cation-selective channels is associated with neurite regeneration in a mammalian central neuron. Neurosci. Lett. 82:21–28.

Marcus DC, Eaton DC (1990). Ca^{2+}-activated nonselective cation channel in apical membrane of vestibular dark cells. Am. J. Physiol. 262:C1423–1429.

McClintock TS, Ache BW (1990). Nonselective cation channel activated by patch excision from lobster olfactory receptor neurons. J. Mem. Biol. 113:115–122.

Partridge LD, Swandulla D, Müller TH (1990). Modulation of calcium-activated non-specific cation currents by cyclic AMP-dependent phosphorylation in neurons of *Helix*. J. Physiol. 429:131–145.

Partridge LD, Swandulla D (1987). Single Ca-activated cation channels in bursting neurons of *Helix*. Pflügers Arch. 410:627–631.

Partridge LD, Swandulla D, Müller TH (1991). Limitations imposed upon channel activation by local Ca^{2+} concentration. Neurosci. Abs. 383.10.

Partridge LD, Swandulla D (1988). Calcium-activated non-specific cation channels. Trends Neurosci. 11:69–72.

Partridge LD (1989). Forskolin depresses CAN current in snail neurons. Neurosci. Abs. 514.7.

Payne R, Corson DW, Fein A (1986). Pressure injection of Ca^{2+} both excites and adapts *Limulus* ventral photoreceptors. J. Gen. Physiol. 88:107–126.

Razani-Boroujerdi S, Partridge LD (1992). Patch clamp study of Ca-activated nonselective cation channels in chick embryo dorsal root ganglion cells. Neurosci. Abs. 363.2.

Roberts WM, Jacobs RA, Hudspeth AJ (1990). Colocalization of ion channels involved in frequency selectivity and synaptic transmission at presynaptic active zones of hair cells. J. Neurosci. 10:3664–3684.

Rod MR, Auer RN (1992). Combination therapy with nimodipine and dizocilpine in a rat model of transient forebrain ischemia. Stroke 23:725–732.

Sawada M, Ichinose M, Maeno T (1990). Activation of a non-specific cation conductance by intracellular injection of inositol 1-3-4-5-tetrakisphosphate into identified neurons of *Aplysia*. Brain Res. 512:333–338.

Schaeffer P, Lazdunski M (1991). K^+ efflux pathways and neurotransmitter release associated to hippocampal ischemia – effects of glucose and of K channel blockers. Brain Res. 539:155–158.

Schwindt PC, Spain WJ, Foehring RC, Chubb MC, Crill WE (1988). Slow conductances in neurons from cat sensorimotor cortex in vitro and their role in slow excitability changes. J. Neurophysiol. 59:450–467.

Siesjo BK, Bengtsson F (1989). Calcium fluxes calcium-antagonists and calcium-related pathology in brain ischemia hypoglycemia and streading depression: a unifying hypothesis. J. Cereb. Blood Flow Metab. 9:127–140.

Simmonneau M, Distasi C, Tauc L, Barbin G (1987). Potassium channels in mouse neonate dorsal root ganglion cells: A patch-clamp study. Brain Res. 412:224–232.

Smith SJ, Augustine GJ (1988). Calcium ions, active zones and synaptic transmitter release. Trends Neurosci. 11:460–464.

Swandulla D, Lux HD (1985). Activation of a nonspecific cation conductance by intracellular Ca^{2+} elevation in bursting pacemaker neurons of *Helix pomatia*. J. Neurophysiol. 54:1430–1443.

Takeuchi S, Marcus DC, Wangemann P (1992). Ca^{2+}-activated nonselective cation, maxi-K^+ and Cl^+ channels in apical membrane of marginal cells of stria vascularis. Hearing Res. 61:86–92.

Wang SS.H, Thompson S (1992). A-type potassium channel clusters revealed using a new statistical analysis of loose patch data. Biophys. J. 63:1018–1025.

Yazejian B, Byerly L (1989). Voltage-independent Ba-permeable channel activated in *Lymnaea* neurons by internal perfusion or patch excision. J. Mem. Biol. 107:63–75.

Yellen G (1982). Single Ca^{2+}-activated nonselective cation channels in neuroblastoma. Nature 296:357–359.

Zucker RS (1988). Intrinsic electrophysiological regulation of firing patterns of bursting neurons in *Aplysia*. In: Neurosecretion. Pickering BT, Wakerly JB, Summerlee AJS, editors. New York: Plenum Pub. Corp., pp. 227–234.

Zucker RS, Fogelson AL (1986). Relationship between transmitter release and presynaptic calcium influx when calcium enters though discrete channels. Proc. Natl. Acad. Sci. 83:3032–3036.

Zufall F, Hatt H, Keil TA (1991). A calcium-activated nonspecific cation channel from olfactory receptor neurons of the silkmoth *Antheraea polyphemus*. J. Exp. Biol. 161:455–468.

Nonselective Cation Channels: Pharmacology, Physiology and Biophysics
ed. by D. Siemen & J. Hescheler

Nonselective Cation Channels in Exocrine Gland Cells

Peter Thorn and Ole H. Petersen

Medical Research Council Secretory Control Research Group, The Physiological Laboratory, Crown Street, P.O. Box 147, University of Liverpool, Liverpool, L69 3BX, U.K.

Summary
The nonselective cation channel has been described in a wide variety of nonexcitable cells. However even in such closely related tissues as the pancreatic acinar cell and the lacrimal acinar cell, which both possess a superficially similar channel, recent work has shown fundamental differences in channel regulation (Sasaki and Gallacher, 1992; Thorn and Petersen, 1992). These differences are a reflection of a diverse function of the nonselective channel in different tissues.

Introduction

Ca^{2+}-activated nonselective cation channels were first described in cultured cardiac cells (Colquhoun et al., 1981). Since then these channels have been found in a wide variety of tissues (review: Partridge and Swandulla, 1988). In neurones the channel is thought to underlie the control of the interburst interval (Partridge and Swandulla, 1987). To date however, the functional significance of the opening of nonselective cation channels in nonexcitable cells is unclear.

Exocrine organs are composed of two main cell types, the acinar cells and the duct cells. The acinar cells are orientated in cell clusters around a central cavity or lumen. These cells are responsible for both regulated exocytotic enzyme secretion and fluid secretion into the lumen. The duct network is arranged in a tree structure that divides and becomes progressively smaller, terminating in the acinar cell clusters. The cells that make up the ducts actively absorb and secrete ions, effectively modifying the primary fluid output of the acinar cells.

Work in a range of exocrine cells has shown the presence of a nonselective cation channel in most cell types studied (guinea pig pancreas, Suzuki and Petersen, 1988; lacrimal gland, Marty et al., 1984; Sasaki and Gallacher, 1990; Vincent, 1992; salivary gland, Maruyama et al., 1983). The nonselective cation channel, recorded in isolated inside-out patches, is activated by an increase in Ca^{2+} (Maruyama and Petersen 1984a; also see Table 1) but blocked by ATP (Sturgess et al., 1986, 1987; Suzuki and Petersen, 1988; Gray and Argent, 1990; Cook et

Table 1. Summary of the data available from the exocrine nonselective cation conductances

Tissue	g (pS)	Intracellular Ca-dep.	Channel regulation	Proposed function	References
ACINAR					
Mouse pancreas	30	yes			Maruyama and Petersen, 1982b pancreas
				Ca influx	Petersen and Maruyama, 1983
			CCK and ACh		Thorn and Petersen, 1992; Maruyama and Petersen, 1982b
Rat pancreas	27	yes			Gögelein et al., 1990; Kasai and Augustine, 1990; Randriamampita et al., 1988
Mouse lacrimal	27	yes		Na influx	Marty et al., 1984
	25		ACh		Marty et al., 1986
	27		ATP	Ca influx	Sasaki and Gallacher, 1990; Sasaki and Gallacher, 1992
	6	no	ATP	Ca influx	Vincent, 1992
DUCTAL					
Rat pancreatic duct	25	yes	secretin	volume regulation	Gray and Argent, 1990
CULTURED					
Mouse mandibular cell line ST885	25	yes		Ca influx	Cook et al., 1990 Poronnik et al., 1991

al., 1990) at the cytoplasmic surface of the patch. Direct measurement of ATP_i in intact cells has shown that millimolar ATP is present and that the concentration of ATP_i remains stable throughout agonist stimulation (Matsumoto et al., 1988). It would appear therefore, that in an intact cell, ATP block would prevent any Ca^{2+}-dependent channel opening. However, cell-attached patch recordings have shown nonselective cation channels activated by a high concentration of cholecystokinin (Maruyama and Petersen, 1982b). Our recent experiments have reexamined the ATP block observed in isolated inside-out patches and we studied possible nonselective channel activation at low and physiological agonist concentrations (Thorn and Petersen, 1992).

We review here, both the channel characteristics (Table 1) and the possible functional significance of the properties of nonselective channels in exocrine gland cells.

Physiological Activation of the Nonselective Channel

Our knowledge of the distribution and regulation of nonselective cation channels has expanded in recent years, but we still lack information on the physiological relevance and the control of these channels.

In the case of mouse pancreatic acinar cells it was shown that cholecystokinin (CCK) could activate this channel (Maruyama and Petersen, 1982b) but it has only been recently demonstrated by us (Thorn and Petersen, 1992) that this activation can occur at the physiological agonist concentrations found in plasma after a meal (5–20 pM CCK8 equivalents) (Walsh, 1987; Forster and Dockray, 1992). In our experiments we employed whole-cell patch-clamp recording techniques and reduced the chloride concentration in the pipette, allowing us to separate the reversal potentials for the nonselective conductance and the chloride conductance (the two major conductances in pancreatic acinar cells (Petersen, 1992)). By stepping the membrane potential of the cell between the two reversal potentials, we were able to gain an independent measure of the chloride and the nonselective cation currents. Similar techniques have previously been employed (Kasai and Augustine, 1990; Randriamampita et al., 1988) but have not been used to study the effects of physiological concentrations of agonist. Figure 1 shows typical current records obtained during the application of the agonists CCK and acetylcholine (ACh). Both the chloride (upper traces) and nonselective cation current (lower traces) are activated in an

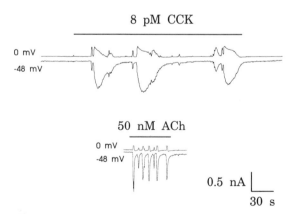

Figure 1. Whole-cell current records obtained in response to 8 pM CCK (upper) and 50 nM ACh (lower). The pipette was filled with a low Cl⁻ solution. The calculated reversal potentials are -48 mV for the Cl⁻ current and 0 mV for the nonselective cation current. The cell membrane potential was clamped at -48 mV and stepped at frequency of 3 Hz to 0 mV. This protocol gives a direct measure of the Cl⁻ current (at 0 mV) and the nonselective current (at -48 mV). It is clear from these records that the Cl⁻ current and the nonselective current are both activated by these low concentrations of agonists in an oscillatory manner. (Figure taken with permission from Thorn and Petersen, 1992.)

oscillatory pattern during the presence of low concentrations of either CCK (upper record) or ACh (lower record). One criticism that could arise from these experiments is that whole-cell recordings, where the internal environment of the cell is dialysed by the pipette solution, could lead artificially to activation of the nonselective channel. Further experiments recording single nonselective channel events using intact cells and cell-attached patches (Figure 2) confirmed activation of the channel even at these low agonist concentrations (Thorn and Petersen, 1992).

Activation of the nonselective cation channel by extracellular ATP has been shown in mouse lacrimal acinar cells (Sasaki and Gallacher, 1990). In these experiments currents activated by extracellular ATP and recorded using whole-cell patch clamp were shown not to be blocked by neomycin, an inhibitor of phospholipase C. In cell-attached patches, the inclusion of ATP into the pipette (extracellular side) dramatically increased the opening of the nonselective channel, and more recently the authors have shown direct, reversible activation of the channel using isolated outside-out patches (Sasaki and Gallacher, 1992). Recent work on the lacrimal gland has shown a nonselective cation current that is activated at micromolar ATP_0 concentrations (Vincent, 1992). Variance analysis of this current gave a conductance of 6 pS, much lower than the 27 pS channel described by Sasaki and Gallacher (1992). It still remains unknown if ATP would be present in effective concentrations at synapses on the lacrimal cell. It is possible that ATP is colocalized with

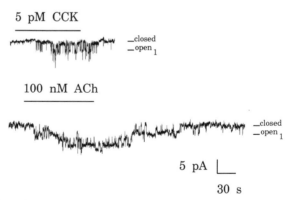

Figure 2. Cell-attached patch recordings obtained from two cells during application of the agonists as shown to the bathing solution surrounding the cell. Na-rich media was used both in the patch pipette and in the bathing solution. The patch potential was +70 mV (Vp) and single-channel current openings are seen as doenward current deflections. Addition of 5 pM CCK to the bathing solution elicited the reversible activation of up to three channels. More channels are seen to be activated by 100 nM ACh and, like CCK, this response was reversible. (Figure taken with permission from Thorn and Petersen, 1992.)

noradrenaline and would be released at the same site (Burnstock, 1990). Vincent (1992) has argued that ACh and noradrenaline are effective on acinar cells at micromolar concentrations and that the expected synaptic concentration of released ATP would be in the same concentration range. Support of this proposition comes from measures of ratios of catecholamines to ATP, found in vesicles of chromaffin cells, of 4:1 (Winkler and Westhead, 1980). The argument for agonist colocalization is further enhanced by evidence from Sasaki and Gallacher (1992) that has demonstrated a synergistic effect between β-adrenoreceptor stimulation and ATP_0 responses.

The nonselective channel in lacrimal acinar cells shown by Marty et al. (1984) was dependent on intracellular Ca^{2+}. The $[Ca^{2+}]_i$ dependence has not been clearly described but the observations of channel opening during stimulation with ACh (2 μM) suggest that it may be activated by micromolar Ca^{2+} concentrations (Marty et al., 1986).

The nonselective cation channel of the rat pancreatic duct cell could only be rarely recorded from cell-attached patches. However prolonged exposure of the cells to cAMP stimulation (0.5 to 6 h) increased the number of patches that showed active channels, suggesting secretin as a possible physiological stimulus for channel opening (Gray and Argent, 1990).

Single-Channel Characteristics

Ca^{2+} Permeability

It can be seen from Table 1 that the nonselective channel is most frequently cited as a possible Ca^{2+} influx pathway. No conclusive evidence has been obtained to show that the channel of the pancreatic acinar cells is permeable to Ca^{2+}. It could be that the Ca^{2+} permeability of the pancreatic acinar cell channel is low. Even a low permeability could exert profound effects on cell function (Petersen and Maruyama, 1983).

It has been shown that the nonselective channel in the mouse mandibular cell line ST885 has a relative permeability to Ca^{2+} only 0.2% that of sodium (Cook et al., 1990), nevertheless, it was apparent that a blocker of the channel (diphenylamine-2-carboxylate, DPC) also effectively blocked Ca^{2+} influx in unstimulated cells (Poronnik et al., 1991). In our laboratory, using acutely dissociated mouse pancreatic acinar cells stimulated with supramaximal ACh concentrations, we have shown that flufenamic acid (related structurally to DPC) blocks channels (see Figure 5) but has relatively little effect on intracellular Ca^{2+} (Thorn, Lawrie, and Schmidt, unpublished observations using fura-2 am loaded cells).

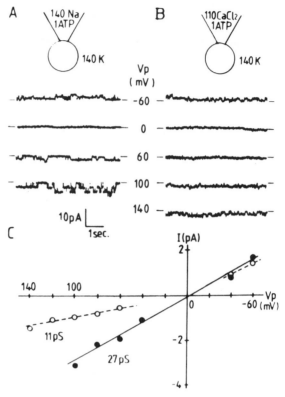

Figure 3. Cell-attached patch recordings obtained from isolated lacrimal acinar cells. Figure 3A shows the normal single-channel openings in the presence of Na-rich media in the patch pipette and in the bathing solution. The single-channel I/V obtained is shown by the filled circles, with a reversal potential of 0 mV and a conductance of 27 pS. Figure 3B shows a separate experiment in a Na-rich bathing media but with substitution of Na^+ with Ca^{2+} in the patch pipette. The I/V derived from this experiment is shown (clear circles) and indicates a similar conductance for the outward currents but a smaller conductance for the inward currents carried by Ca^{2+}. (Figure taken with permission from Sasaki and Gallacher, 1990.)

In lacrimal cells, discrete channel openings with a conductance of 10 pS (compared with a 30 pS conductance when measured with monovalent cations) were observed when Ca^{2+} was used as the only permeant cation (Figure 3, taken with permission from Sasaki and Gallacher, 1990). Similar observations have been made in neutrophils on the nonselective channel (von Tscharner et al., 1986) where channel events of a reduced conductance were observed with Ca^{2+} as the only charge carrier. In another study of the lacrimal acinar cell nonselective current, Vincent (1992) demonstrated a $[Ca^{2+}]_i$ increase associated with the ATP_0 activation of the nonselective current. Using Ca^{2+} imaging techniques ATP_0 was shown to promote a rise in $[Ca^{2+}]_i$ initiated at the secretory pole of the cell (Toescu et al., 1992). This is thought to be due

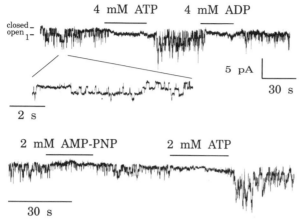

Figure 4. Single-channel current records obtained from an isolated inside-out patch held at a Vp of −70 mV, channel openings downwards. The bathing solution was KCl-rich, 5 μM Ca²⁺, the pipette solution was NaCl-rich, 1 mM Ca²⁺. Nucleotides were added to the bathing solution as indicated. The Ca²⁺/EGTA ratio adjusted to keep the free Ca²⁺ concentration constant. Bath application of 4 mM ATP followed by 4 mM ADP promoted block of channel opening (upper). Bath application of 2 mM AMP-PNP followed by 2 mM ATP (lower), again both promoted a rapid, reversible channel block. After removal of ATP a rebound effect was seen with a transient increase in the number of simultaneously open channels observed. This rebound was never observed after application of ADP or AMP-PNP. (Figure taken with permission from Thorn and Petersen, 1992.)

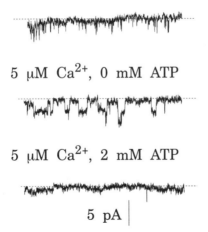

Figure 5. Single-channel current records obtained from an isolated inside-out patch. The patch pipette was held at +70 mV and contained Na-rich media and the bath solution at the cytoplasmic surface of the patch was a K-rich media. The upper record was obtained immediately after patch excision into a solution containing 5 μM Ca²⁺ and 2 mM ATP. Single-channel current openings are seen as downward deflections. The channel open time was seen to increase when the ATP was removed from the solution bathing the intracellular surface of the patch (middle trace). Reintroduction of ATP promoted a rapid block of channel opening. (Figure taken with permission from Thorn and Petersen, 1992.)

to Ca^{2+} entry through nonselective channels promoting Ca^{2+}-induced Ca^{2+} release from Ca^{2+} stores within the cell because the response is abolished in the absence of $[Ca^{2+}]_0$. However in parotid acinar cells ATP_0 evoked $[Ca^{2+}]_i$ increases have been observed in the absence of $[Ca^{2+}]_0$, suggesting that direct, stimulated Ca^{2+} release from intracellular stores is also possible (Soltoff et al., 1990).

Voltage Sensitivity

The Ca^{2+}-dependent nonselective cation channels of lacrimal and pancreatic acinar cells have been shown to be independent of voltage, see table 2. The current/voltage relationships are linear (Marty et al., 1984; Maruyama and Petersen, 1982). The available measurements of mean open time were independent of voltage at 930 ms in the pancreatic acinar cell (Maruyama and Petersen, 1982a) and 225 ms in the lacrimal acinar cell (Marty et al., 1984). The channel open probability in the lacrimal acinar ATP_0-activated channel was shown to be independent of voltage (Sasaki and Gallacher, 1992). Although in contrast, Vincent (1992) does provide evidence that hyperpolarization activated the channels.

In the channel of rat pancreatic duct cells the current/voltage relationship was linear but both the mean open time and the probability of channel opening were strongly dependent on voltage (Gray and Argent, 1990). These authors show an increase in probability of opening from less than 0.1 at membrane potentials of -80 mV to 1.0 at potentials more positive than $+20$ mV. The nonselective channel of the mandibular cell line ST885 also showed a similar voltage-dependent increase in channel open probability at depolarized membrane potentials (Cook et al., 1990). Both channels have open times best described by two exponentials with time constants of 1 and 11 ms (ST885) and 2.7

Table 2. The data available on the voltage sensitivity of exocrine nonselective cation channels

Tissue	g (pS)	Voltage sensitivity	Mean open times (ms) (memb. pot.)	Mean closed time (ms) (memb. pot.)	References
Rat pancreatic duct	25	activated by depolarization	2.7 and 33 at -60 mV	2.2 and 21 at -60mV	Gray and Argent, 1990
Rat mandibular cell line ST885	25	activated by depolarization	1 and 11.4 at -60 mV	1 and 20 at -60 mV	Cook et al., 1990
Mouse pancreatic acinar cell	30	voltage independent	930 at -50 mV	830 at -50 mV	Maruyama and Petersen, 1982a

and 33 ms (pancreatic duct) (Cook et al., 1990; Gray and Argent, 1990). When looking at the closed times the similarity between these two channels is even more striking. Both channel closed times are best fitted by two exponential decays with time constants of 1 and 20 ms (ST885) and 2.2 and 21 ms (pancreatic duct). In addition, the increase in probability of channel opening induced by depolarization is apparently due solely to a decrease of the slow time constant of channel closure; all other time constants remain the same (Cook et al., 1990; Gray and Argent, 1990). It is apparent that these two nonselective channels from rat pancreatic duct and rat mandibular gland, although derived from different tissues, share very similar features that place them in a separate group.

Channel Modulation by Intracellular Messengers

Intracellular Ca^{2+}

One of the first demonstrations of second messenger regulation of channel opening came with experiments carried out on the mouse pancreatic acinar cell nonselective cation channel. In these experiments agonists were applied to the bathing solution while recording from cell-attached patches. It was clearly shown that high concentrations of agonist acutely and reversibly activated the nonselective cation channel (Maruyama and Petersen, 1982b). More recently we have shown that this activation occurs even at much lower, physiological agonist concentrations (Thorn and Petersen, 1992). The major intracellular messenger involved in channel activation in the pancreatic acinar cell was shown to be a rise in intracellular Ca^{2+} (Maruyama and Petersen, 1982a,b). The whole-cell recording experiments (Figure 1) clearly show activation of both the Cl^- and the nonselective currents and these have been correlated with a rise in intracellular Ca^{2+} (Osipchuk et al., 1991). Although other intracellular messengers may modify the channel response, the roughly synchronous activation of both the nonselective and the Cl^- channel suggests that both are primarily dependent on a rise in intracellular Ca^{2+}. In addition, we have conducted experiments using thimerosal, a sulphydryl oxidising agent that promotes a rise in intracellular Ca^{2+} through intracellular Ca^{2+} release and block of the plasma membrane Ca^{2+}-ATPase, both effects are independent of receptor stimulation (Thorn et al., 1992). Whole-cell recordings showed Cl^- and nonselective currents simultaneously activated by thimerosal, and provide further evidence that Ca^{2+} is the primary regulator of the nonselective cation channel.

The Ca^{2+} concentration dependence of the pancreatic acinar cell channel is still unclear. Initial experiments using isolated inside-out patches demonstrated channel activation at greater than 10 μM Ca^{2+}

(Maruyama and Petersen, 1982a). Subsequent experiments used saponin permeabilized cells where the cell is made permeable to small ions but the cell architecture and larger molecules remain intact. Under these conditions channel activation could be observed at a threshold of 5×10^{-8} M Ca^{2+} (Maruyama and Petersen, 1984). It was also shown in the same study that channels recorded from excised patches showed an initial high sensitivity to Ca^{2+} that declined over a period of 1 min. Both of these experiments indicate the possibility that in the intact cell the nonselective channel is maintained in an 'active state' and shows more sensitivity to intracellular Ca^{2+} than that found after isolating isolated inside-out patches. Other studies in rat pancreas have shown a much higher $[Ca^{2+}]_i$ threshold (10^{-5} M) for activation of the nonselective cation conductance in whole-cell recordings (Randriamampita et al., 1988).

The lacrimal nonselective current activated by extracellular ATP was not blocked by the chelation of intracellular Ca^{2+} (10 mM EGTA) (Sasaki and Gallacher, 1992; Vincent, 1992). A different nonselective channel opened by agonists that raise the intracellular Ca^{2+} concentration has also been shown in lacrimal acinar cells (Marty et al., 1984). The evidence for this channel activation by $[Ca^{2+}]_i$ in lacrimal acinar cells is weaker than that in the pancreatic acinar cells; it has been shown that high acetylcholine concentrations (μM) evoked nonselective channel opening in some cell-attached patches (Marty et al., 1986). Evans and Marty (1986) have also shown that the nonselective cation channel was opened when recording from a whole cell perfused with a solution in which the $[Ca^{2+}]_i$ was buffered at between 0.5 and 10 μM. In these experiments the nonselective current was not recorded immediately upon breakthrough to the whole-cell configuration but was observed to develop over a period of time ranging from 10 min at lower Ca^{2+} concentrations to 2 min at 10 μM Ca^{2+} (Evans and Marty, 1986). This later finding suggests that over a period of time the dialysing effect of the pipette solution washed out an inhibitory influence on the nonselective channel which contrasts with nonselective channel 'run down' seen after patch excision in pancreatic acinar cells (Maruyama and Petersen, 1984).

In rat pancreatic duct cells the dependence on intracellular Ca^{2+} has been demonstrated with complete channel closure at $[Ca^{2+}]_i$ concentrations below 10^{-7} M (Gray and Argent, 1990).

Intracellular ATP

The inhibition of nonselective cation channels by intracellular ATP (mM) was demonstrated in an insulinoma cell line (Sturgess et al., 1986, 1987) and subsequently shown in pancreatic acinar cells (Suzuki and Petersen, 1988), the pancreatic β-cell line (Sturgess et al., 1986), kidney ascending Henle's loop (Paulais and Teulon, 1989), rat mandibular cell line (Cook et al., 1990) and in rat pancreatic duct (Gray and Argent,

1990). The normal cellular ATP concentration in pancreatic acinar cells would be in the mM range that blocks channels in isolated inside-out patches. It has been shown that during agonist stimulation the intracellular ATP concentration does not change significantly (Matsumoto et al., 1988). It therefore would seem unlikely in intact cells that conditions would arise where the nonselective cation channel would open. This is not the case; in cell-attached patches nonselective cation channels are observed in response to agonist stimulation (see Figure 2; Maruyama and Petersen, 1982b; Thorn and Petersen, 1992). In addition, whole-cell experiments demonstrate activation of the nonselective cation current in response to high agonist concentrations (Randriamampita et al., 1989; Kasai and Augustine, 1990) and our experiments indicate channel opening at low agonist concentrations even when 2 mM ATP is included in the pipette during whole-cell current recording (Thorn and Petersen, 1992).

Further studies carried out by us have indicated a possible explanation for these discrepant results. We have shown in isolated inside-out patches channel block by ATP, ADP and the nonhydrolysable ATP analogue, AMP-PNP (Thorn and Petersen, 1992). After removal of ATP a rebound activation was observed that was not seen after removal of ADP or AMP-PNP (Figure 4). This suggested the presence of two nucleotide binding sites at the channel. The first site promotes channel block and is unable to discriminate between ATP, ADP and AMP-PNP. The second site, we postulate, involves a phosphorylation event, either at the channel itself or at a site in an adjoining regulatory protein. This second binding site is able to promote channel opening. The transient nature of the channel activation after ATP removal possibly involves the dephosphorylating action of a membrane phosphatase. This model is very similar in many respects to that proposed for the ATP-sensitive K^+-selective channel (Ashcroft, 1989) which is blocked by ATP at the cytoplasmic surface of isolated inside-out patches. ATP is also able to restore channel activity in patches where openings have run down after patch excision (Petersen and Findlay, 1987). However, the block by ATP of the K_{ATP} channel of insulin-secreting cells is partially relieved by the addition of ADP (Dunne and Petersen, 1986; Kakei et al., 1986), an effect not seen in the pancreatic acinar cell nonselective cation channels.

In further experiments we isolated patches into solutions containing 2 mM ATP. Normally, patches are isolated into solutions containing no ATP and we conducted our experiments to test if channel activity would persist in the continuous presence of ATP at the cytoplasmic surface of the patch. Our experiments demonstrated that after patch excision some channel activity was seen (Figure 5) although this activity frequently ran down with time. Channel activity increased in the patch following removal of ATP but was abolished by the reintroduction of ATP (Thorn and Petersen, 1992). These results may be explained by a positive action of ATP maintaining an active state of the channel that is lost when ATP is removed. An alternative explanation for our results is that there is a loss of a regulatory component over time simply due to perfusion of the cytoplasmic surface of the isolated inside-out patch.

Intracellular cAMP

In the rodent pancreatic acinar cells no effect of intracellular cAMP on the nonselective channel has been shown. Two agonists on the pancreatic acinar cells act to elevate cAMP, namely secretin and vasoactive intestinal polypeptide (VIP). Neither agonist has been shown to directly activate the nonselective channels in whole-cell current recordings. It has been shown that VIP and cell permeable cAMP (dibutryl cAMP) promote whole-cell current oscillations but these were blocked by the intracellular perfusion of EGTA, suggesting that it is the $[Ca^{2+}]_i$ elevations that underlie these events (Kase et al., 1991). In contrast, cAMP potentiates the effects of extracellular ATP on the activation of the channel in lacrimal acinar cells (Sasaki and Gallacher, 1992). This effect was seen using three separate prestimulus protocols to elevate intracellular cAMP; direct elevation of cAMP, addition of IBMX to inhibit the phosphodiesterase or stimulation with the β-adrenoreceptor agonist isoprenaline, all possibly acting through a cAMP-dependent kinase (Sasaki and Gallacher, 1992). Although the nonselective cation channel of pancreatic duct cells may be modulated by cAMP concentrations, this has not been shown directly (Gray and Argent, 1990). The evidence the authors present that cAMP may be involved in channel regulation was that, after pretreatment of the cells for 0.5–6 h with a cocktail of cAMP elevating compounds (secretin, forskolin, dibutryl cAMP and IBMX) there was a ninefold increase in the number of patches that contained active channels (Gray and Argent, 1990).

Channel Pharmacology

A major problem in our understanding of the function and activity of the nonselective channel has been the lack of agents that act specifically to block the channel. In an extensive study, DPC derivatives have been shown to be effective blockers of rat acinar cell pancreatic nonselective channels by Gögelein and Pfannmüller (1989). They also demonstrated that the Cl^- channel blockers SITS and DIDS promoted channel opening when applied to the intracellular surface of the patch (Gögelein and Pfannmüller, 1989). Other channel blockers used, such as non-steroidal anti-inflammatory drugs, have a similar chemical structure to DPC, e.g., mefenamic acid and flufenamic acid (Gögelein et al., 1990).

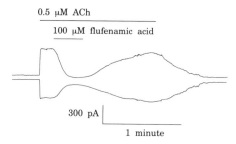

0.5 μM ACh

100 μM flufenamic acid

300 pA

1 minute

Figure 6. Whole-cell current record under the same conditions and voltage protocol as in Figure 1. The upper and lower current traces are the Cl⁻ and nonselective current respectively. The addition of 0.5 μM ACh to the bathing solution produced a rapid increase in both currents and normally gives a sustained response in these cells. At the time shown, we added 100 μM flufenamic acid to the bathing solution, this produced a rapid block of both currents that reversed slowly when the drug was removed. The increase in amplitude of the Cl⁻ and nonselective current was totally reversed by the removal of ACh.

Flufenamic acid and niflumic acid have been shown to block the chloride conductance of the rabbit thick ascending limb of the loop of Henle (Wangemann et al., 1986). In a study of the nonselective channel of the ST885 cell line, Cook et al. (1990) have shown that 4-aminopyridine (on the cytoplasmic surface only), DPC and quinine all blocked single-channel events. They also show that the stilbene, SITS at the cytoplasmic surface, promoted channel opening. In our studies with whole-cell patch-clamp on the pancreatic acinar cell, flufenamic acid has been applied and we have a measure of both the nonselective and the chloride currents. In all our experiments, using whole-cell patch-clamp, 100 μM flufenamic acid promoted a reversible block of both the currents (Figure 6) directly demonstrating the nonspecific action of this drug. It is apparent that further work is required to find pharmacological tools which act specifically on the nonselective channel.

Channel Function

The rodent pancreas is an anomalous exocrine gland in that the acinar cells do not possess a Ca^{2+}-dependent K^+ current (Petersen and Gallacher, 1988). Interestingly, in these cells the nonselective cation channel is present at relatively high densities (up to 10 channels per patch). Current theories of fluid secretion invoke the opening of the Cl⁻ channel located on the luminal membrane and the synchronous opening of a K^+ channel on the basal membrane (Petersen and Gallacher, 1988). The absence of a large K^+ conductance in rodent pancreas suggests two possibilities: firstly, these cells may be capable of fluid secretion in a manner not encompassed by previous theories or secondly, these cells may be poor at fluid secretion. In fact fluid secretion rates of the rat pancreas have indeed been shown to be low, possibly supporting the latter hypotheses (Sewell and Young, 1975).

A recent model of fluid secretion in the rodent pancreas is the so-called 'push-pull' theory proposed by Kasai and Augustine (1990). The crucial element of this model is an observed agonist-induced increase in $[Ca^{2+}]_i$ initiated at the secretory pole of the cell and then spreading to the basal pole (Kasai and Augustine, 1990; Toescu et al., 1992). In separate whole-cell patch-clamp experiments Kasai and Augustine (1990) showed an initial large transient Cl^- current activation followed by activation of the nonselective cation current and a smaller Cl^- current. The major initial Cl^- current is thought to be carried by channels located at the secretory pole of the cell in the luminal membrane. The nonselective channels have been directly observed on the basolateral membrane (Maruyama and Petersen, 1982b). Kasai and Augustine (1990) therefore proposed that the early Cl^- current was correlated with a rise in $[Ca^{2+}]_i$ at the secretory pole and the nonselective current was activated as the $[Ca^{2+}]_i$ signal spread to the rest of the cell. Kasai and Augustine (1990) explain unidirectional flow of Cl^- across the luminal membrane suggesting a 'push' phase at a point where the Cl^- channels are open and the cell membrane potential is more negative than the Cl^- reversal potential leading to outflow of Cl^-. The Ca^{2+} signal then spreads away from the secretory pole; this closes the lumenal Cl^- channels and opens basolateral Cl^- and nonselective cation channels. These nonselective channels depolarize the cell and lead to a 'pull' phase of Cl^- influx. It is suggested that this mechanism of Cl^- flow could provide a functional role for Ca^{2+} oscillations (Kasai and Augustine, 1990).

This model is speculative and it does not extend to secretory acinar cells other than rodent pancreas, however it does offer one possible function for nonselective channel activation in rodent pancreatic acinar cells. We still do not know if the nonselective channel is found on the luminal membrane. If the channels were present on the luminal membrane, as seen in pancreatic ducts (Gray and Argent, 1990), they might participate in water transport (Ussing and Eskesen, 1989) by recycling Na^+ passed into the lumen via the paracellular cation selective junctions (Petersen, 1992).

In the lacrimal gland acinar cell the nonselective channel opened by ATP_0 demonstrates a significant Ca^{2+} permeability. Synaptic activity at synapses on the lacrimal acinar cell possibly co-release ATP and noradrenaline. It has been shown that micromolar concentrations of ATP activate the nonselective conductance (Vincent, 1992) and this would be enhanced by the cooperative stimulation with noradrenaline (Sasaki and Gallacher, 1992). The induced $[Ca^{2+}]_i$ rise as a consequence of Ca^{2+} influx would lead to activation of the Ca^{2+}-dependent Cl^- and K^+ channels, important in fluid secretion (Petersen and Gallacher, 1988).

The functional significance of the nonselective cation channel in the duct cells is less clear. The normal physiological stimuli for bicarbonate secretion in these cells act through an elevation in intracellular cAMP (Case and Argent, 1986). The stimulus for channel activation has yet to be elucidated. Recently, ACh has been shown to promote a rise in $[Ca^{2+}]_i$ in pancreatic duct cells and this leads to some Ca^{2+} influx possibly through nonselective cation channels (Stunkel and Hootman, 1990).

In conclusion, nonselective cation channels are found in the major cell types of exocrine glands. It is clear from all the data presented that the nonselective cation channels of exocrine glands are under multifactoral regulation and more work is required to determine the physiological conditions under which these channels open.

References

Ashcroft FM (1989). Properties and functions of ATP-sensitive K-channels. Ann. Rev. Neurosci. 11:97–118.

Burnstock G (1990). Noradrenaline and ATP as cotransmitters in sympathetic nerves. Neurochem. Int. 17:357–368.

Case RM, Argent BE (1986). Bicarbonate secretion by pancreatic duct cells: mechanisms and control. In: The Exocrine Pancreas: Biology, Pathobiology and Diseases. Go VLW, Gardner JD, Brooks FP, Lebenthal E, Di Magno DP, Sheele GA, editors. New York: Raven Press, pp 213–243.

Colquhoun D, Neher E, Reuter H, Stevens CF (1981). Inward current channels activated by intracellular Ca^{2+} in cultured cardiac cells. Nature 294:752–754.

Cook DI, Poronnik D, Young JA (1990). Characterization of a 25-pS nonselective cation channel on a cultured secretory epithelial cell line. J. Membrane Biol. 114:37–52.

Dunne MJ, Petersen OH (1986). Intracellular ADP activates K^+ channels that are inhibited by ATP in an insulin-secreting cell line. FEBS Lett. 208:59–62.

Evans MG, Marty A (1986). Calcium-dependent choride currents in isolated cells from rat lacrimal glands. J. Physiol. 378:437–460.

Forster ER, Dockray GJ (1992). The role of cholecystokinin inhibition of gastric emptying by peptone in the rat. Exp. Physiol. 77:693–700.

Gögelein H, Dahlem D, Englert HC, Lang HJ (1990). Flufenamic acid, mefanamic acid and niflumic acid inhibit single nonselective cation channels in the rat exocrine pancreas. FEBS Lett. 268(1):79–82.

Gögelein H, Pfannmüller B (1989). The nonselective cation channel in the basoateral membrane of rat exocrine pancreas. Pflügers Arch. 413:287–298.

Gray MA, Argent BE (1990). Non-selective cation channel on pancreatic duct cells. Biochem. Biophys. Acta 1029:33–42.

Kakei M, Kelly RJ, Ashcroft SJH, Ashcroft FM (1986). The ATP-sensitivity of K^+ channels in rat pancreatic B-cells is modulated by ADP. FEBS Lett. 208:63–66.

Kase H, Wakui M, Petersen OH (1991). Stimulatory and inhibitory actions of VIP and cyclic AMP on cytoplasmic Ca^{2+} signal generation in pancreatic acinar cells. Pflügers Arch. 419:668–670.

Kasai H, Augustine GJ (1990). Cytosolic Ca^{2+} gradients triggering unidirectional fluid secretion from exocrine pancreas. Nature 348:735–738.

Marty A, Evans MG, Tan YP, Trautmann A (1986). Muscarinic response in rat lacrimal glands. J. Exper. Biol. 124:15–32.

Marty A, Tan YP, Trautmann A (1984). Three types of calcium-dependent channel in rat lacrimal glands. J. Physiol. 357:293–325.

Maruyama Y, Gallacher DV, Petersen OH (1983). Voltage and Ca^{2+} activated K^+ channel in basolateral acinar cell membranes of mammalian salivary glands. Nature 302:827–829.

Maruyama Y, Petersen OH (1982a). Single channel currents in isolated patches of plasma membrane from basal surface of pancreatic. Nature 299:159–161.

Maruyama Y, Petersen OH (1982b). Cholecystokinin activation of single channel currents is mediated by internal messenger in pancreatic acinar cells. Nature 300:62–63.

Maruyama Y, Petersen OH (1983). Voltage clamp study of stimulant-evoked currents in mouse pancreatic acinar cells. Pflügers Arch. 399:54–62.

Maruyama Y, Petersen OH (1984). Single calcium-dependent cation channels in mouse pancreatic acinar cells. J. Membrane Biol. 81:83–87.

Matsumoto T, Kanno T, Seo Y, Murakami M, Watari H (1988). Phosphorus nuclear magnetic resonance in isolated perfused rat pancreas. Am. J. Physiol. 254:G575–G579.

Osipchuk Y, Wakui M, Yule DI, Gallacher DV, Petersen OH (1990). Cytoplasmic Ca^{2+} oscillations evoked by receptor stimulation, G protein activation, internal application of inositol trisphosphate of Ca^{2+}, simultaneous microfluorimetry and Ca^{2+}-dependent Cl^- current in single pancreatic acinar cells. EMBO Journal 9:697–704.

Partridge LD, Swandulla D (1987). Single Ca-activated cation channels in bursting neurones of Helix. Pflügers Arch. 410:627–637.

Paulais M, Teulon J (1989). A cation channel in the ascending limb of Henle's loop of the mouse kidney: inhibition by adenine nucleotides. J. Physiol. 413:315–327.

Petersen OH, Findlay I (1987). Electrophysiology of the pancreas. Physiol. Rev. 67:1054–1116.

Petersen OH, Gallacher DV (1988). Electrophysiology of pancreatic and salivary acinar cells. Ann. Rev. Physiol. 50:65–80.

Petersen OH, Maruyama Y (1983). What is the mechanism of the calcium influx to pancreatic acinar cells evoked by secretagogues. Pflügers Arch. 396:82–84.

Poronnik R, Cook DI, Allen DG, Young JA (1991). Diphenylamine-2-carboxylate (DPC) reduces calcium influx in a mouse mandibular cell line (ST885). Cell Calcium 12:441–449.

Randriamampita C, Chanson M, Trautmann A (1988). Calcium and secretagogues-induced conductances in rat exocrine pancreas. Pflügers Arch. 411:53–57.

Sasaki T, Gallacher DV (1990). Extracellular ATP activates receptor operated cation channels in mouse lacrimal acinar cells to promote calcium influx in the absence of phophoinositide metabolism. FEBS Lett. 264:130–134.

Sasaki T, Gallacher DV (1992). The ATP-induced inward current in mouse lacrimal acinar cells is potentiated by isoprenaline and GTP. J. Physiol. 447:103–118.

Sewell WA, Young JA (1975). Secretion of electrolytes by the pancreas of the anaesthetized rat. J. Physiol. 252:379–396.

Stunkel EL, Hootman SR (1990). Secretagogue effects on intracellular calcium in pancreatic duct cells. Pflügers Arch. 416:652–658.

Sturgess NC, Hales CN, Ashford MLJ (1986). Inhibition of a calcium-activated, non-selective cation channel, in a rat insulinoma cell line, by adenine derivatives. FEBS Lett. 208:397–400.

Sturgess NC, Hales CN, Ashford MLJ (1987). Calcium and ATP regulate the activity of a non-selective cation channel in a rat insulinoma cell line. Pflügers Arch. 409:607–615.

Soltoff SP, McMillan MK, Cragoe EJ, Cantley LC, Talamo BR (1990). Effects of extracellular ATP on ion transport systems and $[Ca^{2+}]_i$ in rat parotid acinar cells. J. Physiol. 95:319–346.

Suzuki K, Petersen OH (1988). Patch-clamp study of single channel and whole-cell K^+ current in guinea pig pancreatic acinar cells. Am. J. Physiol. 255:G275–285.

Toescu EC, Lawrie AM, Petersen OH, Gallacher DV (1992). Spatial and temporal distribution of agonist-evoked cytoplasmic Ca^{2+} signals in exocrine acinar cells analysed by digital image microscopy. EMBO Journal 11:1623–1629.

Thorn P, Brady P, Gallacher DV, Petersen OH (1992). Cytosolic Ca^{2+} spikes evoked by the thiol reagent thimerosal in both intact and internally perfused single pancreatic acinar cells. Pflügers Arch. 422:173–178.

Thorn P, Petersen OH (1992). Activation of nonselective cation channels by physiological cholecystokinin concentrations in mouse pancreatic acinar cells. J. Gen. Physiol. 100:11–25.

Ussing HH, Eskesen K (1989). Mechanism of isotonic water transport in glands. Acta Physiol. Scand. 136:443–454.

Vincent P (1992). Cationic channels sensitive to extracellular ATP in rat lacrimal cells. J. Physiol. 449:313–331.

Von Tscharner V, Prod'hom B, Baggiolini M, Reuter H (1986). Ion channels in human neutrophils activated by a rise in free cytosolic calcium concentration. Nature 324: 369–372.

Walsh JH (1987). Cholecystokinin. In: Physiology of the Gastrointestinal Tract. (2nd edition) Johnson LR, editor. New York: Raven Press, pp. 195–206.

Wangemann P, Wittner M, Di Stefano A, Englert HC, Lang HJ, Schlatter E, Greger R (1986). Cl^- channel blockers in the thick ascending limb of the loop of Henle. Structure activity relationship. Pflügers Arch. 407:S128–141.

Nonselective Cation Channels: Pharmacology, Physiology and Biophysics
ed. by D. Siemen & J. Hescheler
© 1993 Birkhäuser Verlag Basel/Switzerland

Nonselective Cation Channels in Brown and White Fat Cells

Ari Koivisto*, Elisabeth Dotzler, Ulrich Ruß, Jan Nedergaard* and Detlef Siemen

*Wenner-Gren-Institute, University of Stockholm, S-10691 Stockholm, Sweden; Institut für Zoologie, Universität Regensburg, D-93040 Regensburg, FRG

Introduction

Brown adipose tissue is the main site of nonshivering thermogenesis. Its heat-producing function is utilized during arousal from hibernation, during acclimation to cold, during the neonatal period, and during diet-induced thermogenesis. Chemical energy is dissipated as heat in the mitochondria, where the unique uncoupling protein works as a proton shunt (Cannon and Nedergaard, 1985; Trayhurn and Nicholls, 1986). The organ is well vascularized and innervated; brown adipocytes both activate and maintain their extremely high metabolism due to adrenergic stimulation. This results in a characteristic three-phase change in the cell membrane potential; first, a relatively rapid 25 mV depolarization lasting 10–30 s, then a repolarization with a 5–10 mV hyperpolarization lasting 2–5 min and, finally, a 20–25 mV depolarization for as long as the adrenergic stimulation continues (Connolly et al., 1989; Horwitz et al., 1989; Lucero and Pappone, 1990). The initial depolarization is due mainly to α_1-adrenergic stimulation and is probably caused by Cl^- efflux through 40 pS Cl^--channels (Dasso et al., 1990; Sabanov et al., 1993). The hyperpolarization is activated by mainly α_1- and perhaps by β-adrenergic agonists and is mediated by voltage-gated and Ca^{2+}-activated K^+-channels (Nånberg et al., 1985; Lucero and Pappone, 1989, 1990). The final, long-lasting depolarization is predominantly caused by Na^+ influx, probably through nonselective cation channels (NSC-channels), and is activated by β-adrenergic stimulation (Connolly et al., 1986; Siemen and Reuhl, 1987; Lucero and Pappone, 1990). Temporally, the increased metabolic rate correlates strongly with the last, sustained depolarization, and this suggests a possible relationship between noradrenaline-induced ion fluxes and thermogenesis (Girardier et al., 1968; Schneider-Picard et al., 1985).

The majority of the metabolic responses in the brown fat cells are mediated via β_3-receptors (Arch et al., 1984; Arch, 1989), whereas proliferation seems to occur via β_1-adrenergic receptors (Bronnikov et al., 1992). All these β-responses are mediated via an increase in intracellular cAMP levels. It has been speculated that an increase of intracellular Ca^{2+} together with activation of Na^+ influx could initiate proliferation (Soltoff and Cantley, 1988) – a situation where NSC-channels could also play a significant role.

White adipose tissue is the major energy store in mammals. Its primary function is to store lipid and to release free fatty acids in response to various neural and hormonal stimuli. Similar to the case in brown adipocytes, breakdown of stored triglycerides to free fatty acids (lipolysis) is caused primarily by stimulation with catecholamines. It is well established that β-adrenoceptors induce lipolysis through their coupling to the plasma membrane adenylyl-cyclase. The stimulatory effect on adenylyl-cyclase is mediated via a stimulatory guanine-nucleotide binding protein (G_S-protein). The cyclase catalyzes the formation of cyclic-AMP (cAMP), followed by activation of a hormone-sensitive lipase which breaks down the stored triglycerides (Brooks and Perosio, 1992; Richelsen, 1991).

White fat cells are larger than brown adipocytes (up to 120 μm in diameter), and their cytoplasm is nearly completely displaced to the periphery by a single large lipid droplet. Little is known about the electrical processes at the single-channel level in white adipocytes. Preliminary investigations reveal the presence of a NSC-channel with characteristics almost identical to those described for brown adipocytes (see below).

Electrophysiological Properties of the NSC-Channel in Brown Fat Cells

About one NSC-channel is found per μm^2 in brown adipocyte plasma membranes. This means that there are about 3000 channels in a cell with a diameter of 30 μm. At first glance, the density seems to be low, but consider that even in rat ventricular muscle cells there are only 43 Na^+-channels per μm^2 (Bean and Riós, 1989).

The single-channel conductance of the NSC-channel is 25–30 pS at 25°C (symmetrical solutions). Measurements in the temperature range from 16° to 42°C yielded a Q_{10} of 1.4, a value within the magnitude observed for many ion channels and in good agreement with aqueous diffusion (Siemen and Reuhl, 1987). This Q_{10}-value is also similar to the results of Colquhoun et al. (1981) for the NSC-channel of cultured rat ventricular muscle cells. Histograms of on- and off-times are best fitted by two exponentials of about 100 ms and ≤ 10 ms, respectively. The slow component, especially, shows a steep decrease with increasing temperature (≈ 25 ms at 40°C) (Siemen and Reuhl, 1987).

The open probability is voltage dependent, increasing with depolarization. In excised patches the steepest part of the open probability curve is located in the physiological range between -60 mV and -30 mV. Only in a few experiments was the channel continuously open. A hypothetical gating charge of 3.6 can be calculated from a two-state model of channel gating (Siemen and Weber, 1989). In voltage dependent Na^+-channels the value is about 6.0, expressing a much steeper voltage dependence (Hodgkin and Huxley, 1952).

The selectivity of the pore was determined for the alkali and two earth-alkali metal ions by replacing the ions at the outside of excised patches. This yielded a selectivity sequence of

$$NH^{4+} > Na^+ > Li^+ > K^+ \geq Rb^+ \approx Cs^+ \gg Ca^{2+} \approx Ba^{2+}.$$

The relative permeability for K^+ in comparison to Na^+ is 0.8; the analogous value for the nicotinic acetylcholine receptor is 1.1 (Weber and Siemen, 1989; Adams et al., 1980). Interestingly, the selectivity sequences for the alkali metal ions of these two nonselective channels are almost opposite. The theory of Eisenman suggests an explanation (cf. Hille, 1992): a different field strength of a hypothetical binding site for the permeant ions within a pore could lead to different attraction energies and thus to different selectivity sequences. The sequence of the NSC-channel of brown adipocytes would be closest to the Eisenman sequence X (corresponds to a relatively strong field strength site), while the nAChR equals Eisenman sequence I (a weak field strength site).

In the NSC-channels, as in most ion channels, the permeant ions do not obey the independence principle (Hodgkin and Huxley, 1952); this is most evident by saturation at high ion concentrations ($K_{Na} = 155$ mM). There is obviously competition between the ions for passage through the channel (Weber and Siemen, 1989; cf. J. Dani, this volume). Another indication for the invalidity of the independence principle is derived from the curved shape of the current-voltage relationships at potentials beyond ± 100 mV; at high potentials the curve is bent towards the y-axis (Weber and Siemen, 1989). Both saturation and curved i-E-relations are best explained by a two-barrier/one-binding-site model instead of the Goldman-Hodgkin-Katz model.

It was possible to calculate a mean influx of about 16 fmol Na^+ per second and cell using the data for density, kinetics, and voltage dependence. The cells have to counteract this influx by an increased activity of the Na/K-ATPase which can be calculated to result in an increased oxygen consumption of about 120 nM oxygen per minute and 10^6 cells (Siemen and Weber, 1989). This value agrees remarkably well with the magnitude of the ouabain-blockable part (5–15%) of noradrenaline-induced respiration (Mohell et al., 1987). Thus, there seems to be good correlation between biophysical data from single-channel measurements and biochemical data.

As mentioned above, the presence of the NSC-channel in the plasma membrane of white adipocytes has been indicated by preliminary electrophysiological examinations. The NSC-channel in white and brown fat cells seems to have identical electrophysiological characteristics. For instance, the single-channel slope conductance of 25 pS (24°C) found in white adipose tissue agrees well with data obtained from brown adipose tissue. A further similarity is in the increase of the open probability at depolarizing holding potentials.

Pharmacological Block

Dissection of the macroscopic currents from brown adipocytes has been hampered by the lack of specific pharmacological tools. It was found recently that in inside-out patches it is possible to completely and reversibly block the NSC-channels from brown fat with 0.1 mM flufenamic and mefenamic acid added to the cytoplasmic side (Koivisto et al., 1992). The block seems to be of the slow type as previously described for these kinds of drugs by Gögelein et al. (1990). 0.1 mM mefenamic acid also caused a slow block of the NSC-channel in the plasma membrane of white adipocytes; this occurred rapidly and was completely reversible within a few seconds.

Catecholamine Sensitivity

Lipolysis of both brown and white adipocytes is induced by β-adrenergic stimulation (Bojanic et al., 1985; van Heerden and Oelofsen, 1989; Cawthorne et al., 1992). In brown adipocytes lipolysis is mediated predominantly by β_3-adrenoceptors, whereas typical β_1-adrenoceptors are present but seem less important (Hollenga and Zaagsma, 1989; Hollenga et al., 1991). In white adipocytes the existence of β_3-adrenergic receptors was shown recently by Langin et al. (1991). Experiments with different β-adrenergic agonists were carried out in the cell-attached configuration in order to gain further insight into the regulation of the NSC-channel mediated ion fluxes. 1 μM of extracellularly applied noradrenaline activated the NSC-channel within 1–2 min; the response showed partly burst characteristics, 1 μM of the specific β_1-agonist isoprenaline was more potent than noradrenaline and activation occurred much faster, i.e., within a few seconds. The effects of both agonists were completely reversible. 1 μM BRL 35135A, a potent and selective agonist for the β_3-adrenoceptor, also activated the nonselective cation channels. The effect was achieved after a few minutes and persisted for a long time, during which the open state was interrupted only by a few, brief closures. These data agree well with investigations on stimulation of brown adipocyte respiration via β-agonists (Mohell et al., 1991). In both brown and white adipocytes, BRL 35135A acts as a "slow" agent. One explanation for the time delay of NSC-channel activation could be the fact that BRL 35135A is only active via its deesterified metabolite (Cawthorne et al., 1992).

Ca^{2+} Sensitivity

Coupling of effectors to cell metabolism via Ca^{2+} is a widespread phenomenon. It is therefore interesting to know the Ca^{2+} sensitivity of the NSC-channels in brown fat cells, especially because they are activated by patch excision to a Ca^{2+} containing buffer, and because α_1-adrenergic stimulation has been shown to increase intracellular Ca^{2+} concentration in these cells (Wilcke and Nedergaard, 1989). Concentrations around $10~\mu M$ free Ca^{2+} were necessary to observe channel activity in excised inside-out patches from brown fat cells (Koivisto et al., 1992). This means that the channels thus clearly belong to the group of Ca^{2+}-activated nonselective cation channels (Partridge and Swandulla, 1988; Swandulla and Partridge, 1990). The Ca^{2+} sensitivity of the NSC-channels in brown adipocytes thus resembles that in Schwann cells (Bevan et al., 1984), lacrimal acinar cells (Marty et al., 1984), neutrophils (Tscharner et al., 1986), insulinoma cell line (Sturgess et al., 1987), cultured proximal tubule cells (Merot et al., 1988), mandibular cell line (Cook et al., 1990), corneum endothelial cells (Rae et al., 1990), and colon tumor cells (Champigny et al., 1991). This requirement of intracellular Ca^{2+} for channel activity is also indicated for the NSC-channel of white adipocytes (E. Dotzler, unpublished result).

The NSC-channels in brown fat requires $10~\mu M$ free Ca^{2+} for activation (Kovisto et al., 1992). Such a Ca^{2+} concentration is thought to be unphysiologically high and may reflect the possibility that an important intracellular modulator(s) is lost during the excision of inside-out patches. However, in brown fat cells the cytoplasmic volume is very small; it is possible that local Ca^{2+} concentration close to the channels may be as high, or that phosphorylation of the channel protein could change its Ca^{2+} sensitivity as has been found in Ca^{2+}-activated K^+-channels (Reinhart et al., 1991).

It is still unclear whether Ca^{2+} is released from the endoplasmic reticulum or from the mitochondria in brown fat cells during α_1-adrenergic stimulation (Connolly and Nedergaard, 1988). However, in cell-attached mode it is possible to activate the NSC-channels with $1~\mu M$ Ca^{2+} ionophore (ionomycin or A23187) (Koivisto et al., 1992), a result which agrees with results from pancreatic acinar cells (Thorn and

Petersen, 1992), but disagrees with findings from corneum endothelial cells (Rae et al., 1990).

Purine Nucleotide Sensitivity

It is important to study the nucleotide sensitivity of the ion channels in metabolically active cells because the levels of ATP may change in different metabolic states. A complete and reversible NSC-channel block with 1 mM ATP was found in excised inside-out patches from cultured brown adipocytes (Koivisto et al., 1992). The reduction in open-time probability was dose-dependent and had no effect on the unitary current amplitude. To compensate for the chelating effect of ATP, the free concentrations of Ca^{2+} and Mg^{2+} were kept constant at 1.2 mM in some experiments by increasing their nominal concentrations; this did not reduce the effect of ATP. The possibility that a phosphorylation step in the ATP-blockade is necessary was eliminated by experiments with the nonhydrolyzable ATP-analogue AMP-PCP. There the terminal phosphate group is not available for phosphorylation (Sturgess et al., 1986, 1987) and yet this compound still blocked the channels. ATP in nominally Mg^{2+}-free solutions was also active; this excludes a direct inhibitory action by the MgATP complex (O'Rourke et al., 1992). ADP could also block the channels. This excluded the involvement of a transphosphorylation step and G protein activation by nucleoside diphosphate kinase as found in other systems (Otero et al., 1988; Otero, 1990).

We were able to show that the NSC-channel in the plasma membrane in white adipocytes is also ATP-sensitive. 1 mM ATP causes complete block, whereas 0.1 mM ATP induces only a temporary block of 30–60 s duration.

Adenine nucleotides have been shown to block the NSC-channel of a number of cell types: insulinoma cell line (Sturgess et al., 1986, 1987); cells from the thick ascending limb of Henle's loop (Paulais and Teulon, 1989); a mandibular cell line (Cook et al., 1990); pancreatic duct cells (Gray and Argent, 1990); corneum endothelial cells (Rae et al., 1990); colonic tumor cells (Champigny et al., 1991), and pancreatic acinar cells (Thorn and Petersen, 1992). It is difficult to envisage a signaling function for ATP on the NSC-channels in brown fat cells because ADP seems to work as well. This implication agrees in principle with the results of Connolly et al. (1986), who found that manipulation of the energetic state of the cell did not change Na^+ influx.

Cyclic Nucleotides and NSC-Channels

Manipulations which are thought to increase intracellular cAMP levels were found to increase Na^+ influx in brown fat cells (Connolly et al.,

1986). As this Na^+ influx could be via the NSC-channels it was of interest to study direct effects of cyclic nucleotides on NSC-channels from brown fat. In inside-out patches, 0.1 mM cAMP and cGMP were unexpectedly able to completely and reversibly block the NSC-channel openings. The physiological compound 3'5'-cGMP seemed to be somewhat more potent than the synthetic 2'3'-cGMP (Koivisto et al., 1992).

Paulais and Teulon (1989) have reported that, in the thick ascending limb of Henle's loop cells, 1 mM and 0.1 mM cAMP reduce open-time probability of the NSC-channels by 65% and 21%, respectively; 1 mM cyclic GMP was found to be less potent, decreasing open-time probability by only 7%. Reale et al. (1992) reported that cAMP, cGMP, and cUMP can all regulate NSC-channels from an insulinoma cell line. Low concentrations $(0.1-1 \mu M)$ were found to be stimulatory, whereas higher concentrations were inhibitory.

It is necessary to investigate whether the failure to observe a stimulatory effect of cAMP is due to experimental conditions. Thus, it is of interest to test whether low concentrations of cAMP can stimulate NSC-channel activity in brown fat.

G Protein Connection

Gating of ion channels by G protein subunits seems to be a widespread phenomenon (Brown and Birnbaumer, 1988, 1990). An attempt was, therefore, made to establish a possible G protein connection with the NSC-channels from brown fat. It was found that 0.1 mM of the nonhydrolyzable GTP-analogue GTP-γ-S (which is thought to activate G proteins directly without receptor stimulation) blocked the channel openings completely and persistently in inside-out patches. The nonhydrolyzable GDP-analogue GDP-β-S, which is widely used as a competitive inhibitor of a G protein activation, had no effect, whereas GTP itself also caused a block without any agonist added. However, it has been discussed by Okabe et al. (1991) that even agonist-free receptors could, to some extent, activate or inactivate ion channels by G proteins in the presence of GTP.

It has previously been found that G protein activation opens NSC-channels from renal inner medullary collecting duct cells (Light et al., 1990) and ileum cells (Inoue and Isenberg, 1990). Gating of the NSC-channels from brown fat by G proteins would add flexibility to its hormonal control.

Function

The physiological function of the NSC-channels in brown fat physiology is still far from clear. Their role in thermogenesis, i.e., maintaining respiration by ATP-dependent ion pumping, is still an unsolved problem because mefenamic acid seemed unable to reduce noradrenaline-induced respiration in isolated hamster cells (Koivisto et al., unpublished). However, this result might be caused by the Ca^{2+}-releasing effect of mefenamic acid (Poronnik et al., 1992). Modulation of the membrane potential by NSC-channels could be very important in regulating divalent cation entry to the cells (Mertz et al., 1992). Also, the nature of the Ca^{2+}-channels in brown adipose tissue is an enigma. Therefore, the hypothesis that NSC-channels in brown fat could also function as Ca^{2+}-channels is worth considering, although Ca^{2+} permeability was found to be very low in inside-out patches (Weber and Siemen, 1989). But even low permeabilities may show an effect if the concentration gradient and the Ca^{2+}-sensitivity of the system are high. However, Poronnik et al. (1991) also found low Ca^{2+} permeability in inside-out patches in mandibular cell line, but they, nevertheless, could partly block Ca^{2+} entry by the NSC-channel blocker DPC (diphenyl-amine-2-corboxylate) in whole cells. It has been suggested that NSC-channels could function in conjunction with the Na^+/Ca^{2+}-exchanger (Petersen, 1990). In a model like this a significant Ca^{2+} entry would occur via the Na^+/Ca^{2+}-exchanger.

Whether NSC-channels in brown adipocytes could also play a part in transmission of the mitogenic signal, as has been found in fibroblasts, (Magni et al., 1991); Jung et al. (1992), is a fascinating hypothesis that remains to be tested.

Acknowledgements
We are indebted to Drs. Jenny Kien and W. Vogel for reading the manuscript, to M. Dietl and U. Schmitt for technical assistance. Financial support from the Hasselblad Foundation, the Swedish Natural Science Council (to J.N.), and the Deutsche Forschungsgemeinschaft (to D.S.) is gratefully acknowledged.

References

Adams DJ, Dwyer TM, Hille B (1980). The permeability of endplate channels to monovalent and divalent metal cations. J. Gen. Physiol. 75:493–510.

Arch J, Ainsworth AT, Cawthorne MA, Piercy V, Sennit MV, Thody VE, Wilson C, Wilson S (1984). Atypical β-adrenoceptor on brown adipocytes as target for anti-obesity drugs. Nature 309:163–165.

Arch JRS (1989). The brown adipocyte β-adrenoceptor. Proc. Nutr. Soc. 48:215–223.

Bean BP, Riós E (1989). Nonlinear charge movement in mammalian cardiac ventricular cells. Components from Na and Ca channel gating. J. Gen. Physiol. 94:65–93.

Bevan S, Gray PTA, Ritchie JM (1984). A calcium-activated cation-selective channel in rat cultured Schwann cells. Proc. R. Soc. (London) B 222:349–355.

Bojanic D, Jansen JD, Nahorski SD, Zaagsma J (1985). Atypical characteristics of the β-adrenoceptor mediating cyclic AMP generation and lipolysis in the rat adipocyte. Br. J. Pharmac. 84:131–137.

Bronnikov G, Houstek J, Nedergaard J (1992). β-adrenergic, cAMP-mediated stimulation of proliferation of brown fat cells in primary culture. J. Biol. Chem. 267:2006–2013.

Brooks JJ, Perosio PM (1992). Adipose tissue (Chapter 2). In: Histology for Pathologists. Sternberg SS, editor, New York: Raven Press, pp 33–60.

Brown AM, Birnbaumer L (1988). Direct G protein gating of ion channels. Am. J. Physiol. 254:H401–H410.

Brown AM, Birnbaumer L (1990). Ionic channels and their regulation by G protein subunits. Ann. Rev. Physiol. 52:197–213.

Cannon B, Nedergaard J (1985). Biochemistry of an inefficient tissue: brown adipose tissue. Essays Bichem. 20:110–164.

Cawthorne MA, Sennitt MV, Arch JRS, Smith SA (1992). BRL 35135A, a potent and selective atypical β-adrenoceptor agonist. Am. J. Clin. Nutr. 55:252S–257S.

Champigny G, Verrier B, Lazdunski M (1991). A voltage, calcium, and ATP sensitive nonselective cation channel in human colonic tumor cells. Biochem. Biophys. Res. Comm. 176:1196–1203.

Colquhoun D, Neher E, Reuter H, Stevens CF (1981). Inward current channels activated by intracellular Ca in cultured cardiac cells. Nature 294:752–754.

Connolly E, Dasso L, Nedergaard J (1989). Adrenergic regulation of ion fluxes in brown adipocytes. In: Thermoregulation: Research and Clinical Application. Lomax P, Schönbaum E, editors, Basel: Karger: pp 31–34.

Connolly E, Nedergaard J (1988). β-Adrenergic modulation of Ca^{2+} uptake by isolated brown adipocytes. J. Biol. Chem. 263:10574–10582.

Connolly E, Nånberg E, Nedergaard J (1986). Norepinephrine-induced Na^+ influx in brown adipocytes is cyclic AMP-mediated. J. Biol. Chem. 261:14377–14385.

Cook DI, Poronnik P, Young JA (1990). Characterization of a 25-pS nonselective cation channel in a cultured secretory epithelial cell line. J. Membrane Biol. 114:37–52.

Dasso L, Connolly E, Nedergaard J (1990). Alpha 1-adrenergic stimulation of Cl^--efflux in isolated brown adipocytes. FEBS Lett. 262:25–28.

Girardier L, Seydoux J, Clausen T (1968). Membrane potential of brown adipose tissue. A suggested mechanism for the regulation of thermogenesis. J. Gen. Physiol. 52:925–940.

Gögelein H, Dahlem D, Engelert HC, Lang HJ (1990). Flufenamic acid, mefenamic acid and niflumic acid inhibit single nonselective cation channels in the rat exocrine pancreas. FEBS Lett. 268:79–82.

Gray MA, Argent BE (1990). Non-selective cation channel on pancreatic duct cells. Biochim. Biophys. Acta 1029:33–42.

Heerden van M, Oelofsen W (1989). A comparison of norepinephrine and ACTH-stimulated lipolysis in white and brown adipocytes of female rats. Comp. Biochem. Physiol. 93:275–279.

Hille B (1992). Ionic channels of excitable membranes. Sunderland, Mass.: Sinauer.

Hodgkin AL, Huxley AF (1952). A quantitative description of membrane current and its application to conductance and excitation in nerve. J. Physiol. 117:500–544.

Hollenga C, Brouwer F, Zaagsma J (1991). Relationship between lipolysis and cyclic AMP generation mediated by atypical β-adrenoceptors in rat adipocytes. Br. J. Pharmacol. 102:577–580.

Hollenga C, Zaagsma J (1989). Direct evidence for the atypical nature of functional β-adrenoceptors in rat adipocytes. Br. J. Pharmacol. 98:1420–1424.

Horwitz B, Hamilton JS, Lucero MT, Pappone PA (1989). Catecholamine-induced changes in activated brown adipocytes. In: Living in the Cold. Malan A, Canguilem B, editors. London: John Libbey, pp 377–386.

Inoue R, Isenberg G (1990). Acetylcholine activates nonselective cation channels in guinea pig ileum through a G protein. Am. J. Physiol. 258:C1173–C1178.

Jung F, Selveraj S, Gargus JJ (1992). Blockers of platelet-derived growth factor-activated nonselective cation channel inhibit cell proliferation. Am. J. Physiol. 262:C1464–C1470.

Koivisto A, Russ U, Nedergaard J, Siemen D (1992). Modulation of the nonselective cation channel of rat brown adipocytes by guanosine nucleotides and Ca^{2+}. Pflügers Arch. 420:R82.

Langin D, Portillo MP, Saulnier-Blache J-S, Lafontan M (1991). Coexistence of three β-adrenoceptor subtypes in white fat cells of various mammalian species. Eur. J. Pharmacol. 199:291–301.

Light DB, Corbin JD, Stanton BA (1990). Dual ion-channel regulation by cyclic GMP and cyclic GMP-dependent protein kinase. Nature 344:336–339.

210

Lucero MT, Pappone PA (1989). Voltage-gated potassium channels in brown fat cells. J. Gen. Physiol. 93:451–472.

Lucero MT, Pappone PA (1990). Membrane responses to norepinephrine in cultured brown fat cells. J. Gen. Physiol. 95:523–544.

Magni M, Meldolesi J, Pandiella A (1991). Ionic events induced by epidermal growth factor. J. Biol. Chem. 266:6329–6335.

Marty A, Tan YP, Trautmann A (1984). Three types of calcium-dependent channel in rat lacrimal glands. J. Physiol. 357:293–325.

Merot J, Bidet M, Gachot B, Le Maout S, Tauc M, Poujeol P (1988). Patch clamp study on primary culture of isolated proximal convoluted tubules. Pflügers Arch. 413:51–61.

Mertz LM, Baum BJ, Ambudkar IS (1992). Membrane potential modulates divalent cation entry in rat parotid acini. J. Membrane Biol. 126:183–193.

Mohell N, Siemen D, Cannon B, Nedergaard J (1991). β_3-adrenergic stimulation of thermogenesis in brown adipose tissue. In: Adrenoceptors: Structure, Mechanisms, Function; Advances in Pharmacological Sciences. Basel: Birkhäuser, pp 355–356.

Mohell N, Connolly E, Nedergaard J (1987). Distinction between mechanisms underlying α_1- and β-adrenergic respiratory stimulation in brown fat cells. Am J. Physiol. 253:C301–C308.

Nånberg E, Connolly E, Nedergaard J (1985). Presence of a Ca^{2+}-dependent K^+-channel in brown adipocytes. Possible role in maintenance of α_1-adrenergic stimulation. Biochim. Biophys. Acta 844:42–49.

Okabe K, Yatani A, Brown AM (1991). The nature and origin of spontaneous noise in G protein-gated ion channels. J. Gen. Physiol. 97:1279–1293.

O'Rourke B, Backx P, Marban E (1992). Phosphorylation-independent modulation of L-type calcium channels by magnesium-nucleotide complexes. Science 257:245–248.

Otero ADS (1990). Transphosphorylation and G protein activation. Biochem. Pharmacol. 39:1399–1404.

Otero AS, Breitwieser G, Szabo G (1988). Activation of muscarinic potassium currents by ATPγS in atrial cells. Science 242:443–445.

Partridge LD, Swandulla D (1988). Calcium-activated nonspecific cation channels. Trends Neurosci. 11:69–72.

Paulais M, Teulon J (1989). A cation channel in the thick ascending limb of Henle's loop of the mouse kidney: Inhibition by adenine nucleotides. J. Physiol. 413:315–327.

Petersen OH (1990). Regulation of calcium entry in cells that do not fire action potentials. In: Intracellular Calcium Regulation. Brennan F, editor. New York: Alan R. Liss Inc.: pp 77–96.

Poronnik P, Cook DI, Allen DG, Young JA (1991). Diphenylamine-2-carboxylate (DPC) reduces calcium influx in a mouse mandibular cell line (ST_{885}). Cell Calcium 12:441–447.

Poronnik P, Ward MC, Cook DI (1992). Intracellular Ca^{2+} release by flufenamic acid and other blockers of the nonselective cation channel. FEBS Lett. 296:245–248.

Rae JL, Dewey J, Cooper K, Gates P (1990). A nonselective cation channel in Rabbit corneal endothelium activated by internal calcium and inhibited by internal ATP. Exp. Eye Res. 50:373–384.

Reale V, Hales CN, Ashford MLJ (1992). Cyclic AMP regulates a calcium-activated nonselective cation channel in a rat insulinoma cell line. J. Physiol. 446:312P.

Reinhart PH, Chung S, Martin BL, Brautigan DL, Levitan IB (1991). Modulation of calcium-activated potassium channels from rat brain by protein kinase A and phosphatase 2A. J. Neurosci. 11:1627–1635.

Richelsen B (1991). Prostaglandines in adipose tissue – with special reference to triglyceride metabolism. Danish Medical Bulletin 38:228–244.

Sabanov V, Koivisto A, Nedergaard J (1993). Cl^--channels in cultured mouse brown adipocytes. In preparation.

Schneider-Picard G, Coles JA, Girardier L (1985). α- and β-adrenergic mediation of changes in metabolism and Na/K exchange in rat brown fat. J. Gen. Physiol. 86:169–188.

Siemen D, Reuhl T (1987). Nonselective cationic channel in primary cultured cells of brown adipose tissue. Pflügers Arch. 408:534–536.

Siemen D, Weber A (1989). Voltage-dependent and cation-selective channel in brown adipocytes. In: Thermal Physiology. Mercer J, editor. Amsterdam: Elsevier Science Publishers B.V., pp 241–246.

Soltoff SP, Cantley LC (1988). Mitogens and ion fluxes. Ann. Rev. Physiol. 50:207–223.

Sturgess NC, Hales CN, Ashford MLJ (1986). Inhibition of a calcium-activated, nonselective cation channel, in a rat insulinoma cell line, by adenine derivatives. FEBS Lett. 208:397–400.

Sturgess NC, Hales CN, Ashford MLJ (1987). Calcium and ATP regulate the activity of a nonselective cation channel in a rat insulinoma cell line. Pflügers Arch. 409:607–615.

Swandulla D, Partridge LD (1990). Nonspecific cation channels. In: Potassium Channels; Structure, Classification, Function and Therapeutic Potential. Cook N, editor. Chichester: Ellis Horwood Ltd., pp 167–180.

Thorn P, Petersen OH (1992). Activation of nonselective cation channels by physiological cholecystokinin concentrations in mouse pancreatic acinar cells. J. Gen. Physiol. 100:11–25.

Trayhurn P, Nicholls DG (1986). Brown Adipose Tissue. London: Edward Arnold, pp. 374.

Tscharner von V, Prod'hom B, Baggiolini M, Reuter H (1986). Ion channels in human neurophils activated by a rise in free cytosolic calcium concentration. Nature 324:369–372.

Weber A, Siemen D (1989). Permeability of the nonselective channel in brown adipocytes to small cations. Pflügers Arch. 414:564–570.

Wilcke M, Nedergaard J (1989). Alpha 1- and beta-adrenergic regulation of intracellular Ca^{2+} levels in brown adipocytes. Biochem. Biophys. Res. Comm. 163:292–300.

Nonselective Cation Channels: Pharmacology, Physiology and Biophysics
ed. by D. Siemen & J. Hescheler

Inhibitors of Nonselective Cation Channels in Cells of the Blood-Brain Barrier

R. Popp, H. C. Englert*, H. J. Lang* and H. Gögelein*

*Max-Planck-Institut für Biophysik, Kennedyallee 70, D-60596 Frankfurt; *Hoechst AG, Department of Pharmacology, D-65926 Frankfurt, FRG*

Summary

In the antiluminal membrane of isolated capillaries of rat and porcine brain (blood-brain barrier) nonselective cation channels with g = 31 pS were observed in cell-excised membrane patches. The channel inactivated by decreasing cytosolic Ca^{2+} below 1 μM and was inhibited by 1 mM ATP on the intracellular side. Anions and divalent cations did not pass the channel, but Na^+ and K^+ were equally permeant. Like the nonselective cation channel of rat exocrine pancreas cells, the channel in cerebral capillary endothelial cells was inhibited reversibly by derivatives of diphenylamine-2-carboxylate (DPC), like 3′,5-dichlorodiphenylamine-2-carboxylic acid (DCDPC, $k_i = 1$ μM), and flufenamic acid ($k_i = 4.9$ μM). 4′-methyldiphenylamine-2-carboxylic acid (4-MDPC), 5-chloro-2(3-trifluormethylphenylamino)-3-nitrobenzoic acid, and 5-nitro-2-(3-phenylpropylamino)-2-carboxylic acid (NPPB), as well as the antiinflammatory drug ((Z)-5-chloro2,3-dihydro-3-(hydroxy-2-thienylmethylene)-2-oxo-1H-indole-1-carboxamide (Tenidap) had a relatively low blocking potency ($k_i > 10$ μM). Gadolinium (10 μM), a blocker of stretch-activated channels, inhibited the nonselective cation channel potently.

Introduction

Blood capillaries of the brain possess unique properties. On the one hand, they have characteristics of endothelial cells such as, for example, release of nitric oxide after stimulation by bradykinin (Wiemer et al., 1992), but on the other hand, they show properties of a tight epithelium. In contrast to endothelial cells of the peripheral system, brain capillary endothelial cells are polarized. Na^+/K^+-ATPase is located in the brain-facing membrane, whereas Na^+ influx pathways are present in the luminal membrane (Betz and Goldstein, 1986).

Little is known about ion transport mechanisms in these cells. Patch-clamp studies revealed inward-rectifying K^+ channels (Hoyer et al. 1991), as well as stretch-activated nonselective cation channels (Popp et al., 1992) in the antiluminal membrane. In addition, Ca^{2+}-activated nonselective cation channels were observed in the brain-facing membrane of intact isolated capillaries as well as in isolated cultured cells (Popp and Gögelein, 1992). In this communiction we focus on inhibitors of the latter channel.

General Properties of the Channels

In freshly isolated capillaries of rat brain the antiluminal membrane can be investigated with the patch-clamp method. In cell-attached patches, no other channel types than inward-rectifying K^+ channels and stretch-activated channels could be observed. However, after excision of the membrane patch into a NaCl-solution containing 1.3 mM $CaCl_2$, ion channels with a single-channel conductance of 31 pS appeared. The channel was about equally permeant to Na^+ and K^+ ions, but was not measurably permeant to Cl^- or the divalent ions Ca^{2+} and Ba^{2+}. The channel open propability was independent of the applied potential in the range from -80 mV to $+80$ mV. Lowering bath Ca^{2+} concentration on the cytosolic side to $1 \mu M$ or less inactivated the channel reversibly. The channel was also sensitive to intracellular ATP. Addition of 1 mM ATP to the bath inhibited the channel completely and reversibly. Thus, this nonselective cation channel has many properties in common with channels observed, for example, in the exocrine pancreas (Suzuki and Petersen, 1988), pancreatic duct cells (Gray and Argent, 1990), pancreatic β-cells (Sturgess et al., 1987), thick ascending limb of the mouse kidney (Paulais and Teulon, 1989), and in the cultured secretory epithelial cell line ST_{885} (Cook et al., 1990).

Inhibitors

In contrast to other ion channels, there exist only a few studies with channel blocker for nonselective cation channels. In the insulin-secreting cell line CRI-G1, Sturgess et al. (1987) observed that 4-aminopyridine inhibits the channel in millimolar concentrations and that quinine blocks the channel by inducing flickering. In excised patches of isolated cells of rat distal colon it was reported that quinine inhibits nonselective cation channels by apparently reducing the single-channel current (Gögelein and Capek, 1990). This means that quinine acts as a fast blocker. A different class of substances was discovered to act as blockers for the nonselective cation channel in rat exocrine pancreas when, by chance, related substances of the chloride channel blocker 5-nitro-2-(3-phenylpropylamino-2-carboxylic acid (NPPB, Wangemann et al., 1986) were tested. Although NPPB and diphenylamine-2-carboxylic acid (DPC) were poor blockers (significant effects at about $100 \mu M$), the derivative 3',5-dichlorodiphenylamine-2-carboxylic acid (DCDPC) inhibited the channel with an IC_{50} of about $10 \mu M$ from the cytosolic side (Gögelein and Pfannmüller, 1989). All substances acted by increasing the mean closed-time of the channel (slow block). As these derivatives of DPC have structural similarities with some nonsteroidal antiinflammatory drugs, we tested such compounds on single channels of the rat exocrine pancreas (Gögelein et al., 1990). Indeed, it was observed that flufenamic- and mefenamic acid had blocking properties similar to DCDPC, whereas niflumic acid was somewhat less potent. Some of the compounds used to investigate at the channels of the blood-brain barrier are shown in Figure 1; Figure 2 summarizes the results.

Under control conditions the channels had an open-state probability P_o of 0.65. Addition of 1 μM of DCDPC to the bath decreased P_o to nearly half of maximal, whereas the substances flufenamic acid, 4-methyl-DPC, and NPPB had no significant effects. At a concentration of 10 μM, channels where completely blocked by DCDPC and P_o was markedly reduced by flufenamic acid. 4-methyl-DPC and NPPB decreased P_o slightly, However, at a concentration of 100 μM, also 4-methyl-DPC, NPPB, as well as 5-chloro-2-(3-trifluormethylphenyl-amino)-3-nitrobenzoic acid markedly inhibited channel activity. Thus, DCDPC has a stronger potency of inhibition on nonselective cation channels in cerebral capillary endothelial cells than on channels in the exocrine pancreas.

The novel antiinflammatory drug Tenidap showed a low potency of block, whereas the blocker of stretch-activated channels, gadolinium, inhibited completely at 10 μM (Popp and Gögelein, 1992).

Other substances with antiinflammatory actions, such as indo-methacin, aspirin, diltiazem, ibuprofen or metamizol had no effect on the nonselective cation channel. Thus, it seems that the chemical struc-

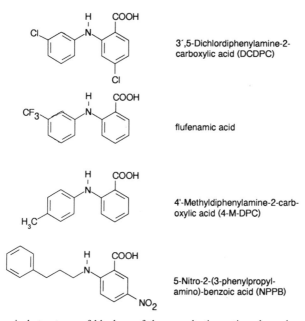

Figure 1. Chemical structures of blockers of the nonselective cation channel.

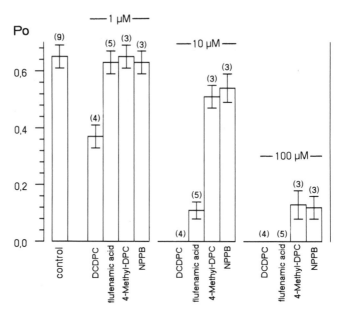

Figure 2. Inhibitors of single-channel open-state probability P_o of nonselective cation channels in the antiluminal membrane of isolated rat cerebral capillaries.

ture of two phenyl rings, linked by an amino group (Figure 1) is important for channel blockage.

Specificity of Channel Blockers

It seems that nonsteroidal antiinflammatory drugs such as flufenamic acid and mefenamic acid are a new class of blockers for nonselective cation channels in a variety of tissues. Thus far, channel inhibition was demonstrated in nonselective channels of the rat exocrine pancreas (Gögelein et al., 1990), the rat distal colon (Siemer and Gögelein, 1992), the blood-brain barrier (Popp and Gögelein, 1992), in mouse mandibular cells ST_{885} (Poronnik et al, 1992) and in cultured fibroblasts (Jung et al., 1992).

Although some DPC derivatives show potent blocking activity at the nonselective cation channels, these substances are far from being selective and should be used only with great care, especially when applied to whole-cell preparations. For example, it was reported that flufenamic acid is a potent inhibitor of anion transport in erythrocytes (Cousin and Motais, 1982), as well as of Ca^{2+}-activated Cl^- channels in *Xenopus* oocytes (White and Aylwin, 1990). Moreover, the drug may inhibit calcium uptake in mitochondria (McDouglass et al., 1988) and may release Ca^{2+} from intracellular stores (Poronnik et al., 1992).

It is well established that nonsteroidal antiinflammatory drugs such as flufenamic acid decrease prostaglandin bioysynthesis (Flower, 1974). On the other hand, not all therapeutic effects of the drug can be explained by this mechanism (Rees et al., 1988; Abramson and Weissmann, 1989). Therefore, one might speculate that some antiinflammatory action of DPC-derivatives is mediated by inhibition of nonselective cation channels.

References

Abramson SB, Weissmann G (1989). The mechanisms of action of nonsteroidal antiinflammatory drugs. Arthritis Rheum. 32:1–9.

Betz AL, Goldstein GW (1986). Specialized properties and solute transport in brain capillaries. Ann. Rev. Physiol. 48:241–250.

Cook DI, Poronnik P, Young JA (1990). Characterization of a 25-pS nonselective cation channel in a cultured secretory epithelial cell line. J. Membrane Biol. 114:37–52.

Cousin JL, Motais R (1982). Inhibition of anion transport in the red blood cell by anionic amphiphilic compounds. I. Determination of the flufenamate-binding site by proteolytic dissection of the band 3 protein. Biochim. Biophys. Acta 687:147–155.

Flower RJ (1974). Drugs which inhibit prostaglandin biosynthesis. Pharmacol. Rev. 26:33–67.

Gögelein H, Capek K (1990). Quinine inhibits chloride and nonselective cation channels in isolated rat distal colon cells. Biochim. Biophys. Acta 1027:191–198.

Gögelein H, Dahlem D, Englert HC, Lang HJ (1990). Flufenamic acid, mefenamic acid and niflumic acid inhibit single nonselective cation channels in the rat exocrine pancreas. FEBS Lett. 268:79–82.

Gögelein H, Pfannmüller B (1989). The nonselective cation channel in the basolateral membrane of rat exocrine pancreas. Inhibition by 3′,5-dichlorodiphenylamine-2-carboxylic acid (DCDPC) and activation by stilbene disulfonates. Pflügers Arch. 413:287–298.

Gray MA, Argent BE (1990). Non-selective cation channel on pancreatic duct cells. Biochim. Biophys. Acta. 1029:33–42.

Hoyer J, Popp R, Meyer J, Galla HJ, Gögelein H (1991). Angiotensin II, vasopressin and GTP[Γ-S] inhibit inward-rectifying K^+ channels in porcine cerebral capillary endothelial cells. J. Membrane Biol. 123:55–62.

Jung F, Selvaraj S, Gargus JJ (1992). Blockers of platelet-derived growth factor-activated nonselective cation channel inhibit cell proliferation. Am. J. Physiol. 262:C1464–C1470.

McDouglass P, Markham A, Cameron I, Sweetman AJ (1988). Action of the nonsteroidal anti-inflammatory agent, flufenamic acid, on calcium movements in isolated mitochondria. Biochem. Pharmacol. 37:1327–1330.

Paulais M, Teulon J (1989). A cation channel in the thick ascending limb of Henle's loop of the mouse kidney: inhibition by adenine nucleotides. J. Physiol. 413:315–327.

Popp R. Gögelein H (1992). A calcium and ATP sensitive nonselective cation channel in the antiluminal membrane of rat cerebral capillary endothelial cells. Biochim. Biophys. Acta 1108:59–66.

Popp R, Hoyer J, Meyer J, Galla HJ, Gögelein H (1992). Stretch-activated non-selective cation channels in the antiluminal membrane of porcine cerebral capillaries. J. Physiol. 454:435–449.

Poronnik P, Ward MC, Cook DI (1992). Intracellular Ca^{2+} release by flufenamic acid and other blockers of the non-selective cation channel. FEBS Lett. 296:245–248.

Rees MCP, Canete-Soler R, Bernal AL, Turnbull AC (1988). Effect of fenamates on prostaglandin E receptor binding. The Lancet 2:541–542.

Siemer C, Gögelein H (1992). Activation of nonselective cation channels in the basolateral membrane of rat distal colon crypt cells by prostaglandin E_2. Pflügers Arch. 420:319–328.

Sturgess NC, Hales CN, Ashford LJ (1987). Calcium and ATP regulate the activity of a non-selective cation channel in a rat insulinoma cell line. Pflügers Arch. 409:607–615.

Suzuki K, Petersen OH (1988). Patch-clamp study of single-channel and whole-cell K^+ currents in guinea pig pancreatic acinar cells. Am. J. Physiol. 255:G275–G285.

Wangemann P, Wittner M, Di Stefano A, Englert HC, Lang HJ, Schlatter E, Greger R (1986). Cl^--channel blockers in the thick ascending limb of the loop of Henle. Structure activity relationship. Pflügers Arch. 407(Suppl2):S128–S141.

White MM, Aylwin M (1990). Niflumic and flufenamic acids are potent reversible blockers of Ca^{2+}-activated Cl^- channels in *Xenopus* oocytes. Mol. Pharmacol. 37:720–724.

Wiemer G, Hock FJ, Popp R, Gögelein H (1992). Effect of CE-inhibition on NO-formation in cultured endothelial cells from bovine aorta and porcine brain arteriols. In: Rythmogenesis in Neurons and Networks, Elsner N, Richter DW, editors. Stuttgart: Thieme Verlag: pp 585 (Abstr.).

Nonselective Cation Channels: Pharmacology, Physiology and Biophysics
ed. by D. Siemen & J. Hescheler
© 1993 Birkhäuser Verlag Basel/Switzerland

Nonselective Cation Channels in Cells of the Crypt-Base of Rat Distal Colon

Christiane Siemer and Heinz Gögelein*

*Max-Planck-Institut für Biophysik, Kennedyallee 70, D-60596 Frankfurt; *Hoechst AG, Department of Pharmacology, H 821, D-65926 Frankfurt, FRG*

Summary
Cells in the base of isolated intact crypts of rat distal colon were investigated with the slow whole-cell patch-clamp technique with nystatin in the patch pipette. Addition of either prostaglandin E_2 or forskolin to the bath depolarized the cell from -74 mV to -27 mV. This depolarization was reversed when bath Na^+ was replaced by N-methyl-D-glucamine ($NMDG^+$), or when flufenamic acid (50 μM) was added to the bath. In cell-attached and cell-excised patches of the basolateral membrane nonselective cation channels ($\gamma = 38$ pS, $35°$C) were recorded. It is concluded that nonselective cation channels are activated by PGE_2 and forskolin. The channels could be involved in cell proliferation.

Introduction

The mammalian distal colon regulates volume and electrolyte composition of the stool by reabsorbing salt and water from the chymus. On the other hand, secretagogues can induce fluid secretion, which occurs under the pathophysiological situation of diarrhea. Fluid secretion can also be induced experimentally by substances such as vasointestinal polypeptide (VIP) or prostaglandin E_2 (PGE_2) and, therefore, the distal colon is an interesting preparation to study the mechanism of salt and water secretion (for review see Binder and Sandle, 1987).

The epithelium of the colon consists of a single layer of cells and shows deep excavations called the crypts of Lieberkühn. Cells in the crypts and at the surface have different degrees of differentiation and function. Morphological studies showed that the crypt base is constituted of stem cells which divide and migrate to the midcrypt where they start to differentiate (Cheng and Leblond, 1974). Electrophysiological studies confirm the view of different functions of cells along the crypt. There exists good experimental evidence that the reabsorptive processes are confined to the surface cells, whereas fluid secretion takes place in the crypt cells (Welsh et al., 1982; Horvart et al. 1986).

Little is known about the function and properties of stem cells in the crypt base. In order to study the function of these cells, we prepared intact crypts from rat distal colon using a nonenzymatic method by incubation in calcium free medium. Interestingly, the basolateral mem-

220

brane of such isolated crypts can be directly investigated with the patch-clamp method (Böhme et al., 1991; Siemer and Gögelein, 1992).

Occurrence and Properties of the Channel

First, the cell potential was recorded in cells at different locations in the crypt. Cells at the crypt-base exhibited a potential of $-74\,mV$, whereas the potential was $-40\,mV$ near the surface. When the secretagogue PGE_2 was applied to the bath, cells at the crypt base showed a persisting depolarization to $-27\,mV$, whereas cells of the midcrypt depolarized only transiently to about $-50\,mV$, and cells near the surface showed no reaction to PGE_2. In order to elucidate the nature of this depolarization, all cations in the bath medium were replaced by the impermeable cation $NMDG^+$.

A typical potential recording of a cell in the crypt base is shown in Figure 1. After exposure to PGE_2 the cell strongly depolarized, but repolarized completely after removal of bath Na^+. This experiment strongly suggests that the PGE_2-induced depolarization was due to the opening of Na^+ permeable channels. As it was shown that flufenamic acid blocks nonselective cation channels (Gögelein et al., 1990), the drug was added to PGE_2 depolarized cells. As shown in Figure 2, $10\,\mu M$ of the drug partially repolarized the cell potential and $50\,\mu M$

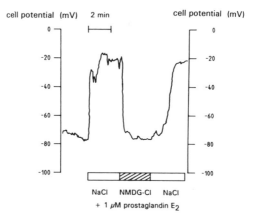

Figure 1. Effect of the Na^+-free bath solution on the cell potential of PGE_2-depolarized cells. NaCl was replaced by N-methyl-D-glucamine chloride (NMDG-Cl). After application of $1\,\mu M$ PGE_2 to the bath the cell potential depolarized from $-78\,mV$ to $-20\,mV$. Subsequent perfusion of the bath with Na^+-free solution repolarized the potential to its control value. The effect of NMDG-Cl solution was reversible. Membrane potential was recorded with the slow whole-cell configuration of the patch-clamp technique. The pipette contained KCl-solution (in mM: 140 KCl, 0.01 $CaCl_2$, 10 HEPES, pH = 7.4) to which $250\,\mu g/ml$ nystatin was added. Crypts were isolated as described previously (Siemer and Gögelein, 1992). The experiments were performed at 35°C. (Reproduced with permission from Siemer and Gögelein, 1992.)

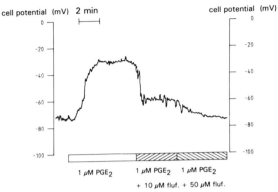

cell potential (mV) 2 min

cell potential (mV)

1 μM PGE₂   1 μM PGE₂   1 μM PGE₂
+ 10 μM fluf. + 50 μM fluf.

Figure 2. Effect of flufenamic acid (fluf.) on the cell potential. The PGE$_2$-induced depolarization was partially reversed by 10 μM of the drug and was completely reversed by 50 μM of flufenamic acid added to the bath. (Reproduced with permission from Siemer and Gögelein, 1992.)

caused complete repolarization. Thus, it can be concluded that PGE$_2$ depolarizes cells in the crypt base by activation of nonselective cation channels.

Further evidence for this idea was obtained by single-channel recordings that revealed the existence of a 38 pS nonselective cation channel in the basolateral membrane of cells in the crypt base. In cell-attached patches of unstimulated cells this channel was never observed. However, upon stimulation with PGE$_2$ channels were evoked. In cell-excised experiments the channel was nearly equally permeable to Na$^+$, Li$^+$, K$^+$, and Rb$^+$, whereas NMDG$^+$ was impermeant. Blockers such as flufenamic acid (50 μM), 3′,5-dichlorodiphenylamine-2-carboxylic acid (DCDPC, 50 μM) (Gögelein and Pfannmüller, 1989) and mefenamic acid (200 μM) inhibited the channel completely and reversibly.

The question remains as to which intracellular messenger activates the channel. Since PGE$_2$ is known to act via adenylyl cyclase, we performed experiments with forskolin, an activator of this enzyme. We observed that forskolin added to the bath medium caused a depolarization similar to that of PGE$_2$. The effect of forskolin was absent in NMDG-Cl solution and after addition of flufenamic acid to the bath medium. These experiments indicate that cAMP is the intracellular messenger which activates the nonselective cation channel.

Next, we investigated the nature of the depolarization in cells at other locations in the crypt. Moving slighty upward in the crypt, the depolarization was only partly caused by the opening of cation channels, but was mostly due to opening of Cl$^-$ channels. This conclusion is mainly based on experiments where Na$^+$ was substituted by NMDG$^+$, showing little effect on the forskolin-induced depolarization. Moreover, flufenamic acid was without effect, whereas the depolarization was inhibited by the chloride channel blocker 5-nitro-2-(3-phenylpropylamino)-benzoate (NPPB).

Function

The physiological function of the channel remains speculative. It is known that the crypt-base consists of undifferentiated stem cells which divide frequently. It was also shown that PGE_2 is involved in the process of cell proliferation. Recently, it was reported that cultured fibroblast L-M(TK$^-$) cells possess a nonselective cation channel which is activated by initiating cell proliferation by platelet-derived growth factor (Frace and Gargus, 1989). Moreover, Jung et al. (1992) showed that proliferation in these cells is inhibited by flufenamic acid. The authors concluded that nonselective cation channels play a crucial role in cell proliferation. Thus, it is also likely that in cells at the base of colon crypts the nonselective cation channels are involved in cell proliferation.

References

Binder HJ, Sandle GI (1987). Electrolyte absorption and secretion in the mammalian colon. In: Physiology of the gastrointestinal tract. Johnson LR, editor. New York: Raven Press: pp 1389–1418.

Böhme M, Diener M, Rummel W (1991). Calcium- and cyclic-AMP-mediated secretory responses in the isolated colonic crypts. Pflügers Arch. 419:144–151.

Cheng H, Leblond CP (1974). Origin, differentiation and renewal of the four main epithelial cell types in the mouse small intestine: V. Unitarian theory of the origin of the four cell types. Am. J. Anat. 141:537–562.

Frace AM, Gargus JJ (1989). Activation of single-channel currents in mouse fibroblasts by platelet-derived growth factor. Proc. Natl. Acad. Sci. USA 86:2511–2515.

Gögelein H, Pfannmüller B (1989). The nonselective cation channel in the basolateral membrane of rat exocrine pancreas. Inhibition by 3′,5-dichlorodiphenylamine-2-carboxylic acid (DCDPC) and activation by stilbene disulfonates. Pflügers Arch. 413:287–298.

Gögelein H, Dahlem D, Englert DC, Lang HJ (1990). Flufenamic acid, mefenamic acid and niflumic acid inhibit single nonselective cation channels in the rat exocrine pancreas. FEBS Lett. 268:79–82.

Horvarth P, Ferriola PC, Weiser MM, Duffey ME (1986). Localization of chloride secretion in rabbit colon: inhibition by anthracene-9-carboxylic acid. Am. J. Physiol. 250:G185–G190.

Jung F, Selvaraj S, Gargus JJ (1992). Blockers of platelet-derived growth factor-activated nonselective cation channel inhibit cell proliferation. Am. J. Physiol. 262:C1464–C1470.

Siemer C, Gögelein H (1992). Activation of nonselective cation channels in the basolateral membrane of rat distal colon crypt cells by prostaglandin E$_2$. Pflügers Arch. 420:319–328.

Welsh MJ, Smith PL, Fromm M, Frizzell RA (1982). Crypts are the site of intestinal fluid and electrolyte secretion. Science 218:1219–1221.

Nonselective Cation Channels as Regulatory Components of Cells from Various Tissues

Nonselective Cation Channels: Pharmacology, Physiology and Biophysics
ed. by D. Siemen & J. Hescheler
© 1993 Birkhäuser Verlag Basel/Switzerland

Poorly Selective Cation Channels in Apical Membranes of Epithelia

Willy Van Driessche, Luc Desmedt, Patrick De Smet and
Jeannine Simaels

Laboratory for Physiology, KULeuven, Campus Gasthuisberg, B-3000 Leuven, Belgium

Summary
The apical membrane of frog skin contains two types of pathways which allow the passage of
several monovalent cations in the absence of external Ca^{2+}. Differences between the two
pathways concern their open-close kinetics, selectivity, and the affinity for several blocking
agents. Type S channels open and close relatively *s*lowly, whereas type F channels display *f*ast
open-close kinetics. Both channel types allow the passage of Na^+, K^+, and Rb^+ currents
which are blocked by divalent cations and La^{3+} added to the extracellular side. Type F
channels are permeable for Cs^+ which is, however, excluded from type S channels. Shifts in
open-close kinetics induced by Mg^{2+} occur at concentrations below $5\ \mu M$ for type F channels,
whereas more than a tenfold higher dose is required for the type S pathway. UO_2^{2+}
concentrations up to $100\ \mu M$ only occlude type S channels while $100\ \mu M$ tetracaine selectively
blocks type F channels.
  Apical membranes of toad urinary bladder, cultured amphibian renal epithelia (A6), and
toad colon contain only type F channels. In toad bladder and A6 cells volume expansion
strongly activates this pathway. Macroscopic currents carried by Ba^{2+} and Ca^{2+} could be
recorded after activation of toad bladders with oxytocin and treatment of the apical surface
with nanomolar concentrations of Ag^+, which seems to interact with a site located at the
channel interior.

Two Types of Poorly Selective Cation Channels in Frog Skin

Previous studies from our laboratory demonstrated the presence of
poorly selective cation channels in the apical membrane of frog skin.
After removing Ca^{2+} from the mucosal side, this pathway is permeable
for several monovalent cations (Na^+, K^+, Rb^+, Cs^+, NH_4^+, Tl^+).
Ca^{2+} and other divalent cations added to the mucosal bath inhibit the
monovalent cation currents rapidly and reversibly.

  In our experiments with frog skin (Desmedt et al., 1993a), we found
two types of Ca^{2+}-blockable channels with different open-close kinetics.
Figure 1 illustrates power density spectra (PDS) recorded during an
experiment with frog skin (*Rana temporaria*) perfused with Na^+ solutions on the mucosal and serosal sides. Initially, Ca^{2+} was removed
from the mucosal solution and $60\ \mu M$ amiloride was added to block the
highly selective Na^+ channels in the apical membrane. In a recent paper
(Desmedt et al., 1991), we demonstrated that amiloride is able to exert

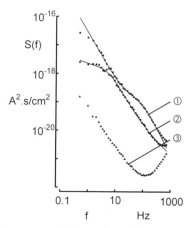

Figure 1. Occurrence of two types of Lorentzian noise components in the power density spectrum (PDS) of the fluctuation in current. PDSs were recorded from short-circuited skin of *Rana temporaria* bathed with Na_2SO_4 solutions on both sides. As in all following figures, Ca^{2+} free conditions were established by removal of Ca^{2+} from the mucosal solution and adding 0.5 mM EGTA. The serosal solutions always contained 1 mM Ca^{2+}. During the initial phase of the experiments (spectrum 1) a high-frequency Lorentzian ($S(f) = S_0/(1 + (f/f_c)^2)$) with $f_c = 90.4$ Hz and $S_0 = 131 \times 10^{-21} A^2 \cdot s/cm^2$) appeared at higher frequencies and $1/f$ noise ($S(f) = A/f^x$) dominated at the lower frequency end of the spectrum. After 2 h of incubation time, the high-frequency Lorentzian disappeared, $1/f$ noise was reduced, and a low-frequency Lorentzian emerged in the PDS (spectrum 2; $f_c = 6$ Hz; $S_0 = 1640 \times 10^{-21} A^2 \cdot s/cm^2$). The addition of 1.5 mM mucosal Ca^{2+} completely depressed the Lorentzian noise (spectrum 3).

its inhibitory effect in the absence of Ca^{2+}. The latter finding contradicts the generally accepted view of the necessity of Ca^{2+} for amiloride binding (Cuthbert and Wong, 1972). Consequently, the inhibition of I_{sc} by Ca^{2+} occurred solely by the occlusion of the Ca^{2+}-blockable pathway and was not biased by indirect effects on the amiloride-inhibitable pathway resulting from a putative requirement of Ca^{2+} for amiloride binding. The PDS recorded in the absence of Ca^{2+} contained a Lorentzian component with corner frequency $f_c = 90.4$ Hz and plateau value $S_0 = 131 \times 10^{-21} A^2 \cdot s/cm^2$. The relative high corner frequency (f_c) suggests fast open-close kinetics of the cation-conducting channels defined as type F channels. In 49 out of 205 experiments we found this type of high frequency Lorentzian. In 58 out of 205 skins we recorded power density spectra which contained a Lorentzian component with markedly lower corner frequency (mean value = 12.3 Hz). In many experiments we found that in the early phase of the experiment the high-frequency Lorentzian was present and that it disappeared slowly while the tissue remained exposed to Ca^{2+} free solutions, as illustrated in Figure 1. The channels which pass the current associated with the low-frequency Lorentzian component are designated type S channels.

The Lorentzian component of both channel types is completely abolished upon addition of 1 mM Ca^{2+} to the mucosal side.

We have found type S and F channels in epithelial cells of the skin of different frog species *Rana temporaria*, *Rana catesbeiana*, *Rana esculenta*, and *Rana pipiens*. However, in toad bladder (Aelvoet et al., 1988; Van Driessche et al., 1987), A6 epithelia (Van Driessche and De Smet, 1992), and in toad colon (Krattenmacher et al., 1990) only type F channels are present. The absence of type S channels in these epithelia might suggest that they belong to the same family as the apical K^+ channels in the skin, since the latter also do not occur in the apical membranes of the other epithelia. However, this hypothesis is in contradiction with results obtained from experiments with skins of *Rana pipiens*. Although the type S channels are present in this frog species, it lacks the apical K^+ channels. Another striking difference of the skin with the other epithelia resides in the continuous replacement of the outermost cell layer. During moulting the K^+ channels disappear from the cell membrane facing the outer side and amiloride-sensitive Na^+ channels are expressed. In this way, a K^+-permeable basolateral membrane is converted into a Na^+-permeable apical membrane. It is conceivable that the Ca^{2+}-blockable channels are precursors of the highly selective Na^+ channels. This hypothesis has been raised previously in relation to the occurrence of the poorly selective pathway in tadpole skin (Hillyard and Van Driessche, 1989) during stages prior to XX where the Na^+-selective channels are not yet expressed. During this stage of development the sensitivity of the channel to external Ca^{2+} was much smaller than in the skin of the adult frog. Indeed, monovalent cation currents were not completely blocked with physiological Ca^{2+} concentrations. As Na^+ transport was further developed during subsequent stages of the larval frog the poorly selective channels disappeared.

Localization of the Poorly Selective Channels

The apical as well as basolateral membranes of epithelial cells contain ion-conductive channels (Civan, 1983). As far as the epithelium of frog skin is concerned, Na^+ and K^+ channels have been described in the apical membranes, whereas the basolateral membrane is predominantly permeable for K^+ (Van Driessche and Zeiske, 1985b). Also, the paracellular pathway allows the passage of ions with some selectivity (Civan, 1983). Moreover, it is conceivable that tight junctions also contain ion channels that open and close randomly, thus giving rise to the Lorentzian noise observed in the absence of external Ca^{2+}. However, there is no transepithelial driving force for Na^+ movements through the paracellular pathway under our experimental conditions (short-circuited tissues incubated with identical Na^+ concentrations on both sides). On the other

228

hand, the existence of concentration gradients between the bulk solution and the lateral interspace cannot be excluded. This could find its origin in the presence of a diffusion barrier between the lateral interspace and the basolateral bath. Ion transport into or out of the cellular compartment could then build up concentration gradients between bulk solution and interspace which could drive currents through the tight junctions.

Microelectrodes were utilized to decide whether the Ca^{2+}-blockable channels are localized in the apical or basolateral membrane or in structures of the tight junctions. Figure 2 illustrates such an experiment in which short-circuit current (I_{sc}), transepithelial conductance (G_t), intracellular potential (V_0), and fractional resistance (fR_0) were recorded. The latter parameter is assumed to be equal to the voltage divider ratio (fV_0) defined as the fraction of voltage drop across the apical membrane (ΔV_0) during the application of a transepithelial voltage pulse (ΔV_t): $fV_0 = \Delta V_0/\Delta V_t$. Agents which interfere with the transcellular conductive pathways will change this parameter, whereas effects on the paracellular path will leave fR_0 unaltered. For example, a complete block of the apical conductances will increase fR_0 to 100%. The experiment depicted in Figure 2 demonstrates that the addition of amiloride or Ca^{2+} alone to the mucosal side did not evoke such a complete block of the apical conductance. The fact that amiloride or Ca^{2+} alone did not cause large changes in the transepithelial or the intracellular parameters can be

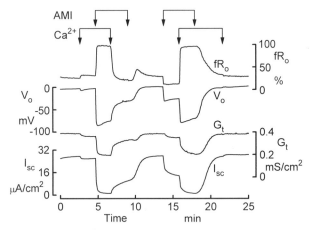

Figure 2. Localization of the Ca^{2+}-blockable pathways. Microelectrode experiment with the epithelium of *Rana temporaria* exposed to Na_2SO_4 solutions. During the initial phase of the experiment Ca^{2+} was absent in the mucosal bath. Cell potential (V_0) was strongly depolarized and the fractional resistance (fR_0) was small. The addition of mucosal Ca^{2+} (1.5 mM) hardly affected V_0 and fR_0 or the short-circuit current (I_{sc}) and transepithelial conductance (G_t). Blocking of the highly selective Na^+ channels with 60 μM amiloride raised fR_0 to values near 100%, demonstrating the validity of the microelectrode impalement (Nagel, 1978). Concomitantly, V_0 hyperpolarized, whereas I_{sc} and G_t were depressed. Similar responses were recorded upon reversing the order of application of the inhibitors.

explained by the interference between both pathways. Indeed, the block of one of these pathways will hyperpolarize the cell and increase the driving force for Na^+ entry across the apical membrane through the unaffected parallel pathway. Moreover, the cellular hyperpolarization might alter the conductance of the remaining conductive pathway. This makes it impossible to predict the effect of the occlusion of one pathway only. With both agents together in the mucosal bath fR_0 increased to 100%. The increase of fR_0 to 100%, accompanied with a complete block of I_{sc}, a strong hyperpolarization of V_0, and a marked decrease of G_t provides strong evidence for an apical localization of the amiloride-sensitive and Ca^{2+}-blockable pathway.

Properties of Ca^{2+}-Blockable Channels

The study of the properties of types S or F channels separately requires tools that selectively block one of the pathways. Several divalent cations and La^{3+} block both pathways and do not provide a tool to discriminate between the two channel types. Quinidine, which has been reported to inhibit cation-selective channels in toad urinary bladder (Das and Palmer, 1989) also blocked type S channels in the same concentration of $100\ \mu M$. We refrained from using this compound because it is well documented that it has considerable effects on intracellular Ca^{2+} stores. The release of Ca^{2+} from these compartments could interfere with both channel types. We found four compounds which displayed some specific affinity for either one of both Ca^{2+}-blockable pathways: UO_2^{2+}, Be^{2+}, Mg^{2+}, and tetracaine. Before discussing the effects of these agents, we will first present the dose-dependent interaction of Ca^{2+} with both channel types.

A) Ca^{2+} Block and Noise

Ca^{2+} and several other divalent cations, as well as La^{3+}, reversibly block both channel types. The random binding of these blockers to the channel gives rise to additional fluctuation in current and to a reduction of the open probability of the channel, causing a decrease in macroscopic current passing through this pathway. An example of the dose-dependent inhibition of the Na^+ current is displayed in Figure 3A. Increasing doses of Ca^{2+} added to the amiloride-containing Na^+ solutions progressively depressed I_{sc}. $[Ca^{2+}]_m = 500\ nM$ already inhibited 96% of the total Ca^{2+}-blockable I_{sc} (I_{sc}^{Ca}). Blocker-induced noise was recorded at concentrations between 30 and 500 nM (Van Driessche et al., 1991). The corner frequency increased with $[Ca^{2+}]_m$ (Figure 3B). Up to $[Ca^{2+}]_m = 150\ nM$, f_c shifted linearly to higher frequencies, as expected for first-order kinetics. At larger concentrations f_c displayed a superlinear increase. This superlinear increase of f_c could be caused by

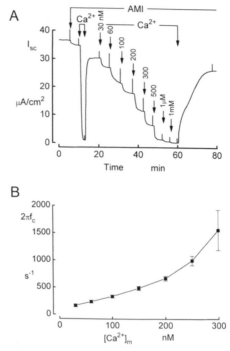

Figure 3. Dose-dependent inhibition of I_{sc} by mucosal Ca^{2+}. Experiment with the skin of *Rana temporaria* exposed to Na_2SO_4 solutions. Na^+ transport through the highly selective Na^+ channels was inhibited with amiloride ($6 \mu M$). Panel A: Inhibition of I_{sc} by mucosal Ca^{2+}. The current changes resulting in the vertical deflections on the current trace are caused by transepithelial voltage pulses of $10 \, mV$, used to measure the transepithelial conductance. Panel B: Shift of the corner frequency of the blocker-induced Lorentzian noise with $[Ca^{2+}]_m$.

different mechanisms: 1) a higher order reaction scheme for Ca^{2+} binding; 2) interference of binding Ca^{2+} with an intracellular factor which depends on extracellular Ca^{2+}; 3) an effect of voltage changes due to the block of the apical membrane conductance. An increase of the potential across the apical membrane caused by the Ca^{2+} block itself will indeed augment the affinity of Ca^{2+} for the receptor located at the channel interior (De Wolf and Van Driessche, 1986). However, microelectrode experiments (Van Driessche et al., 1991) demonstrated that most of the increase in intracellular potential occurs at $[Ca^{2+}]_m$ below $150 \, nM$. This observation makes the involvement of changes in cell voltage unlikely because the superlinear increase occurred at $[Ca^{2+}]_m$ above $150 \, nM$. Nevertheless, it cannot be excluded that at higher Ca^{2+} concentrations screening of or binding to negative surface charges near the channel mouth results in alterations of the transmembranal potential which are not detected with microelectrodes. Also, interference with an intracellular factor cannot be excluded because the addition of Ca^{2+} to the apical solution could affect the cellular milieu by, for example, an increase of cell Ca^{2+}. Modifications of cellular Ca^{2+} could then, directly or indirectly, alter the binding of external Ca^{2+}. Such effects could explain the discrepancies between results obtained with and without chelating agents discussed below. Finally, we do not have any evidence for the possibility that Ca^{2+} binding occurs through a higher order reaction scheme. Such multistate binding reaction should give rise to additional relaxation noise components in the PDS which are not observed. These noise components could, however, be located outside of the frequency range of our analysis.

In our earlier experiments with frog skin and toad urinary bladder, we avoided the use of Ca^{2+} chelating agents because of possible interferences with the paracellular conductive pathway. Indeed, doses of 5 mM EGTA added to the mucosal perfusate markedly increased the transepithelial conductance by opening of the paracellular pathway, at least for toad urinary bladder. After a series of preliminary experiments we chose to utilize 0.5 mM of the chelator, which appeared to have minor effects on the transepithelial conductance. The free Ca^{2+} concentration was estimated after determining the total amount of Ca^{2+} ($[Ca^{2+}]_{tot}$) by flame absorption spectrophotometric analysis of our solutions. We found $[Ca^{2+}]_{tot}$ values of less than 5 μM, for the Cl^- as well as SO_4^{2-} solutions. Cl^- solutions contained (in mM): 120 Na^+, 115 Cl^-, 5 Hepes, 0.5 EGTA, pH = 7.5. SO_4^{2-} solutions had the following composition (in mM): 120 Na^+, 57.5 SO_4^{2-}, 5 Hepes, 0.5 EGTA, pH = 7.5. Ca^{2+} activities calculated according to methods described by Fabiato and Fabiato (1979) with binding constants obtained from Martell and Smith (1974) were less than 1 nM.

Because of the lack of Ca^{2+} chelating agents in our earlier studies (Van Driessche and Zeiske, 1985a), Ca^{2+}-blockable currents and conductances were underestimated and channel kinetics were modified by the interaction of Ca^{2+} with the channel receptors. Indeed, as described above, the reversible interaction of Ca^{2+} with the channel induces additional noise. According to these results, Ca^{2+} concentrations of $1-5$ μM as present in the absence of Ca^{2+} chelators should reduce blocker noise below background levels. Moreover, extrapolation of $2\pi f_c$-$[Ca^{2+}]_m$ relations recorded in EGTA-containing solutions predict corner frequencies which are outside the frequency range of our analysis. These discrepancies are difficult to understand. Different possibilities can be examined: 1) direct effects of EGTA on the channels; 2) effects of external and/or internal Ca^{2+} chelation on intracellular Ca^{2+} together with possible effects of cellular Ca^{2+} on the channel; 3) differences in Ca^{2+} activity between the bulk solution and layer adjacent to the apical membrane.

232

B) Effect of UO_2^{2+} and Be^{2+}: Blockage of Type S Channels

UO_2^{2+} in concentrations up to 100 μM interferes mainly with type S channels. An example of the inhibitory effect of UO_2^{2+} is depicted in Figure 4. Initially, while the mucosal side was perfused with amiloride-containing Ca^{2+} free Na^+ solutions, a low-frequency component dominated in the PDS (Figure 4B), indicating the presence of type S channels. The addition of 100 μM UO_2^{2+} to the mucosal bath depressed the low-frequency Lorentzian component in the PDS and reduced I_{sc} (Figure 4A). Concomitantly, a high-frequency Lorentzian appeared in the PDS. In an extensive study of the UO_2^{2+} effect (Desmedt et al., 1993a), we demonstrated that this Lorentzian was not induced by UO_2^{2+} by randomly interrupting the currents through channel type S, but that it was associated with currents passing through channel type F. In control, this high-frequency Lorentzian was not visible and should, therefore, have been masked by the low-frequency Lorentzian noise which has generally much higher amplitudes. Close inspection of Figure 4B, however, shows that in the higher frequency range the noise levels

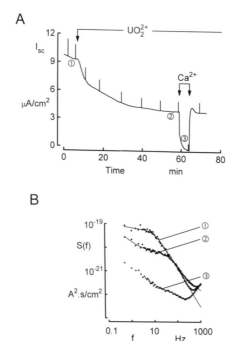

Figure 4. 100 μM UO_2^{2+} blocks type S channels. Skin of *Rana temporaria* was mucosally perfused with amiloride-containing Na_2SO_4 solution. Panel A: UO_2^{2+} inhibits I_{sc} partly. The addition of 1.5 mM Ca^{2+} reduces I_{sc} to zero. Panel B: Power density spectra recorded at times indicated at the current trace in panel A. UO_2^{2+} inhibits the low-frequency Lorentzian noise and reveals the existence of a high-frequency Lorentzian. Spectrum 3 was recorded in the presence of mucosal Ca^{2+}.

of spectrum 2 are larger than or at least of magnitude equal to those of spectrum 1. Therefore, the high-frequency Lorentzian should have been visible in spectrum 1, unless the addition of UO_2^{2+} had elevated the noise of currents through channel type F. This could result from an increased driving force caused by the blocking of type S channels, which would imply that both channels are located in the same cell type. Microelectrode experiments (Desmedt et al., 1993b) confirmed this hypothesis. On the average, we found that UO_2^{2+} blocked 80% of the current through channel types S. The addition of mucosal Ca^{2+} blocked I_{sc} completely and depressed the high frequency Lorentzian noise to levels close to background. This type of experiment clearly demonstrates that UO_2^{2+} interferes with channel type S. Similar experiments were done with Be^{2+}. Here, 5 μM of the divalent also inhibited the low-frequency Lorentzian and unmasked the high-frequency Lorentzian.

C) Effect of Tetracaine: Blockage of Type F Channels

In our search for agents that specifically block channel type F, we found that concentrations up to 100 μM of tetracaine mainly affected the high-frequency Lorentzian and left the low-frequency Lorentzian almost unaltered. Figure 5 illustrates the effect of tetracaine in an experiment where, initially, a high-frequency Lorentzian was clearly expressed in the PDS. After removal of Ca^{2+} from the mucosal bath, we added amiloride to block the highly selective Na^+ channels. In the experiment shown, amiloride depressed 60% of I_{sc}. The addition of 100 μM tetracaine to the mucosal bath completely inhibited the high-frequency Lorentzian noise and the Ca^{2+}-blockable I_{sc} component. In experiments as displayed in Figure 4 where a low-frequency Lorentzian dominated, we tested tetracaine after adding UO_2^{2+}. Also in such experiments, tetracaine depressed the high-frequency Lorentzian recorded after adding UO_2^{2+}. On the other hand, we tested the effect of tetracaine on the low-frequency Lorentzian and found no significant inhibition of this noise component. Therefore, we conclude that tetracaine in concentrations up to 100 μM selectively interacts with type F channels. Lidocaine also inhibited type F channels. However, concentrations larger than 200 μM were required which also affected the type S pathway. Moreover, it is well known that in this concentration range of the local anaesthetic also the basolateral K^+ channels are blocked (Van Driessche, 1986). On the other hand, we demonstrated (Desmedt et al., 1993b) that tetracaine did not influence transepithelial K^+ movements by interfering with the basolateral K^+ conductance.

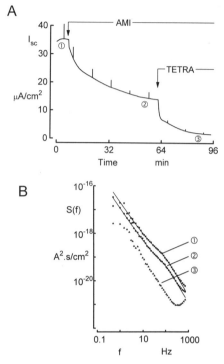

Figure 5. Tetracaine blocks type F channels in the skin of *Rana temporaria* incubated with Na$_2$SO$_4$ solutions. Panel A: Tetracaine (100 μM) was added to the mucosal perfusate after amiloride had depressed 60% of I$_{sc}$. Panel B: PDSs recorded at times indicated in Panel A. Before and during amiloride treatment, the PDS contained a high-frequency Lorentzian component (before adding tetracaine: f$_c$ = 98 Hz; S$_0$ = 50 × 10^{-21} A^2 · s/cm^2). The Lorentzian was abolished by tetracaine.

D) Effect of Mg^{2+}

Among the divalent cations, Mg^{2+} is easier to use in experiments with Ca^{2+} chelating agents than other divalent species. Its affinity for EGTA and BAPTA is negligible (Martell and Smith, 1974). Therefore, the addition of Mg^{2+} will not cause dissociation of Ca^{2+} from EGTA or BAPTA leading to increases of the free Ca^{2+} concentration. Doses of Mg^{2+} smaller than 5 μM mainly influenced the high-frequency noise. Higher concentrations were needed to modify the kinetics of type S channels. The interaction with both channels was completely reversible and gave rise to blocker-induced noise as demonstrated for Ca^{2+} in Figure 6 in Van Driessche et al. (1991). In contrast to Figure 3B, the 2πf$_c$-[Mg^{2+}]$_m$ relation did not display a superlinear increase.

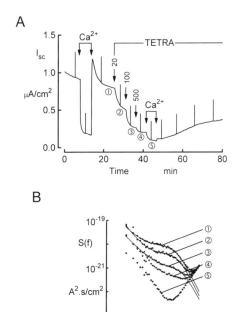

Figure 6. Tetracaine blocks the Cs^+ current in the skin of *Rana temporaria* perfused with Na_2SO_4 and Cs_2SO_4 (Ca^{2+} free) solutions on the serosal and mucosal side, respectively. Panel A: I_{sc} recording. Ca^{2+} was added to the mucosal side during the indicated periods. 500 μM tetracaine blocked I_{sc} almost completely. Panel B: PDSs recorded during periods indicated at the I_{sc} trace in Panel A. Tetracaine progressively depresses the Lorentzian plateau of the high-frequency Lorentzian.

E) Selectivity of Ca^{2+}-Blockable Channels

The properties of the Ca^{2+}-blockable channels discussed so far concerned Na^+ currents across the short-circuited epithelium of frog skin. Na^+ uptake across the apical membrane is driven by the electrochemical driving force which is maintained by the Na^+-K^+ pump at the basolateral membrane. Movements of other cations across the epithelium can be evoked by applying a chemical gradient. As far as K^+ and Rb^+ are concerned, these cations can easily move across the basolateral membrane through the K^+ channels. Very little is known about the permeability of the basolateral membrane for Cs^+. Our results demonstrated that Cs^+ currents diminished slowly, indicating that Cs^+ accumulated in the cell.

Differences between both channel types not only concern the affinity of channel blockers, but also the permeability for different monovalent cations. We found that Na^+, K^+ and Rb^+ permeate through both channel types, whereas Cs^+ can pass through type F channels only. The latter statement is based on the effect of UO_2^{2+} and tetracaine on the currents and noise recorded with mucosal Cs^+ solutions and on the presence of a high-frequency Lorentzian in the PDS. Figure 6 illustrates such an experiment with Cs^+ in which we tested the effect of tetracaine. The control spectrum recorded after we tested the sensitivity of the current to Ca^{2+} contained a high-frequency Lorentzian (Figure 6B), suggesting that Cs^+ mainly flows through the type F channels. This was confirmed by the inhibitory effect exerted by tetracaine on I_{sc} and the Lorentzian noise. Tetracaine depressed I_{sc} and the Lorentzian plateau in a dose-dependent way. Similar experiments with Cs^+ in which we tested the effect of UO_2^{2+} demonstrated that Cs^+ currents and noise were not inhibited by this agent. These findings thus demonstrate that Cs^+ does not pass through the type S channels which are sensitive to UO_2^{2+}.

Activation of Type F Channels in Toad Urinary Bladder by cAMP

Several studies from our laboratory (Aelvoet et al., 1988; Van Driessche et al., 1988) concerned the poorly selective cation channels in the urinary bladder of the toad (*Bufo marinus*). A typical example of the experiments with the bladder is illustrated in Figure 7. In this experiment the tissue was exposed to mucosal K^+ solutions. The PDS of the fluctuation in current contained a Lorentzian component with a high f_c value, indicating that this pathway belongs to the type F channels. This is confirmed by the fact that the current and noise are insensitive to UO_2^{2+} but inhibited by tetracaine (Desmedt et al., 1993b). Ca^{2+} reversibly blocks the current and its fluctuation. Agents which elevate cellular cAMP (oxytocin, theophyline, IBMX, exogenous cAMP analogues) considerably increase the Lorentzian noise as well as Ca^{2+}-blockable current. In the example depicted in Figure 7, we used oxytocin. From these results we concluded that this channel is activated by cAMP (Aelvoet et al., 1988), although data obtained with A6 epithelia necessitate reconsideration of this issue. In this cell type, as well as in toad bladder, we found that the Ca^{2+}-blockable pathway is drastically elevated by volume expansion. Cell swelling can be evoked by serosal hypotonicity or replacement of sucrose by organic compounds with reflection coefficients smaller than 1. Part of the increase in current and noise might therefore be caused by cell volume expansion which could be evoked by agents increasing cellular cAMP. Indeed, in urinary bladder these agents drastically increase the conductance of several ionic pathways as well as the water permeability (Φ_{H_2O}) of the apical membranes. The increases in water and ion permeabilities might then alter cell volume and activate the channel. With the data presently available the question of the involvement of cell volume expansion or of

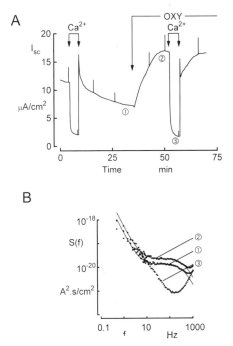

Figure 7. Ca^{2+}-blockable current and Lorentzian noise in toad urinary bladder (*Bufo marinus*) perfused with Na_2SO_4 and K_2SO_4 (Ca^{2+} free) at the serosal and mucosal side, respectively. Ca^{2+} was added to the mucosal side. Oxytocin (0.1 U/ml) stimulated I_{sc} and the Lorentzian noise. Ca^{2+} completely blocked I_{sc} and the Lorentzian noise.

the cAMP pathway remains open and we are unable to univocally exclude one of these putative affectors of the cation pathway.

Activation of Type F Channels in Toad Bladder and A6 Cells by Cell Volume Expansion

Epithelia grown from a cell line derived from a distal part of the nephron of *Xenopus laevis* have been used to study Na^+ and Cl^- transport systems (Wills and Millinoff, 1990) and its hormonal regulation. During the early stage of development (4 to 10 days of epithelial growth) of these epithelia also the Ca^{2+}-blockable pathway is present, although in an inactivated state. With isotonic Cl^- or SO_4^{2-} solution (215 mOsm/kg) on both sides the tissues do not have a Ca^{2+}-blockable conductance and the PDS does not contain a Lorentzian component. However, a Ca^{2+}-blockable current and associated Lorentzian noise can be activated by reducing the tonicity of the basolateral solution (Figure 8). We lowered the tonicity at the basolateral side because only this membrane is water permeable (Candia and Yorio, 1992). In control, I_{sc}

238

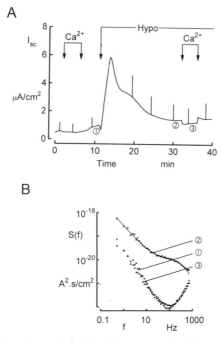

Figure 8. Hypotonicity activates Ca^{2+}-blockable current in A6 epithelia. The tissue used after 4 days of epithelial growth was exposed to NaCl and KCl (Ca^{2+} free) to the serosal and mucosal side, respectively. In control, I_{sc} (Panel A) was insensitive to Ca^{2+} and no Lorentzian noise component was detectable in the PDS (Panel B). Reduction of the serosal osmolality from 215 to 142 mOsm/kg stimulated I_{sc} transiently and reached a plateau after 20 min. Lorentzian noise, blockable by Ca^{2+}, was recorded during hypotonicity.

was completely insensitive to Ca^{2+} and the PDS did not contain Lorentzian noise. Upon reduction of serosal tonicity to 142 mOsm/kg, I_{sc} increased transiently and reached a steady state after approximately 15 min. Concomitantly, the Lorentzian noise appeared in the PDS. The value of f_c (320 Hz) indicates that we are dealing with type F channels. The activated I_{sc} and Lorentzian noise were inhibited by mucosal Ca^{2+}. It is well known that cells exposed to hypotonic media will swell and then shrink back to volumes near their resting volume (Völkl et al., 1988). This regulatory volume decrease (RVD) is accompanied by the loss of cellular K^+ and Cl^-. Ca^{2+} plays an important role in the control of the RVD since this response seems to be Ca^{2+} dependent (Montrose-Rafizadeh and Guggino, 1991). Cellular Ca^{2+} is regulated by different transporters and depends on the extracellular Ca^{2+} concentration. Consequently, if Ca^{2+} entry occurs through the apical membranes (McCarthy and O'Neil, 1991), the time-course and level of the RVD might be markedly altered by removal of Ca^{2+} from the apical solution. Moreover, if tissues are exposed to mucosal KCl solutions, the loss of K^+ and Cl^- might be considerably diminished by the opening of the K^+-permeable Ca^{2+}-blockable channels and the activation of an apical Cl^- pathway. Under these conditions, we observed inward instead of outward movements of both ion species which counteract RVD. Recent data of Crowe and Wills (1991) reporting volume measurements in A6 cells incubated in Cl^- solutions demonstrated the lack of a RVD. This observation could agree with the hypothesis that the cation channel is activated by volume expansion. In the absence of a RVD the cell will maintain larger volumes which keep the cation channels in the activated state. Moreover, removal of external Ca^{2+} has been described to inhibit the ability of renal cells to regulate their volume (Montrose-Rafizadeh and Guggino, 1991).

Cell volume can be challenged with a hypotonic incubation medium. However cell volume will also change upon replacement of sucrose, which has a high reflection coefficient, by glycerol or ureum, which easily pass the cell membrane (Stein, 1967). The accumulation of the cell permeable compounds will be followed by water and results in an expansion of the cell. This was confirmed by our experimental results as exemplified by the serosal sucrose-glycerol replacement (Figure 9). In control (isotonic sucrose-containing solutions) the Ca^{2+}-blockable pathway was completely inactivated, as can be concluded from the lack of sensitivity of I_{sc} to Ca^{2+} and the absence of a Lorentzian noise component in the PDS. Upon substitution of glycerol for sucrose, I_{sc} increased transiently and reached a plateau after approximately 10 min. Concomitantly, a Lorentzian noise component, sensitive to Ca^{2+}, appeared in the PDS. As hypotonicity as well as the sucrose-glycerol replacements induce the Ca^{2+}-blockable I_{sc} and associated Lorentzian noise, we conclude that the Ca^{2+}-blockable pathway in A6 cells seems to be activated by cell volume expansion. Similarly, we found that serosal hypotonicity and sucrose-glycerol replacements strongly activated the poorly selective pathway in toad urinary bladder. As discussed above, this makes a direct involvement of cAMP pathways questionable.

Activation of Ca^{2+} Currents by Ag^+

So far, our overview of the properties of the Ca^{2+}-blockable pathway dealt with properties of monovalent cation currents and associated noise. The currents and noise are inhibited by Ca^{2+} and other divalent cations, but a finite permeability for divalent cations has not yet been discussed. Several types of nonselective channels found in other tissues are permeable for monovalent cations and are modulated by cellular Ca^{2+} (Marcus et al., 1992). Effects of external Ca^{2+}, as discussed above,

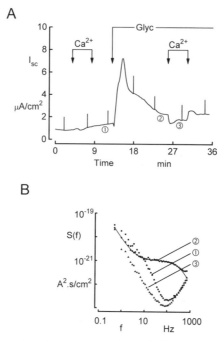

Figure 9. Sucrose-glycerol replacement activates Ca^{2+}-blockable current in A6 epithelia. Epithelial growth: 5 days. Mucosal solution: Ca^{2+} free KCl. Serosal solution: 70 mM NaCl + 70 mM sucrose. In control, I_{sc} (Panel A) was insensitive to Ca^{2+} and no Lorentzian noise component was detectable in the PDS (Panel B). The substitution of glycerol for sucrose activates the current (Panel A) and noise (Panel B).

were mostly not investigated in other systems. However, a finite permeability could be demonstrated for some nonselective channels (Bear, 1990). In intact epithelial tissues, as used in our studies, it is unlikely that cells can cope with a measurable Ca^{2+} influx. Such Ca^{2+} currents might occur with high Ca^{2+}-containing solutions while activating the pathway with hormones or cell volume expansion. In some experiments with toad urinary bladder we were able to record currents carried by divalent cations. Such an experiment was shown in a previous study (Figure 4B in Van Driessche, 1987). We exposed the apical surface to Ba^{2+} solutions lacking any monovalent cation. In control, I_{sc} was close to zero and did not change upon addition of 20 μM La^{3+} to the mucosal side. The activation of the tissue with oxytocin however caused a very small current increase which was inhibited by mucosal La^{3+}. This observation strengthened our hypothesis that the activation of the nonselective channels might allow the influx of Ca^{2+}. As already suggested above, the epithelial nonselective channel seems to be modulated by intracellular Ca^{2+} which could result in a rapid inactivation during Ca^{2+} influx elicited by a hormonal challenge or volume expansion. This hypothesis is supported by much evidence showing interferences of Ca^{2+} with the activation of water and Na^+ transport across the toad urinary bladder by antidiuretic hormone (Levine and Schlondorff, 1984). A Ca^{2+} efflux into the mucosal compartment (Cuthbert and Wong, 1974) and changes in intracellular Ca^{2+} concentrations (Wong and Chase, 1988) have been demonstrated during hormonal treatment. Interferences of Ca^{2+} channel blockers (Levine and Schlondorff, 1984) and alterations of intracellular Ca^{2+} with the hydroosmotic response to antidiuretic hormone have been described.

The mechanism which protects epithelial cells for a steady Ca^{2+} influx could be overridden by treatment of the apical surface with nanomolar concentrations of Ag^+. Similar effects were reported for heavy metal ion (Ag^+, Hg^{2+}) induced Ca^{2+} release from sarcoplasmic reticulum of skeletal (Salama and Abramson, 1984) and cardiac (Prabhu and Salama, 1990) muscle by binding to an accessible free sulfhydryl group on or adjacent to the Ca^{2+} release channel. In an earlier study, we demonstrated that treatment of the apical surface with nanomolar concentrations (50 nM) of Ag^+ drastically increased the permeability of type F channels in toad bladder for monovalent cations (Na^+, K^+, Rb^+, Cs^+). The monovalent currents were blocked by $20 \mu M La^{3+}$ or $500 \mu M Cd^{2+}$. We used these blockers instead of Ca^{2+}, because it turned out that the latter divalent permeates through the Ag^+ stimulated pathway, which makes it useless as blocking agent. The currents could only be stimulated after activation of the pathway with oxytocin. Therefore, it seems that the hormone acts via insertion or activation of channels, whereas Ag^+ increases the permeability of the individual channel. Effects of Ag^+ on intracellular compartments or affectors were excluded on the basis of the lack of an effect of serosal Ag^+ in concentrations up to 500 nM. At Ag^+ concentrations larger than $1 \mu M$ unspecific increases of the conductances were observed, most likely related to opening of the paracellular shunt, by effects on intracellular compartments and/or composition.

Ag^+ added to the mucosal bath not only increased the permeability of the poorly selective pathway for monovalent cations, but it also enabled the passage of divalents $(Ca^{2+}, Ba^{2+}, Mg^{2+})$. This fact supports the concept that the augmentation of the monovalent cation current occurs through the interaction of Ag^+ with the individual channel by increasing its permeability. An example of the stimulation of divalent currents by Ag^+ is depicted in Figure 10. The apical side was perfused with $Ca(NO_3)_2$ solutions free of monovalent inorganic cations. Before, as well as during oxytocin treatment, the La^{3+}-sensitive current was not measurable. However, upon addition of Ag^+ the current inhibited by La^{3+} reached about $0.4 \mu A/cm^2$. This result suggests that Ag^+ disrupts the selectivity filter which modulates cation influx under physiological conditions. Therefore, it seems unlikely that the binding site is located

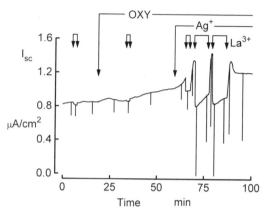

Figure 10. Mucosal Ag^+ (50 nM) enables the passage of divalent cations through the poorly selective cation channel in toad urinary bladder. Mucosal solution (in mM): 50 $Ca(NO_3)_2$, 5 TEA/Hepes, serosal solution (in mM): 115 $NaNO_3$, 2.5 $KHCO_3$, 1 $Ca(NO_3)_2$. 50 nM Ag^+ added to the mucosal side of the oxytocin treated tissue increased I_{sc} which could be blocked by 20 μM La^{3+}.

at the outer side of the channel mouth. This agrees with the finding that the poorly selective channels in epithelia are not stimulated by organic SH reagents because the site is not accessible. This contrasts the hypothesis raised for Ag^+ binding at the outside or adjacent to the Ca^{2+} channel of the sarcoplasmic reticulum (Salama and Abramson, 1984).

The opening of the poorly selective channel for movements of Ca^{2+}, Ba^{2+}, and Mg^{2+} provides a tool to elevate intracellular Ca^{2+} without using Ca^{2+} ionophores. Therefore, we expose the apical side to Ca^{2+} solutions as used in the above experiment. After challenging the epithelium with oxytocin, we treat the membrane with Ag^+ which will open the pathway for Ca^{2+} entry. We used this procedure to test the effect of intracellular Ca^{2+} on osmotic water flow (Φ_{H_2O}) across toad urinary bladder (Figure 11), which, according to several reports in the literature (Parisi et al., 1987), should be depressed by a Ca^{2+} influx. The bladder was exposed to diluted Ca^{2+} solutions. After activating Φ_{H_2O} with oxytocin, 100 nM Ag^+ was added to the mucosal bath, which depressed Φ_{H_2O}. The latter observation is in agreement with the hypothesis that Ag^+ opens a pathway for Ca^{2+} entry and enables a Ca^{2+} influx. This was confirmed by blocking Ca^{2+} entry with 100 μM La^{3+}, which partly reversed the inhibition of Φ_{H_2O}. The hormonal activation of Φ_{H_2O} was also studied in bladders exposed to Ca^{2+} free mucosal solutions (Van Driessche et al., 1990). Similar increases in Φ_{H_2O} were recorded as in Ca^{2+}-containing solutions. However, the addition of Ca^{2+} to the mucosal side reduced Φ_{H_2O} significantly. This observation suggests that also in the absence of Ag^+ treatment a finite influx of Ca^{2+} occurs which elevates cellular Ca^{2+} modulating Φ_{H_2O}.

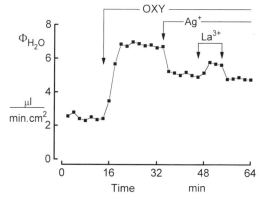

Figure 11. Effect of Ca^{2+} entry on the osmotic water flow (Φ_{H_2O}) through toad urinary bladder. Mucosal solution (in mM): 9 mM $Ca(NO_3)_2$, 3 TEA/Hepes. Φ_{H_2O} was activated with oxytocin (0.1 U/ml) and Ca^{2+} entry was enabled by adding 100 nM Ag^+ to the mucosal side. 100 μM La^{3+} blocked the Ca^{2+} influx and elevated Φ_{H_2O}.

Conclusion

The poorly selective channels in the apical membranes of epithelia constitute a special category because of their extreme sensitivity to external Ca^{2+} and other divalent cations. Many other nonselective channels reported in this volume are permeable for monovalent cations in the presence of physiological Ca^{2+} concentrations. The differences might find their origin in intracellular factors which modulate the sensitivity to external Ca^{2+}. This assumption seems reasonable, because even with physiological Ca^{2+} concentrations this pathway is also conducting monovalent cations in the epithelium of the larval frog skin (Cox and Alvarado, 1979). In skins of the adult frog the affinity for external Ca^{2+} is in the range of physiological intracellular concentrations. This could find its origin in the fact that Ca^{2+} can reach its binding site from the external side as well as from the cell interior. Therefore, it seems likely that Ca^{2+} can cross the channel. The Ca^{2+} transfer seems however to be modulated by intracellular affectors, protecting the cell for Ca^{2+} overload. After activation with oxytocin we could disrupt this protection mechanism by treatment of the apical surface with nanomolar concentrations of Ag^+ which opens the channels for divalent cation movement. This was confirmed by measurements of osmotic water flow which is reduced by elevation of intracellular Ca^{2+} concentration. Also, without treating the bladders with Ag^+, we could depress the osmotic water flow by addition of Ca^{2+} to the mucosal solution, suggesting Ca^{2+} entry through the apical membrane. Taken together, it seems that, with normal physiological Ca^{2+} concentrations, the poorly selective cation channels participate in the regulation of intracellular Ca^{2+}.

244

References

Aelvoet I, Erlij D, Van Driessche W (1988). Activation and blockage of a calcium-sensitive cation-selective pathway in the apical membrane of toad urinary bladder. J. Physiol. (London) 398:555–574.

Bear CE (1990). A nonselective cation channel in rat liver cells is activated by membrane stretch. Am. J. Physiol. 268:C421–C428.

Candia OA, Yorio T (1992). Water permeability response to vasopressin (AVP) of A6 cells grown on Anocell filter inserts. FASEB J. 6:A1194.

Civan MM (1983). Epithelial ions and transport. Application of biophysical techniques. In: Transport in Life Sciences. Bittar EE, editor. New York: John Wiley & Sons, vol. 4.

Cox TC, Alvarado RH (1979). Electrical and transport characteristics of skin of larval *Rana catesbeiana*. Am. J. Physiol. 237:R74–R79.

Crowe WE, Wills NK (1991). A simple method for monitoring changes in cell height using fluorescent microbeads and an Ussing-type chamber for the inverted microscope. Pflügers Arch. 419:349–357.

Cuthbert AW, Wong PYD (1972). The role of Ca^{2+} ions in the interaction of amiloride with membrane receptors. Mol. Pharmacol. 8:222–229.

Cuthbert AW, Wong PYD (1974). Calcium release in relation to permeability changes in toad bladder epithelium following antidiuretic hormone. J. Physiol. 241:407–422.

Das S, Palmer LG (1989). Extracellular Ca^{2+} controls outward rectification by apical cation channels in toad urinary bladder: patch-clamp and whole-bladder studies. J. Membrane Biol. 107:157–168.

Desmedt L, Simaels J, Van Driessche W (1991). Amiloride blockage of Na^+ channels in amphibian epithelia does not require external Ca^{2+}. Pflügers Arch. 419:632–638.

Desmedt L, Simaels J, Van Driessche W (1993a). Ca^{2+}-blockable poorly selective cation channels in the apical membrane of amphibian epithelia. I. UO_2^{2+} reveals two channel types. J. Gen. Physiol. 101:85–102.

Desmedt L, Simaels J, Van Driessche W (1993b). Ca^{2+}-blockable poorly selective cation channels in the apical membrane of amphibian epithelia. II. Tetracaine blocks the UO_2^{2+}-insensitive pathway. J. Gen. Physiol. 101:103–116.

De Wolf I, Van Driessche W (1986). Voltage dependent Ba^{2+} block of K^+ channels in the apical membrane of frog skin. Am. J. Physiol. 251:C696–C706.

Fabiato A, Fabiato F (1979). Calculator programs for computing the composition of solutions containing multiple metals and ligands used for experiments in skinned muscle cells. J. Physiol. 75:463–505.

Hillyard SD, Van Driessche W (1989). Effect of amiloride on the poorly selective cation channel of larval bullfrog skin. Am. J. Physiol. 256:C168–C174.

Krattenmacher R, Voigt R, Clauß W (1990). Ca-sensitive sodium absorption in the colon of *Xenopus laevis*. J. Comp. Physiol. B 160:161–165.

Levine SD, Schlondorff D (1984). The role of calcium in the action of vasopressin. Seminars in Nephrol. 4:144–158.

Marcus DC, Takeuchi S, Wangemann P (1992). Ca^{2+}-activated nonselective cation channel in apical membrane of vestibular dark cells. Am. J. Physiol. 262:C1423–C1429.

Martell AE, Smith RM (1974). Critical stability constants. In: Amino Acids. New York: Plenum, vol. 1.

McCarthy NA, O'Neil RG (1991). Calcium-dependent control of volume regulation in renal proximal tubule cells: II. Roles of dihydropyridine-sensitive and -insensitive Ca^{2+} entry pathways. J. Membrane Biol. 123:161–170.

Montrose-Rafizadeh C, Guggino WB (1991). Role of intracellular calcium in volume regulation by rabbit medullary thick ascending limb cells. Am. J. Physiol. 260:F402–F409.

Nagel W (1978). Effects of ADH upon electrical potential and resistance of apical and basolateral membranes of frog skin. J. Membrane Biol. 42:99–122.

Parisi M, Ibarra C, Porta M (1987). Intracellular Ca^{2+} concentration and antidiuretic hormone-induced increase in water permeability: effects of ionophore A23187 and quinidine. Biochem. Biophys. Acta 905:399–408.

Prabhu SD, Salama G (1990). The heavy metal ions Ag^+ and Hg^{2+} trigger calcium release from cardiac sarcoplasmic reticulum. Arch. Biochem. Biophys. 277:47–55.

Salama G, Abramson J (1984). Silver ions trigger Ca^{2+} release by acting at the apparent physiological release site in sarcoplasmic reticulum. J. Biol. Chem. 259:13363–13369.

Stein WD (1967). The movement of molecules across cell membranes. New York, London: Academic Press.

Van Driessche W (1986). Lidocaine blockage of basolateral potassium channels in the amphibian urinary bladder. J. Physiol. (London) 381:575–593.

Van Driessche W (1987). Ca^{2+} channels in the apical membrane of the toad urinary bladder. Pflügers Arch. 410:243–249.

Van Driessche W, Aelvoet I, Erlij D (1987). Oxytocin and cAMP stimulate monovalent cation movements through a Ca^{2+}-sensitive, amiloride-insensitive channel in the apical membrane of toad urinary bladder. Proc. Natl. Acad. Sci. USA 84:313–317.

Van Driessche W, Desmedt L, Simaels J (1991). Blockage of Na^+ currents through poorly-selective cation channels in the apical membrane of frog skin and toad urinary bladder. Pflügers Arch. 418:193–203.

Van Driessche W, De Smet P (1992). Cell volume expansion activates poorly-selective cation channels in the apical membrane of A6 cells. FASEB J. 6:A1195.

Van Driessche W, Erlij D, Aelvoet I (1990). Ca^{2+} entry through the apical membrane reduces antidiuretic-induced hydroosmotic response in toad urinary bladder. Pflügers Arch. 417:342–348.

Van Driessche W, Simaels J, Aelvoet I, Erlij D (1988). Cation-selective channels in amphibian epithelia: electrophysiological properties and activation. Comp. Biochem. Physiol. 90A: 693–699.

Van Driessche W, Zeiske W (1985a). Ca^{2+}-sensitive, spontaneously fluctuating, cation channels in the apical membrane of the adult frog skin epithelium. Pflügers Arch. 405:250–259.

Van Driessche W, Zeiske W (1985b). Ionic channels in epithelial cell membranes. Physiol. Rev. 65:833–903.

Völkl H, Paulmichl M, Lang F (1988). Cell volume regulation in renal cortical cells. Renal Physiol. Biochem. 3–5:158–173.

Wills NK, Millinoff LP (1990). Amiloride-sensitive Na^+ transport across cultured renal (A6) epithelium: evidence for large currents and high Na:K selectivity. Pflügers Arch. 416:481–492.

Wong SME, Chase HS (1988). Effect of vasopressin on intracellular [Ca] and Na transport in cultured toad bladder cells. Am. J. Physiol. 255:F1015–F1024.

Nonselective Cation Channels: Pharmacology, Physiology and Biophysics
ed. by D. Siemen & J. Hescheler
© 1993 Birkhäuser Verlag Basel/Switzerland

Nonselective Cation Channels in Cardiac and Smooth Muscle Cells

G. Isenberg

Department of Physiology, University of Cologne, Robert-Koch-Str. 39, D-50931 Köln, FRG

Summary
In cardiac and smooth muscle cells, nonselective cation channels can be activated by hormones and neurotransmitters, by cell stretch, and by changes in membrane potential. Activation of nonselective cation channels can depolarize the cell membrane, induce Ca^{2+} influx through voltage-gated Ca^{2+} channels and contraction. Activation of nonselective cation channels may trigger contraction even when membrane depolarization is absent or when voltage-gated Ca^{2+} channels are blocked, provided the Ca^{2+} permeability of these channels is sufficiently high.

Introduction

Nonselective cation channels are channels through which extracellular Na^+ and intracellular K^+ can permeate equally well. However, it is not only the permeability but also the electrochemical gradient that determines the ion flux. Since most cardiac and vascular myocytes have membrane potentials between -45 and -80 mV, there is more driving force for Na^+ influx (Na^+ equilibrium potential approx. $+70$ mV) than for K^+ efflux (K^+ equilibrium potential -90 mV). As a result, the net flux is inwardly directed at the resting potential. More positive potentials increase the driving force for K^+ efflux and reduce the one for Na^+ influx, and the current may reverse polarity. The reversal potential is thought to be close to zero mV for a permeability ratio $pNa^+:pK^+ \approx 1:1$; measurements of this value, or of its change upon substitution, e.g., Na^+ by impermeable n-methyl-glucosamine$^+$, can evaluate the permeability ratio for the channel of consideration.

The importance of nonselective cation channels for regulating contractile activity of cardiac and smooth muscle cells depends on their permeability for Ca^{2+} ions. The question of Ca^{2+} permeability has been addressed by experiments where all extracellular Na^+ ions were substituted by Ca^{2+} ions; the result was that some nonselective cation channels are Ca^{2+} permeable with a permeability ratio $pCa^+:pNa^+ \approx 1:1$ and other are not. In vivo, 150 mM $[Na^+]_0$ and 1.5 mM $[Ca^{2+}]_0$ are present simultaneously, and Ca^{2+} influx through channels that are not highly selective for Ca^{2+} seems to be low. However, recent evidence suggests (Wellner and Isenberg, this volume) that divalent Ca^{2+} ions bind inside the pore, thereby increasing the channel selectivity for Ca^{2+}. If that result is confirmed, Ca^{2+} influx through nonselective cation channels may be greater than the values extrapolated from the present measurements in isotonic $CaCl_2$ solutions.

The importance of Ca^{2+} influx through nonselective cation channels is suggested by measurements of ^{45}Ca influx or by measurements of the increase in the free cytosolic $[Ca^{2+}]_c$. In preparations that are not voltage-clamped (as in the $^{45}Ca^{2+}$ flux studies), the activation of nonselective cation channels depolarizes the membrane and voltage-gated Ca^{2+} channels (L-type) are activated when the potential of -35 mV is exceeded. This "secondary effect" by which nonselective cation channels can activate Ca^{2+} influx is suppressed when the L-type Ca^{2+} channels are blocked by Ca^{2+} channel antagonists like the dihydropyridine nifedipine. Vice versa, the nifedipine-insensitive Ca^{2+} influx can be attributed to the "direct" Ca^{2+} permeation through the nonselective channels (cf. van Breemen and Saida, 1989).

The efficacy of nonselective cation channels in depolarizing the membrane depends on many factors. For example, the amplitude of an excitatory junction potential (EJP) on the postsynaptic muscle membrane will be attenuated if the transmitter activates not only nonselective cation channels, but also K^+ channels in parallel. In this report, I address only those factors that arise from voltage-dependence of the ion flux. In some nonselective cation channels the channel activity (product of number of channels and open probability, NP_0) is independent of the membrane potential. Nevertheless, the current amplitude becomes disproportionally smaller when the membrane potential is changed from -50 to 0 mV and at, for example, $+50$ mV the current is of smaller amplitude than at -50 mV. The effect is termed "inward rectification" and may be attributed to a channel block by intracellular Mg^{2+} ions. Another class of nonselective cation channels changes its activity under the influence of membrane potential, e.g., hyperpolarization can both decrease (deactivate) and increase (activate) NP_0. Hyperpolarization-activated nonselective channels are involved in the generation of the pacemaker depolarization as it occurs in a variety of smooth muscle cells as well as in the pacemaker regions of the heart (SA-node, AV-node, Purkinje fibers). For some smooth muscle cells, pacemaker activity requires the convergence of several gating parameters, for example, in urinary bladder myocytes, nonselective cation channels can be modulated by membrane hyperpolarization in the presence of stretch.

In the field of cardiac and smooth muscle cells, studies of nonselective cation channels with whole-cell and single-channel recordings started only several years ago. There is a wide heterogeneity between the channels analyzed in different species (e.g., frog versus mammals) and tissues (e.g., mycocytes from bladder cells versus myocytes from coronary arteries), and the understanding of the channel function is far from being complete. Hence, this short review can give only an outline of what I consider to be important.

ATP-Gated Nonselective Channels (P_{2x}-Receptors)

In most mammalian systemic arteries, and some smooth muscles as in the vas deferens, stimulation of sympathetic nerves evokes excitatory junction potentials (EJPs) that can lead to action potentials and contractions (cf. Bolton, 1979). The EJPs have two components (cf. Burnstock, 1990), the slow depolarization is due to the release of noradrenaline (NA, see below) and the rapid one is gated by co-released ATP.

ATP is co-released from adrenergic nerve terminals. At the post-synaptic membrane of the smooth muscle cell ATP interacts with receptors that may either induce IP_3-formation and Ca^{2+} release from SR (P_{2y}-receptors) or open cation channels (P_{2x}-receptors; cf. Pearson and Gordon, 1989; Benham, 1990) that generate the fast component of the EJPs. These currents clearly are not due to activation of voltage-gated L-type Ca^{2+} channels because they can be induced at the resting potential (-60 mV), in the absence of $[Ca^{2+}]_0$, and in the presence of 10 μM nifedipine (Benham et al., 1987; Schneider et al., 1991). The amplitude of the current increases with $[ATP]_0$, the $K_{0.5}$ value is 2.3 μM, and the Hill-coefficient 1.8 (Inoue and Brading, 1990). The ATP-gated current peaks within less than 18 ms and, subsequently, desensitization causes the current to decay within <1 s, the decay accelerated with increasing $[ATP]_0$ (Inoue and Brading, 1990).

ATP-gated whole-cell currents depend on membrane potential with moderate inward rectification, i.e., the current amplitude became disproportionally smaller at potentials positive to -50 mV, and at positive potentials the outward current had a very small amplitude (Schneider et al., 1991). At physiological 150 mM $[Na^+]_0$ and 2 mM $[Ca^{2+}]_0$ the current reversal potential was close to zero mV, and it changed only slightly when extracellular Na^+ was substituted with Li^+, Cs^+, K^+, Ba^{2+} or Ca^{2+}. Therefore, it was concluded that ATP opens a nonselective cation channel (Benham et al., 1987; Schneider et al., 1991). The small changes in reversal potential upon ion substitution suggest that ATP-gated channels are permeable to Ca^{2+} with a permeability ratio $pCa:pNa = 3:1$ (Benham and Tsien, 1987; Schneider et al., 1991). Due to this low Ca^{2+} selectivity and different extracellular ion concentration (150 mM $[Na^+]_0$, 2 mM $[Ca^{2+}]_0$) less than 10% of the total ATP-gated current is carried by Ca^{2+} influx. Addition of 1.5 mM Ca^{2+} to a Ca^{2+} free 130 mM NaCl medium reduced the unitary current amplitude, probably by binding of Ca^{2+} ions inside the pore (Benham and Tsien, 1987). The result suggests that the assumptions (independence principle) used for the calculation of permeability ratios may have been incorrect. Hence, it was important that the Ca^{2+} permeability of the ATP-gated channel was confirmed by results where the inward current and the resulting increment in $[Ca^{2+}]_c$ were recorded simultaneously; the comparison of the transported charge with the approx. 1 μM rise in $[Ca^{2+}]_c$ yielded a permeability ratio $pNa^+:pCa^{2+}$ of approx. 1:2 (Benham, 1990; Schneider et al., 1991). In cells that are not held under voltage-clamp, activation of ATP-gated channels will depolarize the membrane, which in turn can activate Ca^{2+} influx through voltage-gated Ca^{2+} channels. However, the rise in $[Ca^{2+}]_c$ can activate BK-channels to such an extent that the expected depolarization is almost prevented (Schneider et al., 1991). In nonclamped vascular preparations ATP stimulated a $^{45}Ca^{2+}$ influx that was insensitive to Ca^{2+} channel antagonists, but was inhibited by the "antagonist" α,β-methylene-ATP, or by the Sandoz compound NCDC (Wallnover et al., 1989; Rüegg et al., 1989).

The ATP-gated channels were one of the first "receptor-operated Ca^{2+} channels" in smooth muscle cells that were analyzed on the single-channel level (rabbit ear artery, Benham and Tsien, 1987). Usually, one single patch contained a large number of ATP-gated channels that carried both mono- and divalent cations. Upon application of ATP to outside-out patches, single-channel currents rapidly appeared; they lasted hundreds of milliseconds and were interrupted by brief, flickering events. A detailed kinetic scheme is still missing because the rapid desensitization hampers the analysis. Since ATP activated the channels in isolated outside-out patches, the receptor channel coupling seems to occur without the involvement of readily diffusible second messengers. It is difficult, however, to exclude the possibility that a membrane-bound cofactor such as a G-protein is involved (Benham, 1990; Xiong et al., 1991).

ATP induces inward currents through nonselective channels in bullfrog atrial cells (Friel and Bean, 1988); the concentration-effect curves, the fast desensitization, and the permeability ratios were similar as described above for the smooth muscle cells. An ATP-gated current was also reported for cells from pregnant rat myometrium (Honoré et al., 1989). Different from the above reports, this ATP-gated current did not desensitize and did not rectify; Ca^{2+} and other divalent cations were impermeable and induced channel block. These different properties suggest that there might be more than one class of ATP-gated nonselective channels.

Noradrenaline-Activated Nonselective Channels (α_1-Receptors)

The slow component of the EJP has its counterpart in a nonselective inward current. Single vascular myocytes, voltage-clamped to -50 mV, respond to an 0.2 s pulse of NA with a 20-s long, noisy inward current, one part of which is due to a Ca^{2+}-activated chloride efflux, the other due to an influx of cations, mostly Na^+ ions (rabbit portal vein, Byrne and Large, 1988). Since the cation current was insensitive to 10 μM nifedipine the NA-activated channels are distinct from the voltage-gated L-type Ca^{2+} channels. Experiments on multicellular vascular preparations suggest that activation of contraction is related to binding of NA to α_1 receptors which leads to both Ca^{2+} influx and IP_3-induced Ca^{2+} release from intracellular stores. The Ca^{2+} from both sources has multiple secondary effects on channels such as activation of chloride and BK channels, modulation of nonselective channels and inactivation L-type Ca^{2+} channels.

On the single-channel level, NA-activated channels have not been studied up to now. The changes of whole-cell currents upon ion substitution (external NaCl replaced by $BaCl_2$) suggest that NA-activated channels have a higher permeability for Ba^{2+} or Ca^{2+} than for Na^+ ions. The permeability for Ca^{2+} is thought to mediate the Ca^{2+} influx at resting membrane potential of -65 mV (myocytes of the ear artery, cf. Suzuki 1989). The NA-activated, whole-cell current rectified inwardly, i.e., the current amplitude became disproportionately smaller at potentials positive to -50 mV, and at positive potentials the outward current had a very small amplitude (Wang and Large, 1991; Amédée et al., 1990). NA also activated the current in cells dialyzed with 10 mM EGTA, therefore, NA-induced Ca^{2+} release and increase in $[Ca^{2+}]_c$ is unlikely to gate the channel activation. However, 10 min of EGTA-dialysis abolished the current thus, a permissive role for $[Ca^{2+}]_c$ in the generation of the cation current was suggested (Wang and Large, 1991).

Muscarinic ACh-Receptors Activate Nonselective Channels Through a G-Protein

In a variety of mammalian intestinal muscles and some vessels acetylcholine (ACh) and carbachol induce graded depolarizations, action potentials, and contractions. The depolarization was attributed to an increase in Na^+ conductance (Bülbring and Kuriyama, 1963). With the development of the single-cell voltage-clamp experiments, ACh was discovered to activate an inward current through nonselective cation channels that desensitizes slowly with time (Benham et al., 1985). Since ACh-activated channels have a low density, single-channel currents could be recorded only in whole-cell configuration. The analysis yielded a 25 pS single-channel conductance for both Na^+ and K^+ ions (pNa:pK $\approx$ 1:0.3; Inoue et al., 1987; myocytes from jejunum). For stomach myocytes a 30 pS single-channel conductance and a permeability ratio pNa:pK $\approx$ 1:1 was evaluated (Vogalis and Sanders, 1990). The

Ca^{2+} permeability of the channel seems to be small but remains to be evaluated.

Voltage Dependence

The ACh-gated nonselective current is deactivated by membrane hyper-polarization (Benham et al., 1988; Inoue and Isenberg, 1990a). When the whole cell is clamped from -20 to -80 mV an instantaneous jump to a negative current peak (increased driving force) is followed by a monoexpoential decay; the steady current is approx. 20% of the peak. The deactivation became faster and more complete when the potential was more negative. Deactivation reduced the currents in such a way that the linear instantaneous i-v curve (reversal potential 0 mV) became a bell-shaped i-v curve with maximal inward currents at -20 mV. The deactivation at negative potentials makes the membrane potential an effective regulator of the ACh-induced depolarization, similar as in other types of receptor-operated currents.

Dependence on $[Ca^{2+}]_c$

In the absence of ACh, elevation of $[Ca^{2+}]_c$ by dialysis of 40 mM Ca-EGTA did not change the membrane conductance. Once the current was activated by ACh, it was augmented by $[Ca^{2+}]_c$ with a half maximal effect at 200 nM $[Ca^{2+}]_c$ (Inoue and Isenberg, 1990b). The $[Ca^{2+}]_c$ dependence has been used as an explanation for the facilitation of the ACh-gated current; preceding action potentials or depolarizing clamp steps increase the ACh-activated current several times, and the extent of facilitation correlates with the amount of preceding Ca^{2+} influx (Inoue and Isenberg, 1990b). The dependence of the muscarinic current on $[Ca^{2+}]_c$ has been proven for other types of smooth muscle cells (e.g., Pacaud and Bolton, 1991; Sims, 1992).

Coupling Through a G-Protein

When ACh binds to the muscarinic receptor the information is trans-ferred to the channel protein through a pertussis toxin-sensitive GTP-binding protein. Accordingly, the response is blocked by cell dialysis of 0.1 mM GDPβS, or vice versa, in the absence of the agonist thio-phos-phorylation with GTPγS induces a sustained inward current that has properties similar to the ACh-gated nonselective cation current. These results of Inoue and Isenberg (1991c) on myocytes from guinea-pig jejunum have been confirmed (Komori et al., 1992), however, for other tissues and other species the coupling through G-proteins has been questioned (Komori and Bolton, 1990). Experiments reconstituting the G-protein system in inside-out patches have not been done because of the low channel density.

Probably, the same muscarinic ACh receptor also passes information via a pertussis toxin-insensitive G-protein to a phospholipase C that activates formation of IP_3 and Ca^{2+} release from SR (e.g., Pacaud and Bolton 1991; Wang et al., 1992). The interaction of the two pathways constitute a positive feedback: once Ca^{2+} is released it facilitates the activity of ACh-gated channels and Ca^{2+} influx; augmented Ca^{2+} influx may re-load the SR with Ca^{2+} for the subsequent Ca^{2+} release. The ACh-activated Ca^{2+} release and the ACh activation of the nonselective currents are tightly linked; depletion of the stores of releasable Ca^{2+} nearly abolished the efficacy of ACh to induce the current (Sims, 1992).

Nonselective Channels Activated by Peptides

The peptides endothelin (ET) and arginine-vasopressin (AVP) are potent vasoconstrictors. In cultured aortic cells (line A7r5) AVP modulated the spontaneous electrical activity by IP_3-induced Ca^{2+} release (Ca^{2+} activation of BK channels, inhibition of L-type Ca^{2+} channels) and by the activation of an inward current (van Renterghem et al., 1988a). The channel was impermeable for anions but permeable for Na^+, K^+, Cs^+ and Ca^{2+}, i.e., it was a nonselective cation channel. The channel was not sensitive to changes in $[Ca^{2+}]_c$ or to Ca^{2+} channel antagonists.

Endothelin exerted effects similar to those described above for AVP; it triggered IP_3-induced Ca^{2+} release that secondarily inhibited L-type Ca^{2+} channels and activated BK channels. In addition, endothelin opened Ca^{2+} permeable, nonselective cation channels (van Renterghem et al., 1988b). A later study of the endothelin effect (aortic cells cultured for 2 days, Chen and Wagoner, 1991) suggested that the peptides endothelin-1, endothelin-3, AVP, and sarafotoxin S6b activated the same nonselective cation channel, e.g., sarafotoxin desensitized the cell to a subsequent application of endothelin. The endothelin-induced current could be carried by Na^+, K^+, Cs^+ (but not by $Tris^+$) with a current reversal potential near zero mV. The current was blocked by extracellular Co^{2+} or Ni^{2+}, but not by nifedipine. Removal of the extracellular Ca^{2+} abolished the endothelin-induced current, and neither Ba^{2+} nor Sr^{2+} could substitute for the Ca^{2+} effect. The results suggest that the endothelin-activated channel is modulated by $[Ca^{2+}]_0$.

Stretch-Activated Nonselective Channels (SACs)

SACs have been described for a variety of cell types (Guruhary and Sachs, 1984) including smooth muscle (Kirber et al., 1988; Davis et al., 1992; Wellner and Isenberg, this volume) and cardiac cells (Sigurdson et al., 1992). In the usual experiment, SACs are activated by application of a negative pressure (-2 to -4 kPa) to the open end of the patch electrode; the suction may stress the cytoskeleton that activates the channel by some unknown interaction (Sachs, 1986).

SACs in Cardiac Myocytes

This and other laboratories have experienced that ventricular myocytes hardly respond with activation of SACs when negative pressure is applied. One can speculate that SACs are suppressed with cell differentiation since SACs were regularly found in neonatal mammalian and in chick embryonic heart cells (cf. Sigurdson et al., 1992). In the latter preparation, a comparison of single-channel currents through SACs carried by either 150 mM NaCl or 110 mM $CaCl_2$ suggested a moderate Ca^{2+} permeability. Evidence for Ca^{2+} influx through SACs came also from the increase in $[Ca^{2+}]_c$ that was induced by stretch (Sigurdson et al., 1992). Cells of the SA-node resemble the embryonic cardiac tissue in many respects, and in SA-nodal cells SACs-like inward currents could be induced by either negative or positive pressure (Irisawa and Hagiwara, 1991).

In smooth muscle stretch is known to cause depolarizations, action potentials, and contractions (myogenic response, Bayliss, 1902; Harder, 1984). On the single-channel level, SACs were first described by Kirber et al. (1988) for myocytes from the stomach of *Bufo marinus*. SACs occurred at the density of 1 channel per 3 μm^2. In both cell-attached and inside-out patches the negative pressure was shown to increase channel activity, mainly by a decrease of the closed time durations. The SACs of smooth muscle cells are hyperpolarization-activated (see below). The open channel i-v curve follows a modest inward rectification and only changes silghtly when K^+ is substituted by Na^+ as the charge-carrying cation. The single current reversal potentials and their changes suggest that the permeabilities for Na^+, K^+, Ca^{2+}, and Ba^{2+} are very similar. Addition of 2 mM Ca^{2+} to 140 mM Na^+ significantly decreased the slope conductance and 10 μM Gd^{3+} blocked the channel.

Mammalian smooth muscle cells bear SACs with similar properties (Wellner and Isenberg, this volume). In guinea-pig urinary bladder myocytes SACs appeared at a density of 1 per 2 μm^2, and they were permeable for mono- and divalent cations, but not for anions. Once the channels were activated, activity could be enhanced by hyperpolarization. Differing from the situation with stomach myocytes (Hisada et al., 1991), hyperpolarization per se did not activate the channel. With K^+ ions as charge carrier, SACs had a conductance of 82 pS in absence and of 40 pS in presence of 2 mM Ca^{2+}. 20 μM Gd^{3+} blocked the channels. It was discussed that the stretch, exerted by the filling of the urinary bladder, would modulate the membrane potential through activation of SACs. Stretch activation of SACs could increase $[Ca^{2+}]_c$ and activate contraction in a nifedipine-insensitive way; Ca^{2+} influx through SACs leads to Ca^{2+} induced Ca^{2+} release. The nifedipine-sensitive pathway would be depolarization of the membrane leading to activation of Ca^{2+} influx through voltage-gated Ca^{2+} channels.

Recently, SACs of coronary myocytes were activated, not only by application of negative pressure, but also by stretching the cells by 5–30% above their slack length (Davis et al., 1992). Under current clamp, stretch elicited depolarizations of 5 to 40 mV and sometimes action potentials. Under whole-cell voltage clamp, with the cells in physiological salt solution, stretch induced an inward current that reversed at about -15 mV. The whole-cell stretch could significantly increase $[Ca^{2+}]_c$, even when the L-type Ca^{2+} channels were blocked by 5 μM nifedipine.

Hyperpolarization-Activated Nonselective Cation Channels

Currents activated by membrane hyperpolarization are known for a variety of smooth muscle and cardiac tissues; probably, they are based on more than one channel type. In *cardiac SA-nodal cells and Purkinje fibers* the current was initially described as a pacemaker K^+ current that deactivates with hyperpolarization (i_{K2}, cf. Noble, 1979). In 1981, DiFrancesco interpreted the current more correctly as a Na^+ current (i_F) that is activated by hyperpolarization (i_H by Yanagihara and Irisawa, 1980). The current may belong to the class of nonselective cation channels because measurements with ion-sensitive electrodes demonstrated a cation permeability in the rank of $K^+ > Na^+ > Cs^+$ ions (Glitsch et al., 1986). Under physiological conditions, however, when the potential is close to the K^+ equilibrium potential, the current is mostly due to Na^+ influx. Cs^+ blocks the i_H current (Isenberg, 1976; DiFrancesco, 1982). Recent single-channel analysis yielded the low conductance of 1 pS, a long lifetime of the open state, and a channel density between 0 and 12 per μm^2 (DiFrancesco and Tromba, 1989).

Voltage Dependence

The gating of i_H can be described with a simple two-state model. Hyperpolarization activates with a forward rate constant of 0.2 s^{-1} at

$-50\,\text{mV}$ and $1\,\text{s}^{-1}$ at $-80\,\text{mV}$. Deactivation occurs with $0.5\,\text{s}^{-1}$ at $-50\,\text{mV}$ and $0.1\,\text{s}^{-1}$ at $-80\,\text{mV}$. As a result, pacemaker depolarizations starting from $-80\,\text{mV}$ (as in Purkinje fibers) are much more dependent on i_H than those starting from $-60\,\text{mV}$ (as in the SA-node). In SA-nodal cells elevation of $[\text{Ca}^{2+}]_c$ augments i_H, the amplitude is increased, and the activation threshold shifted by $+13\,\text{mV}$ (Hagiwara and Irisawa, 1989), confirming previous results on the $[\text{Ca}^{2+}]_c$ modulation of i_{K2} (Isenberg, 1977) within the framework of the new i_H interpretation. Similar shifts are produced also by β-agonists (isoproterenol) or by second messenger (cAMP), with the result that pacemaker depolarization becomes faster and sinus rhythm more frequent. Acetylcholine may effect i_H by reduction of cAMP ("antiadrenergic effect").

A hyperpolarization-activated pacemaker current was also reported for *some smooth muscle cells*. In myocytes of the rabbit ileum, i_H was activated in the voltage range between -60 and $-110\,\text{mV}$. At physiological ion concentrations, the current had a reversal potential of $-25\,\text{mV}$, and ion-substitution experiments suggested that K^+ and Na^+ ions can permeate equally well. In *Bufo marinus* stomach myocytes, analysis of single-channel currents suggested a single-channel conductance of $40\,\text{pS}$ when both $145\,\text{mM}$ NaCl and $1.8\,\text{mM}$ CaCl_2 were present (Hisada et al., 1991). Many properties of these channels resembled those reported for the stretch-activated channels as there are the permeability ratios, the single-channel conductance or the reduction of conductance for monovalent cations by Ca^{2+} or Mg^{2+}. Vice versa, the activity of stretch-activated channels was increased by hyperpolarization (see Wellner and Isenberg, this volume). Hence, it has been suggested that the channel proteins may be identical (Hisada et al., 1991). As already discussed for the stretch-activated channels, the hyperpolarization-activated channels contribute to the resting potential and the pacemaker depolarization that finally determines the frequency of the action potentials.

Sodium Channels in Smooth Muscle Cells

For cardiac myocytes, the existence and functional role of TTX-sensitive sodium channels is well established. In vascular and intestinal smooth muscle cells, Na^+ channels were thought to be absent because the action potentials depended on Ca^{2+} influx and were insensitive to tetrodotoxin. More recently, however, TTX-sensitive Na^+ inward currents have been described for myocytes from ureter, jejunum (Smirnov et al., 1992) or from myometrium. In the latter tissue, the density of TTX-sensitive Na^+ channels increases with the stage of gestation (Ohya and Sperelakis, 1991). It has been speculated that the fast Na^+ inward current could be important for the spread of the excitation that synchronizes the myometrial contraction during delivery.

In A7r5 cells, a muscle cell line from rat aorta, an epithelial-like Na channel has been described on the single-channel level (van Renterghem et al., 1991). The channel has a density of approx. 1 per 2 μm^2, has a relatively high selectivity for Na (pNa:pK > 11), a 10 pS conductance, and is voltage-independent. It is insensitive to TTX or amiloride, but can be blocked by > 10 μM phenamil. The channel was suggested to stay under metabolic control and to play a role in the balance of [Na$^+$] and membrane potential.

Leakage Channels and Background Current

Usually, the pacemaker-depolarization is attributed to a decline in K$^+$ conductance as well as to an increase of hyperpolarization-activated nonselective cation and Ca^{2+} conductances. However, these time-dependent ionic currents superimpose on a time-independent background current (leakage current) that provides the net current negative (cf. Noble, 1979). In SA-nodal cells the background current was measured after all other currents through channels or electrogenic transporters were blocked (Irisawa and Hagiwara, 1991). The amplitude of the background current increased with [Na$^+$]$_0$, but it was not sensitive to TTX. Ni^{2+}, Ba^{2+} or Ca^{2+} reduced the current and 10 μM Gd^{3+} suppressed it by 70%. With Na$^+$ as the main cation in the patch pipette, the single-channel conductance was approx. 20 pS. Surprisingly, Tris$^+$ was able to carry current through these channels with approx. 20% of the Na$^+$ permeability (Irisawa and Hagiwara, 1991).

Ca^{2+}-Activated Nonselective Cation Channels

These channels were discovered in primary-cultures from neonatal heart cells (Colquhoun et al., 1981). The channels had a conductance between 30 and 40 pS, they excluded anions and conducted mono- and divalent cations with poor selectivity. The channel gating was not appreciably affected by the membrane potential and activation required a rise of [Ca^{2+}]$_c$. In adult ventricular myocytes the Ca^{2+}-activated nonselective channels have been described recently (Ehara et al., 1988). Due to their low density of 0.04 per μm^2 they were difficult to find and to investigate. The single-channel conductance was 15 pS and the reversal potential approx. 0 mV; substitution of Na$^+$ by K$^+$, Li$^+$ or Cs$^+$ on either side of the membrane did not change these parameters, suggesting a similar permeability for these ions. The channel did not show voltage-dependent gating. However, the channel activity increased with [Ca^{2+}]$_c$ up to a maximal open probability at 10 μM [Ca^{2+}]$_c$. The [Ca^{2+}]$_c$-activation curve had a Hill-cofficient of 3 and a K$_{0.5}$ of 1.2 μM. These properties

258

suggest that the channel starts to activate at 0.3 μM $[Ca^{2+}]_c$. Hence, it can be responsible for some of the Ca^{2+} activated currents that can be recorded in the whole-cell clamp, and it may contribute to the plateau of the cardiac action potential. Ca^{2+}-activated nonselective cation channels have also been reported for a variety of other mammalian cells (see this volume), however, not in smooth muscle cells up to now.

References

Amédée T, Benham CD, Bolton TB, Byrne NG, Large WA (1990). Potassium, chloride and nonselective cation conductances opened by noradrenaline in rabbit ear artery cells. J. Physiol. (London) 423:551–568.

Bayliss WM (1902). On the local reaction of the arterial wall to changes in arterial pressure. J. Physiol. (London) 28:220–231.

Benham CD (1989). ATP-activated channels gate calcium entry in single smooth muscle cells dissociated from rabbit ear artery. J. Physiol. (London) 419:689–701.

Benham CD (1990). ATP-gated channels in vascular smooth muscle cells. Ann. N.Y. Acad. Sci. 603:275–286.

Benham CD, Bolton TB, Byrne NG, Large WA (1987). Action of externally applied adenosine triphosphate on single smooth muscle cells dispersed from rabbit ear artery. J. Physiol. (London) 387:473–488.

Benham CD, Bolton TB, Lang RJ (1985). Acetylcholine activates an inward current in single mammalian smooth muscle cells. Nature 316:345–347.

Benham CD, Tsien RW (1987). A novel receptor-operated Ca^{2+}-permeable channel activated by ATP in smooth muscle. Nature 328:275–278.

Bolton TB (1979). Mechanisms of action of transmitters and other substances on smooth muscle. Physiol. Rev. 59:606–718.

Bülbring E, Kuriyama H (1963). Effects of changes in ionic environment on the action of acetylcholine and adrenaline on the smooth muscle cells of the guinea-pig taenia coli. J. Physiol. (London) 166:59–74.

Burnstock G (1990). Purinergic mechanisms, an overview. Ann. N.Y. Acad. Sci. 603:1–18.

Byrne NG, Large WA (1988). Membrane ionic mechanisms activated by noradrenaline in cells isolated from the rabbit portal vein. J. Physiol. (London) 404:557–573.

Chen C, Wagoner PK (1991). Endothelin induces a nonselective cation current in vascular smooth muscle cells. Circ. Res. 69:447–454.

Colquhoun D, Neher E, Reuter H, Stevens CF (1981). Inward current channels activated by intracellular Ca in cultured cardiac cells. Nature 294:752–754.

Davis MJ, Donovitz JA, Hood JD (1992). Stretch-activated single-channel and whole cell currents in vascular smooth muscle cells. Am. J. Physiol. Cell. Physiol. 262:C1083–C1088.

DiFrancesco D (1981). The contribution of the 'pacemaker' current (i_f) to generation of spontaneous activity in rabbit sino-atrial node myocytes. J. Physiol. (London) 434:23–40.

DiFrancesco D (1982). Block and activation of the pace-maker channel in calf purkinje fibres: effects of potassium, caesium and rubidium. J. Physiol. (London) 329:485–507.

DiFrancesco D, Tromba C (1989). Channel activity related to pacemaking. In: Isolated Adult Cardiomyocytes (Vol II). Piper HM, Isenberg G, editors. Boca Raton, Florida: CRC, pp 97–115.

Ehara T, Noma A, Ono K (1988). Calcium-activated non-selective cation channel in ventricular cells isolated from adult guinea-pig hearts. J. Physiol. (London) 403:117–133.

Friel DD, Bean BP (1988). Two ATP-activated conductances in bullfrog atrial cells. J. Gen. Physiol. 91:1–27.

Glitsch HG, Pusch H, Verdonck F (1986). The contribution of Na and K ions to the pacemaker current in sheep cardiac Purkinje fibres. Pflügers Arch. 406:464–471.

Guharay F, Sachs F (1984). Stretch-activated single ion channel currents in tissue-cultured embryonic chick skeletal muscle. J. Physiol. (London) 352:685–701.

Hagiwara N, Irisawa H (1989). Modulation by intracellular Ca^{2+} of the hyperpolarization-activated inward current in rabbit single sino-atrial node cells. J. Physiol. (London) 409:121–141.

Harder DR (1984). Pressure-dependent membrane depolarization in cat middle cerebral artery. Circ. Res. 55:197–202.

Hisada T, Ordway RW, Kirber MT, Singer JJ, Walsh JV, Jr. (1991). Hyperpolarization-activated cationic channels in smooth muscle cells are stretch sensitive. Pflügers Arch. 417:493–499.

Honoré E, Martin C, Mironneau C, Mironneau J (1989). An ATP-sensitive conductance in cultured smooth muscle cells from pregnant rat myometrium. Am. J. Physiol. 257:C297–C305.

Inoue R (1991). Effect of external Cd^{2+} and other divalent cations on carbachol-activated non-selective cation channels in guinea-pig ileum. J. Physiol. (London) 442:447–463.

Inoue R, Brading AF (1991). Human, pig and guinea-pig bladder smooth muscle cells generate similar inward currents in response to purinoceptor activation. Br. J. Pharmacol. 103:1840–1841.

Inoue R, Isenberg G (1990). Effect of membrane potential on acetylcholine-induced inward current in guinea-pig ileum. J. Physiol. (London) 424:57–71.

Inoue R, Isenberg G (1990). Intracellular calcium ions modulate acetylcholine-induced inward current in guinea-pig ileum. J. Physiol. (London) 424:73–92.

Inoue R, Isenberg G (1990). Acetylcholine activates nonselective cation channels in guinea pig ileum through a G protein. Am. J. Physiol. Cell. Physiol. 258:C1173–C1178.

Inoue R, Kitamura K, Kuriyama H (1987). Acetylcholine activates single sodium channels in smooth muscle cells. Pflügers Arch. 410:69–74.

Irisawa I, Hagiwara N (1991). Pacemaker mechanism in the isolated rabbit sinoatrial node cells. Presence and role of a background current. In: The Proceedings of the 18th International Symposium of Cardiovascular Electrophysiology, February 19, 1991, Seoul, Korea, pp 57–64.

Isenberg G (1976). Cardiac purkinje fibres. Caesium as a tool to block inward rectifying potassium currents. Pflügers Arch. 365:99–106.

Isenberg G (1977). Cardiac purkinje fibres. $[Ca^{2+}]_i$ controls the potassium permeability via the conductance components g_{K1} and g_{K2}. Pflügers Arch. 371:77–85.

Kirber MT, Walsh Jr JV, Singer JJ (1988). Stretch-activated ion channels in smooth muscle: a mechanism for the initiation of stretch-induced contraction. Pflügers Arch. 412:339–345.

Knot HJ, De Ree MM, Gähwiler BH, Rüegg UT (1991). Modulation of electrical activity and of intracellular calcium oscillations of smooth muscle cells by calcium antagonists, agonists, and vasopressin. J. Cardiovasc. Pharmacol. 18(Suppl. 10):S7–S14.

Komori S, Bolton TB (1990). Role of G-proteins in muscarinic receptor inward and outward currents in rabbit jejunal smooth muscle. J. Physiol. (London) 427:395–419.

Komori S, Kawai M, Takewaki T, Ohashi H (1992). GTP-binding protein involvement in membrane currents evoked by carbachol and histamine in guinea-pig ileal muscle. J. Physiol. (London) 540:105–126.

Noble D (1979). The Initiation of the Heartbeat, 2nd edn. Oxford: Clarendon Press.

Ohya Y, Sperelakis N (1989). Fast Na^+ and slow Ca^{2+} channels in single uterine muscle cells from pregnant rats. Am. J. Physiol. 257:C408–C412.

Pacaud P, Bolton TB (1991). Relation between muscarinic receptor cationic current and internal calcium in guinea-pig jejunal smooth muscle cells. J. Physiol. (London) 441:477–499.

Pearson JD, Gordon JL (1989). P_2 purinoceptors in the blood vessel wall. Biochem. Pharmacol. 38:4157–4163.

Rüegg UT, Wallnöfer A, Weir S, Cauvin C (1989). Receptor-operated calcium-permeable channels in vascular smooth muscle. J. Cardiovasc. Pharmacol. 14(Suppl. 6):S49–S58.

Sachs F (1986). Biophysics of mechanoreception. Membr. Bichem. 6:173–195.

Schneider P, Hopp HH, Isenberg G (1991). Ca^{2+} influx through ATP-gated channels increments $[Ca^{2+}]_i$ and inactivates I_{Ca} in myocytes from guinea-pig urinary bladder. J. Physiol. (London) 440:479–496.

Sigurdson W, Ruknudin A, Sachs F (1992). Calcium imaging of mechanically induced fluxes in tissue-cultured chick heart: Role of stretch-activated ion channels. Am. J. Physiol. Heart Circ. Physiol. 262:H1110–H1115.

Sims SM (1992). Cholinergic activation of a non-selective cation current in canine gastric smooth muscle is associated with contraction. J. Physiol. (London) 449:377–398.

Smernov SV, Sholos AV, Shuba MF (1992). Potential-dependent inward currents in single isolated smooth muscle cells of the rat ileum. J. Physiol. (London) 454:549–571.

Suzuki H (1989). Electrical activities of vascular smooth muscles in response to acetylcholine. Asia Pacific J. Pharmacol. 4:141–150.

Van Breemen C, Saida K (1989). Cellular mechanisms regulating $[Ca^{2+}]_i$ smooth muscle. Annu. Rev. Physiol. 51:315–329.

Van Renterghem C, Lazdunski M (1991). A new non-voltage-dependent, epithelial-like Na^+ channel in vascular smooth muscle cells. Pflügers Arch. 419:401–408.

Van Renterghem C, Romey G, Lazdunski M (1988a). Vasopressin modulates the spontaneous electrical activity in aortic cells (line A7r5) by acting on three different types of ionic channels. Proc. Nat. Acad. Sci. USA 85:9365–9369.

Van Renterghem C, Vigne P, Barhanin J, Schmid-Alliana A, Frelin C, Lazdunski M (1988b). Molecular mechanism of action of the vasoconstrictor peptide endothelin. Biochem. Biophys. Res. Commun. 157(3):977–985.

Vigne P, Breittmayer J-P, Lazdunski M, Frelin C (1988). The regulation of the cytoplasmic free Ca2+ concentration in arotic smooth muscle cells (A7r5 line) after stimulation by vasopressin and bombesin. Eur. J. Biochem. 176:47–52.

Vogalis F, Sanders KM (1990). Cholinergic stimulation activates a non-selective cation current in canine pyloric circular muscle cells. J. Physiol. (London) 429:223–236.

Wallnöfer A, Cauvin C, Lategan TW, Rüegg UT (1989). Differential blockade of agonist- and depolarization-induced $^{45}Ca^{2+}$ influx in smooth muscle cells. Am. J. Physiol. 257:C607–C611.

Wang Q, Large WA (1991). Noradrenaline-evoked cation conductance recorded with the nystatin whole-cell method in rabbit portal vein cells. J. Physiol. (London) 435:21–39.

Wang X-B, Osugi T, Uchida S (1992). Different pathways for Ca^{2+} influx and intracellular release of Ca^{2+} mediated by muscarinic receptors in ileal longitudinal smooth muscle. Jpn. J. Pharmacol. 58:407–415.

Xiong Z, Kitamura K, Kuriyama H (1991). ATP activates cationic currents and modulates the calcium current through GTP-binding protein in rabbit portal vein. J. Physiol. (London) 440:143–165.

Yanagihara K, Irisawa H (1980). Inward current activated during hyperpolarization in the rabbit sinoatrial node cell. Pflügers Arch. 385:11–19.

Nonselective Cation Channels: Pharmacology, Physiology and Biophysics
ed. by D. Siemen & J. Hescheler
© 1993 Birkhäuser Verlag Basel/Switzerland

Physiology of Muscarinic Receptor-Operated Nonselective Cation Channels in Guinea-Pig Ileal Smooth Muscle

Ryuji Inoue and Shan Chen

Department of Pharmacology, Faculty of Medicine, Kyushu University, Fukuoka 812, Japan

Summary
Stimulations of autonomic nerves in smooth muscle often evoke both fast and slow excitatory junction potentials (EJPs), which are thought to involve activations of several distinct types of nonselective cation channels (NSC channels). The ACh-activated NSC channel in guinea-pig ileum ($I_{ns,ACh}$) is one probably responsible for the slow EJP and seems to undergo various regulations. This short paper will review the physiology of $I_{ns,ACh}$, with particular emphasis on its dynamic interactions with other physiologically important factors such as the membrane potential, $[Ca^{2+}]_i$ and pH.

It is broadly agreed that ACh is the principal excitatory neurotransmitter released from the cholinergic nerves in the gastrointestinal tract and that it participates in the regulation of gut motility. This is thought to occur by activating muscarinic receptors present on smooth muscle cells as well as presumably on other types of cells, in particular the interstitial cells of Cajal which probably contribute to the pacemaking of electrical activity in the gut muscle. By use of patch-clamp experiments, muscarinic receptor activations in this type of smooth muscle have been found to induce several distinct changes in the membrane permeability, including voltage-dependent Ca channels (Russel and Aaronson, 1990), Ca-dependent K channels (Cole et al., 1989), M-current (Sims et al., 1985), and NSC channels (Benham et al., 1985; Inoue et al., 1987; Vogalis and Sanders, 1990; Sims, 1992). Since maintained contraction to exogenously applied ACh in gut muscle depends exclusively on extracellular Ca and is very sensitive to blockers for voltage-dependent Ca channels (e.g., Brading and Sneddon, 1980), NSC channels seem to have particular functional importance in this muscle. This is because they are capable of depolarizing the membrane and increasing the frequency of spike discharge, and also because a close correlation has been found between the frequency of spike discharge and the extent of contraction in this muscle.

Early evidence that ACh increases transmembrane Na permeability of the longitudinal muscle of guinea-pig ileum came from experiments using flux measurements and membrane potential recordings with the microelectrode technique. As Bolton described in his review (1979), low concentrations of ACh produce small depolarizations with superimposed increased spike discharge, the amplitude and duration of which are reduced and prolonged, while higher concentrations lead to pronounced depolarization and cessation of spike discharge (depolarization block). The apparent reversal potential of this depolarization, which was calculated from the membrane potential and changes in the electrotonic potentials, does not agree with any of the equilibrium potentials estimated for K, Na, Ca, and Cl, thus suggesting the nonselective ionic nature of the depolarization. Later, it turned out by applying the patch-clamp technique to single smooth muscle cells dissociated from the rabbit jejunum that this depolarization occurs through induction of a sustained inward current upon muscarinic receptor activation (Benham et al., 1985). Experiments changing the holding potential revealed two important properties of this inward current: It can be carried by Na and K as well as possibly Ca, and it is voltage-dependent. These properties were further confirmed in terms of single-channel recording under a whole-cell clamped condition in which the input resistance exceeded ca. 5 GΩ and thereby allowed individual openings of unitary currents to be resolved (Inoue et al., 1987). The unit conductance of the ACh-activated NSC channel ($I_{ns,ACh}$) calculated in this way was 20–25 pS and its relative open probability measured as NP_o was potential-dependent: the reconstructed current as a product of the relative open probability (NP_o) and unit current amplitude (i) gave a similar inverted bell-shaped current-voltage relationship to that obtained from the whole-cell experiment (Figure 5C in Inoue et al., 1987). A similar experiment was attempted later in the canine stomach and confirmed this conclusion (Vogalis et al., 1990).

50% of the $I_{ns,ACh}$ Channels are Available Near the Resting Membrane Potential

By loading Cs-aspartate into the cell, the channel properties can be more unequivocally demonstrated, since other conductances which may be activated as a consequence of muscarinic receptor activation, e.g., Ca-dependent K and Cl currents, can be eliminated. Figure 1Aa shows a typical current-voltage relationship of the ACh-induced cationic inward current evaluated for a wide range of membrane potentials (for method see the figure legend). It is of an inverted bell-shape and reverses in polarity between 0 and 10 mV. The availability of the current is a function of the membrane potential which can be evaluated in a number of different ways (short pulse analysis, tail current analysis; Inoue and Isenberg, 1990b; slow ramp potential, this figure) and is a sigmoid curve of Boltzmann type. Half maximal activation occurs at about − 50 mV

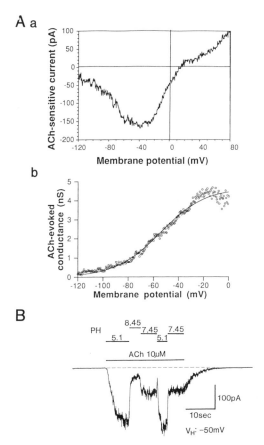

Figure 1. Aa: A typical current-voltage relationship of the ACh-evoked sustained inward current evaluated by slow rising ramp potentials. After the membrane had been held at −120 mV for 2 s, ramps of −120 mV to 80 mV were applied at a rate of 0.1 V/s. The curent-voltage relationship of the ACh-evoked inward current is defined as the difference between net. currents in the presence and absence of ACh 100 μM. Ab: The steady-state activation curve obtained from a. The best fit with a Boltzmann equation p = 4.6/ (1 + exp((Vm−Vh)/k)) gave in this case − 54 mV and − 17 mV for Vh and k, respectively. B: Prompt changes in external pH using a concentration jump technique caused a rapid and reversible change in the ACh-evoked current.

and is close to the resting membrane potential of this muscle (ca. − 55 mV). This suggests that about half of the $I_{ns,ACh}$ channels are available at this potential, where the influence of potential changes must be greatest. Indeed, a small hyperpolarization from the resting membrane potential has been demonstrated to strongly retard the time-course of ACh-evoked depolarization under current-clamped conditions (Figure 7Bb in Inoue and Isenberg, 1990b). It is therefore conceivable that in multicellular preparations slow oscillations in the membrane potential (slow waves) interact with this voltage-dependent mechanism and strongly alter the extent of ACh-evoked depolarization.

Both Static and Dynamic Changes in $[Ca^{2+}]_i$ Effectively Regulate the $I_{ns,Ach}$ Channel

The ACh-induced cationic inward currents have been found to be sensitive to changes in $[Ca^{2+}]_i$ (Inoue and Isenberg, 1990c). This means that although the currents are not primarily activated by a rise in $[Ca^{2+}]_i$ in the physiological range (ca. $0.1–1 \mu M$), the extent of their activation by ACh appears to be greatly enhanced by a rise in $[Ca^{2+}]_i$. The procedures which elevate $[Ca^{2+}]_i$ are all able to induce this effect. For example, activation of voltage-dependent Ca^{2+} influx by depolarizing the membrane is followed by a large transient potentiation of the ACh-evoked inward current, if it occurs before or during the application of ACh (Figures 2 and 3 in Inoue and Isenberg, 1990c). The main part of this effect is not due to the depolarization but to the Ca^{2+} influx itself, since organic and inorganic blockers for voltage-dependent Ca channels and internal dialysis with high concentrations of EGTA strongly attenuate the potentiation. In fact, a recent experiment using a fluorescent dye fura-2 demonstrated that the time-course of this potentiation coincides well with that of a transient increase in $[Ca^{2+}]_i$ caused by depolarization (Pacaud and Bolton, 1991). The potentiating effect is specific to Ca ions (Ba^{2+} and Sr^{2+} are ineffective) and can be graded by changing the duration of depolarizations or the amount of Ca^{2+} influx. The extent of potentiation is a nonlinear function of the amount of Ca charge entry, half maximal and submaximal potentiating effect being observed for $2–4$ and 10 pC, respectively. It is interesting that the former amount of Ca can be carried by a single action potential if we assume a mean input capacitance of 50 pF and a mean spike amplitude of $60–80$ mV. This suggests that whatever mechanism is involved dynamic $[Ca^{2+}]_i$ changes due to spike discharges strongly affect the extent of ACh-evoked depolarization. The prolonged duration of an action potential in the presence of low concentrations of ACh may partly be ascribed to this mechanism. It has also been pointed out that the release of stored Ca upon muscarinic receptor activation may be involved in potentiating the ACh-evoked depolarization indirectly via this mechanism. Thus, in ryanodine or caffeine-pretreated cells the size of the cationic inward current was significantly reduced (Inoue and Isenberg, 1990c), and inclusion of heparin in the patch pipette, which is known to prevent inositol 1,4,5-trisphosphate-triggered Ca release from the sarcoplasmic reticulum, resulted in only a small rise in $[Ca^{2+}]_i$ on exposure to ACh (Pacaud and Bolton, 1991).

When the $[Ca^{2+}]_i$ was clamped by internally dialyzing with Ca/EGTA mixtures via the patch pipette, the size of the ACh-evoked inward current was found to be a sigmoidal function of $[Ca^{2+}]_i$. The $[Ca^{2+}]_i$–$I_{ns,ACh}$ availability relationship spans the dynamic range of physiological $[Ca^{2+}]_i$: approximate $[Ca^{2+}]_i$ values for half maximal and submaximal activations are found near 100–200 nM and 1 μM, respectively, and these are very close to the resting and peak values of $[Ca^{2+}]_i$, indicating that change in $[Ca^{2+}]_i$ is a critical determinant of the ACh-evoked conductance or depolarization. However, prolonged dialysis over 10 min always reversed the effect to strong attenuation of the ACh-induced responses. This might suggest that more than one Ca-dependent process is involved in the regulation of $I_{ns,ACh}$ channel. It will be intriguing to investigate the mechanism responsible for such Ca-dependent regulations as well as to determine more precisely the difference in the proposed "Ca-activated" and "Ca-modulated" mechanisms, since ACh and other agonists are also proposed to increase the sensitivity of the contractile machinery to $[Ca^{2+}]_i$ by modifying intracellular enzymic activities (e.g., Kitazawa and Somlyo, 1990).

Extracellular pH and Ca^{2+} Are Also Very Effective Modulators of the I$_{ns,ACh}$ Channel Activity

In addition to the profound effects of the membrane potential and $[Ca^{2+}]_i$, the $I_{ns,ACh}$ channel appears to have various modulatory sites which are sensitive to extracellular protons and polyvalent cations. Figure 1B shows an example of extracellular pH effect on the $I_{ns,ACh}$ channel using a rapid soluton switching technique. Most noteworthy here is that the extracellular acidification augments the ACh-evoked inward current while alkalinization reduces it. This pH effect is unique to $I_{ns,ACh}$ channels and is quite the reverse of that usually observed for other types of channel (e.g., voltage-dependent Ca channels; Kaibara and Kameyama, 1988), implying that it is not due simply to a blocking action nor to titration of negatively charged residues distributed on the membrane (i.e., reduction in the surface potential), but presumably to a kind of allosteric modification of the channel conformation. The apparent pK for this effect is found around 7.4 and a ± 0.1 change in pH unit around the pK produces more than a 10% change in the ACh-evoked cationic conductance (R. Inoue and S. Chen, unpublished observation). Good agreement between the dynamic range of physiological pHs and the effective range of pH on the $I_{ns,ACh}$ channel emphasizes the significant role of protons in the physiological situation.

Divalent cations Zn^{2+}, Cd^{2+}, Ni^{2+}, Mn^{2+}, and Co^{2+}, which are known to block voltage-dependent Ca channels, have been found to reduce the ACh-evoked cationic conductance (Inoue, 1991; R. Inoue and S. Chen, unpublished data for Zn^{2+} action). The mode of this inhibitory action is almost voltage-independent with little change in reversal potential or sensitivity to ACh, and the apparent sequence of potency is Zn^{2+} ($IC_{50} = 38 \mu M$) > Cd^{2+} ($98 \mu M$) > Ni^{2+} ($131 \mu M$) $\gg$ Co^{2+} ($700 \mu M$) > Mn^{2+} ($1000 \mu M$). In contrast, extracellular Ca^{2+} and La^{3+}, chemical properties of which resemble Ca in many respects, augmented the ACh-evoked cationic conductance (Inoue, 1990; 1991). This effect is related to a shift of the steady-state activation curve toward more negative potentials and the most effective range of the extracellular Ca^{2+} action seems to be between 0.1 and 10 mM.

The target sites of pH and polyvalent cations are likely to exist, not on the muscarinic receptor, but on the external side of the separate channel protein. This is because $GTP\gamma S$-induced inward cationic currents, which has been suggested to be activated via a PTX-sensitive G-protein, and they possess identical properties to the currents flowing through the $I_{ns,ACh}$ channels (Inoue and Isenberg, 1990a; see also G. Isenberg, this volume), and are also sensitive to the modulatory effects of these cations (Inoue and Chen, unpublished observation).

Conclusions

Figure 2 schematically illustrates what could occur during the first few seconds after ACh or other spasmogenic substances have reached the

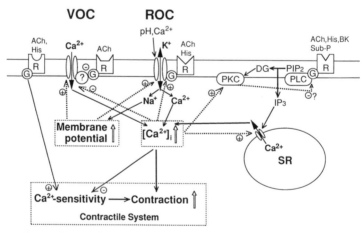

Figure 2. A simplified model of interactions between transmembrane channels and the sarcoplasmic reticulum on muscarinic receptor activation in guinea-pig ileum. VOC: voltage-dependent Ca channels, ROC: receptor-operated NSC channels, R: receptors, G: GTP-binding proteins, PLC: phospholipase C, PKC: protein kinase C, PIP_2: phosphatidylinositol 4,5-bisphosphate, IP_3; inositol 1,4,5-trisphosphate, DG: diacylglycerol, ACh: acetylcholine, His: histamine, Sub-P: substance P, BK: bradykinin.

muscarinic receptors on the smooth muscle membrane of guinea-pig ileum. The scheme is incomplete in that it does not incorporate a full set of elements regulating the contractile state of this muscle, but only some which contribute to increase $[Ca^{2+}]_i$ and develop tension, i.e., voltage-dependent Ca channels, NSC channels, and the sarcoplasmic reticulum. Nevertheless, it is readily understood that instead of operating independently the elements involved seem to constitute a complicated network of numerous mutual interactions via, for example, changes in the membrane potential and $[Ca^{2+}]_i$. The extent and contribution of each element may vary at each moment as the fraction of stimulants occupying the receptors changes with time and the effector systems involved are activated and desensitized with different kinetics. Although we are now far from fully understanding the mechanism of cholinergic control of the tone of this muscle, future experiments designed to consider such dynamic aspects of interactions promise to advance our knowledge.

Acknowledgement
We are grateful to Dr. A. F. Brading, University Department of Pharmacology, Oxford, for improving our manuscript.

References

Benham CD, Bolton TB, Lang RJ (1985). Acetylcholine activates an inward current in single mammalian smooth muscle cell. Nature 316:345–347.

Bolton TB (1979). Mechanisms of action of transmitter and other substances on smooth muscle. Physiol. Rev. 59:606–718.

Brading AF, Sneddon P (1980). Evidence for multiple sources of calcium for activation of the contractile mechanism of guinea-pig taenia coli on stimulation with carbachol. Br. J. Pharmacol. 70:229–240.

Cole WC, Carl A, Sanders KM (1989). Muscarinic suppression of Ca^{2+}-dependent K current in colonic smooth muscle. Am. J. Physiol. 257:C481–C487.

Inoue R (1990). Lanthanum augments ACh-induced inward current through modification of voltage-dependent gating in isolated guinea-pig ileum. J. Physiol. 430:119P.

Inoue R (1991). Effect of external Cd^{2+} and other divalent cations on carbachol-activated nonselective cation channels in guinea-pig ileum. J. Physiol. 442:447–463.

Inoue R, Kitamura K, Kuriyama H (1987). ACh activates single sodium channels in smooth muscle cells. Pflügers Arch. 410:69–74.

Inoue R, Isenberg G (1990a). ACh activates nonslective cation channels in guinea pig ileum through a G-protein. Am. J. Physiol. 258:C1173–C1178.

Inoue R, Isenberg G (1990b). Effect of membrane potential on ACh-induced inward current in guinea-pig ileum. J. Physiol. 424:57–71.

Inoue R, Isenberg G (1990c). Intracellular Ca ions modulate ACh-induced inward current in guinea-pig ileum. J. Physiol. 424:73–92.

Kaibara M, Kameyama M (1988). Inhibition of the calcium channel by intracellular protons in single ventricular myocytes of the guinea-pig. J. Physiol. 403:621–640.

Kitazawa T, Somlyo AP (1990). Desensitization and muscarinic re-sensitization of force and myosin light chain phosphorylation to cytoplasmic Ca^{2+} in smooth muscle. Biochem. Biophys. Res. Comm. 172:1291–1297.

Pacaud P, Bolton TB (1991). Relation between muscarinic receptor cationic current and internal calcium in guinea-pig jejunal smooth muscle cells. J. Physiol. 441:477–499.

Russel SN, Aaronson PI (1990). Carbachol inhibits the voltage-gated calcium current in smooth muscle cells isolated from the longitudinal muscle of the rabbit jejunum. J. Physiol. 426:23P.

Sims SM, Singer JJ, Walsh JV (1985). Cholinergic agonists suppress a potassium current in freshly dissociated smooth muscle cells of the toad. J. Physiol. 367:503–529.

Sims SM (1992). Cholinergic activation of a non-selective cation current in canine gastric smooth muscle is associated with contraction. J. Physiol. 449:377–398.

Vogalis F, Sanders KM (1990). Cholinergic stimulation activates a non-selective cation current in canine pyloric circular muscle cells. J. Physiol. 429:223–236.

Nonselective Cation Channels: Pharmacology, Physiology and Biophysics
ed. by D. Siemen & J. Hescheler
© 1993 Birkhäuser Verlag Basel/Switzerland

Nonselective Ion Pathways in Human Endothelial Cells

Bernd Nilius, Guy Droogmans, Marion Gericke, and Gero Schwarz

KU Leuven, Campus Gasthuisberg, Department of Physiology, B-3000 Leuven, Belgium

Summary
Four probably different transmembrane pathways are described in human endothelial (EN) cells that are all nonselective for cations.

i) A nonselective cation channel that is more permeable for Na^+ and K^+ than for Ca^{2+} can be gated by agonists such as histamine. This channel provides an agonist-gated entry route for Ca^{2+} into EN cells with a single-channel conductance of 25 pS for Na^+, K^+, and approximately 4 pS for Ca^{2+} (110 mM).

ii) Another Ca^{2+}-permeable pathway can be activated by shear stress. This supposedly mechanically activated channel is more permeable for divalent than for monovalent cations and provides mechano-sensing properties to EN cells.

iii) A third ionic current, activated by the selective Ca^{2+}-ATPase blocker thapsigargin, seems to be related to Ca^{2+}-release from Ca^{2+}-stores in the endoplasmic reticulum. In EN cells, this Ca^{2+}-entry route is cation selective, but cannot differentiate between Na^+ and K^+. Activation of this nonselective current is associated with an increase in intracellular Ca^{2+}. We therefore assume a Ca^{2+}-entry through this thapsigargin-activated pathway.

iv) A nickel-blockable, Ca^{2+}-permeable, nonselective leak is described that is present in nonstimulated EN cells.

It will be discussed whether agonist-gated channels and leak channels might be related to the Ca^{2+}-release activated Ca^{2+}-entry mechanism.

Introduction

Endothelial cells lining the inner surface of blood vessels act as a transducing surface for many physiological functions such as control of the contractile state of the underlying smooth muscle cells, proliferation of these cells, modulation of the function of white blood cells, platelets or constituents of plasma. They synthesize and release endothelial active molecules such as proteins, e.g., platelet-derived growth factor and interleukins, peptides, e.g., endothelins, as well as smaller products, e.g., prostaglandins and especially endothelium-derived relaxation factor (EDRF). Other secreted vasoactive compounds (endothelium-derived contracting factors different from endothelins and endothelium-derived hyperpolarization factor) are not yet chemically characterized. Most of the endothelial functions are controlled by intracellular Ca^{2+} (Newby and Henderson, 1991, for review), for example: EDRF is synthesized

270

from the precursor L-arginine by a Ca^{2+}-dependent NO-synthase. Another Ca^{2+}-dependent mechanism has been described for the Ca^{2+}-dependent enzyme lyso-PAF-acetyl-transferase which triggers the synthesis of platelet-activating factor, PAF.

Therefore, transmembrane pathways for Ca^{2+} or activation of ion channels that modulate the driving force for Ca^{2+} are critically involved in all these biological important mechanisms. Ion channels in endothelium are difficult to study, however, and until now no detailed picture of these channels could be given. In this article, we will summarize nonselective transmembrane ion pathways that are activated by different stimuli and that could play a crucial role in controlling Ca^{2+}-entry pathways.

Agonist-Induced Nonselective Cation Channels in Endothelium

It is well known that vasoactive compounds such as histamine induce intracellular Ca^{2+}-signals in endothelial cells that trigger different cellular responses in neighboring cells. Figure 1 shows a measurement in which membrane currents and intracellular Ca^{2+} (Ca^{2+}_i) were simultaneously recorded. A voltage clamped single endothelial cell from human

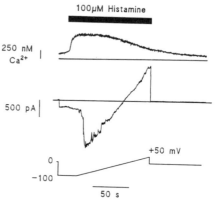

Figure 1. Application of a 2-min voltage ramp from -100 to $+50$ mV during application of 100 μM histamine. Only a small current response can be seen after stepping from 0 to -100 mV. An increase in the driving force induced only a very small increase in Ca^{2+}_i. A Ca^{2+} signal appears immediately after application of histamine that declines again by depolarizing the endothelial cell to $+50$ mV. With a latency of some seconds, histamine induced a large inward current at negative potentials. The current reversed near 0 mV. (Cells: cultured endothelial cells from human umbilical cord veins; for details see Jaffe et al., 1973; solutions: bath-(in mM): 140 NaCl, 1.5 or 10 $CaCl_2$, 5.9 KCl, 1.2 $MgCl_2$, 11.5 Hepes-NaOH, 10 glucose, titrated to pH 7.3, pipette: 100 Cs^+ or K^+ aspartate, 40 mM CsCl, 510 Na_2ATP, 5.5 $MgCl_2$, 10 Hepes, 0.1 EGTA buffered at pH 7.2 with CsOH, KOH; room temperature. Ca^{2+}_i-measurements: FURA II/AM loaded cells, slow sampling at 3 s^{-1}, whole-cell patch-clamp technique; see Neher, 1989. Electrophysiology: patch-clamp technique, fast sampling between 1 and 20 ms, 200–500 Hz filter, slow sampling 3 s^{-1}).

umbilical cord vein was stimulated by a linear voltage ramp. Five seconds after the voltage step from 0 mV to -100 mV histamine was applied to induce a Ca^{2+}-signal. With a latency of approximately 10 s after application of histamine, a current was activated that reversed near 0 mV and amounted to approximately 1 nA at -100 mV. The Ca^{2+}-transient was decreased by changing the membrane potential towards more positive values.

Figure 2A shows the ionic currents activated by histamine. A single EN cell was clamped at different potentials. At the time indicated, 100 μM histamine were applied. After a latency of some seconds a current appeared that reversed near 20 mV. In cell-attached patches, application of histamine to the bath induced single-channel activity. The unitary currents reversed near 0 mV. Under the conditions used (see legend of Figure 1, 140 K^+ inside, 6 K^+ outside, 140 Na^+ outside, 20 Na^+ inside, 40 Cl^- inside, 140 Cl^- outside), we measured a reversal potential of 3.2 ± 3.1 mV (n = 8) for whole-cell currents, 0.3 ± 0.4 mV (n = 4) in excised patches.

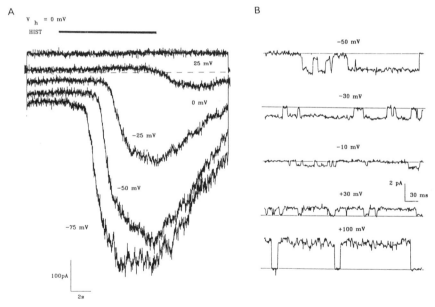

Figure 2. Histamine induces activity of a nonselective cation channel. A: A single endothelial cell is clamped at a holding potential of 0 mV. A 20 s step to different holding potentials (starting at -75 mV, 25 mV spaced) is applied. After 2 s, 100 μM histamine are applied to the surface of the cell through a multibarreled perfusion pipette. Development of an inward current can be seen depending on the membrane potential. The size of the current is 650 pA at -75 mV. This current is not a chloride current that would reverse near -40 mV. The reversal potential is at approximately $+20$ mV (whole-cell patch-clamp mode, fast sampling with 2 ms interval, 250 Hz filter). B: Single-channel recordings after application of histamine to the bath (cell-attached patch, bath solution in the pipette, 1 kHz sampling, 500 Hz filter). The single-channel conductance in this experiment is 26 pS. The reversal potential is 0 mV.

This current cannot be carried by Cl^- because the expected reversal potential would be close to -40 mV. Thus, histamine activated a nonselective cation channel. The permeation ratio of this channel was $P_{NA}:P_K:P_{Ca} = 1:0.9:0.2$. The single-channel conductance was approximately 25 pS for Na^+ and K^+, 4 pS for Ca^{2+} (110 mM Ca^{2+}, for details see Nilius and Riemann, 1990; Nilius, 1990; Nilius, 1991a, b). This agonist- or receptor-activated channel is Ca^{2+}-permeable (but less than reported by Yamamoto et al., 1992). Single-channel currents can be described by a two barrier – one binding site model (Weber and Siemen, 1989). From this description, we calculated an affinity to bind at a site within the channel of 35 mM, 40 mM, and 17 mM for Na^+, K^+, Ca^{2+}, respectively. The current is decreased by elevation of extracellular Ca^{2+}. This decrease can be explained by the much smaller conductance for Ca^{2+} than for monovalent cations. The channel can be down-regulated by phorbol esters such as PMA, TPA that all activate PKC.

The macroscopic currents can amount to the order of nano-amperes. However, the small conductance for Ca^{2+} and the low extracellular Ca^{2+}-concentration (Ca^{2+}_e) would only permit a small Ca^{2+}-influx via this nonselective cation channel. This influx could be in the range of 5 to 10 pA which is still enough to explain the Ca^{2+}_e-dependent plateau of the Ca^{2+}_i-signal after application of histamine.

Mechanically Gated Nonselective Ion Currents

Vascular endothelial cells are continuously exposed to the flow-induced changes in shear stress. It has already been demonstrated that these flow-induced changes affect a number of endothelial cell functions, such as i) synthesis of prostacyclin, ii) activation of a potassium channel, iii) production of tissue plasminogen activator, iv) changes in cytoskeleton and morphology, v) changes in pinocytosis, vi) modulation of the response of endothelial cells to vasoactive agonists, vii) changes in the cell proliferation, and viii) especially an increase in the synthesis of EDRF (for a review see Nollert et al., 1991). A direct influence of mechanical events on Ca^{2+} signaling could be an intriguing explanation for the modulation of different cell functions by shear stress.

We have measured, for the first time, membrane currents induced by shear stress together with intracellular calcium signals in human endothelial cells from umbilical cord veins. These transients are modulated by extracellular Ca^{2+} and are obviously different from the Ca^{2+} transients induced by vasoactive agonists. The shear-stress sensitive channel is obviously different from receptor-operated channels as discussed for the channel activated by histamine (Schwarz et al., 1992a).

In the presence of extracellular calcium (Ca^{2+}_e), shear stress induced an inward current at 0 mV holding potential that is associated with an intracellular Ca^{2+} transient. In the absence of extracellular calcium, shear stress was unable to evoke a calcium signal but still activates a membrane current.

Figure 3A shows the first simultaneous measurement of shear-stress-induced membrane currents and intracellular Ca^{2+}-signals in a single

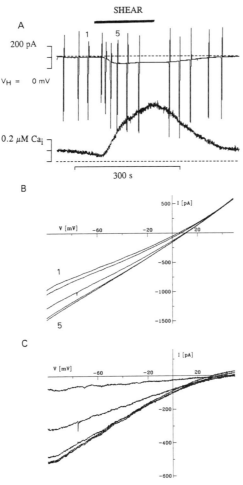

Figure 3. Shear stress induces Ca^{2+}-transients and inward currents in human vascular endothelial cells at 10 mM extracellular calcium. A: Simultaneous measurement of intracellular calcium and membrane currents. The bar indicates the application of shear stress to the surface of a single endothelial cell (approximately 10 dyn/cm^2). A holding potential of 0 mV was applied throughout the experiment (slow sampling rate of 3 Hz). B: Application of 500 ms linear voltage ramps from -100 to 50 mV (sampling rate 4 kHz). C: Shear-stress-induced currents were obtained by subtracting an averaged "control" ramp current (averaged over the 5 ramp currents before application of shear stress) from the 2nd to the 6th current during shear stress. The development of the shear stress induced current can be seen. The reversal potential obtained is close to 30 mV.

endothelial cell. At a holding potential of 0 mV, shear stress (approximately 10 dyn/cm^2) evoked an inward current together with a slowly developing Ca^{2+} transient. Shear stress was applied in the presence of 10 mM Ca$^{2+}_e$. Figure 3A shows slowly sampled records (3/s). During the experiment, 500 ms linear voltage ramps from -100 to $+50$ mV were applied. In Figure 3B membrane currents are shown which were evoked by voltage ramps at different periods before and during the mechanical stimulation (indicated by the bar, the fast blips in the slow sampled traces correspond to ramp currents and are successively numbered). From the trace before stimulation, a control current-voltage (IV) relation was constructed. This IV-curve was then subtracted from the currents during shear stress to obtain the mechanically induced currents. Figure 3C depicts these inward currents. The shear stress induced currents reversed at a potential E_{rev}, near to $+30$ mV. E_{rev} was significantly shifted from -2.3 ± 0.8 mV (n = 4) at 0 mM Ca$_e$, to $+1.5 \pm 1.6$ mV at 1.5 Ca$_e$ (n = 4), and to $+21.9 \pm 4.4$ mV (n = 7) at 10 mM Ca$_e$.

From these data it can be concluded that shear stress opened a nonselective cation channel that is, however, 12.5 ± 2.9 (n = 7) times more permeable for calcium than for sodium or cesium. We have shown that this pathway was not blocked by PKC activators. It was also permeable for Ni^{2+} and Ba^{2+}, but was reversibly blocked by La^{3+} and nonsteroid inflammation inhibitors such as mefenamic acid. Incubation of EN cells with modulators of the cytoskeleton such as cytochalasin B, increased the sensitivity of the cells to shear stress (Schwarz et al., 1992a, b).

Ionic Currents Activated by Intracellular Release of Ca^{2+}

Intracellular Ca^{2+} ions play a fundamental role in linking information from receptors or mechanosensors in the plasma membrane with various distinct cellular functions of EN cells. The initial rise in intracellular Ca^{2+} concentration upon receptor stimulation originates from the release of Ca^{2+} ions from intracellular stores sensitive to Ins(1, 4, 5)P$_3$. The activator Ca^{2+}, responsible for the more long-lasting effects on the different cell functions comes from the extracellular space, however. Nonexcitable cell types lack voltage-operated Ca^{2+} channels, but have developed a Ca^{2+}-entry mechanism, Ca^{2+}-release-activated Ca^{2+}-permeable channels (CRAC, Hoth and Penner, 1992), which appear to be coupled to the level of filling with Ca^{2+} of the internal stores: the mere emptying of these intracellular stores seems to increase the Ca^{2+} permeability of the plasma membrane (Putney, 1990; Jacob, 1990; Meldolesi et al., 1991). This novel mechanism must be distinguished from classical receptor-operated channels. A CRAC-related Ca^{2+}-entry mechanism is also present in EN cells. Figure 4A shows an experiment from which evidence for the existence of a CRAC-mechanism can be obtained: superfusion of a single EN cell with Ca^{2+}-free solution resulted in a decrease of Ca^{2+}_i. Under these conditions, histamine can still release Ca^{2+}. The Ca^{2+}-signal was short and showed no long-lasting plateau. The absence of extracellular Ca^{2+} prevented refilling of the intracellular stores. Reapplication of a high concentration of extracellular Ca^{2+} induced an increase in $[Ca^+]_i$ without any further application of a stimulus. $[Ca^{2+}]_i$ declined again when the intracellular stores were refilled in the presence of extracellular Ca^{2+}.

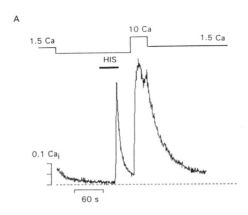

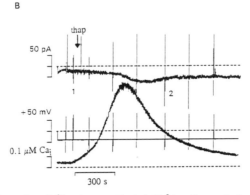

Figure 4. Evaluation of a Ca^{2+}-release activated Ca^{2+}-pathway in human EN cells. A: Discharge of Ca^{2+}-stores by histamine in Ca^{2+}-free medium permitted a Ca^{2+}-influx in human EN cells after increase of extracellular Ca^{2+}. Refilling of the stores in Ca^{2+}-containing extracellular medium decreases the Ca^{2+}-signal. B: Simultaneous measurement of the effects of $2\,\mu M$ thapsigargin on $[Ca^{2+}_i]$ and on the transmembrane currents. The top panel shows the transmembrane current measured at a holding potential of -40 mV. The artefacts in this recording are due to currents induced by repetitive applications of short voltage ramps. After application of thapsigargin (thap) an inward current is activated. Analysis of the voltage ramps (not shown) reveals a reversal potential of the currents activated by thapsigargin near 0 mV (perforated patch-clamp technique with application of the pore-forming nystatin).

The following approach was used to study CRAC activation in EN cells. Ca^{2+}-stores are refilled by action of a Ca-ATPase that can be specifically blocked by a tumor-promoting agent, the sesquiterpene lactone thapsigargin, extracted from the umbelliferous plant *Thapsia garcanica* (Takemura et al., 1989; Thastrup et al., 1991). Thapsigargin has been used as a pharmacological tool to study intracellular calcium pools that participate in the generation of agonist-induced Ca^{2+}-signals in nonexcitable cells. Thapsigargin increases Ca^{2+}_i by preventing refilling of normally leaking Ca^{2+}-stores in EN cells. Depletion of the stores can be directly monitored by application of agonists such as histamine.

We measured thapsigargin-induced changes in Ca^{2+}_i simultaneously with activation of a current in the same EC cell using the perforated patch whole-cell configuration (Figure 4B). The onset of a thapsigargin-induced current seemed to match the time-course of intracellular Ca^{2+}_i changes induced by thapsigargin. The current was measured at a holding potential of -40 mV which is close to the Cl^--equilibrium potential. From Figure 4A it is obvious that the development of the thapsigargin-induced inward current lags behind the changes in Ca^{2+}_i. Linear voltage ramps were applied in the absence and in the presence of thapsigargin to determine the voltage-dependence of this thapsigargin-activated current (shown as deflections in the current trace). The difference current (not shown), which represents the current activated by thapsigargin, reversed near 0 mV, i.e., -8 ± 3 mV (n = 5). Thus, thapsigargin activates a nonselective transmembrane pathway in human EN cells from umbilical vein. In EN cells these CRAC-related currents vary considerably in size. They range between a few to several hundreds of pA. At present, we do not have any detailed data on permeation properties or on mechanisms of modulation of this nonselective pathway. Activation of a very similar nonselective cation channel by bradykinin has been described in bovine endothelial cells (Mendelowitz et al., 1992). This pathway is permeable for Ca^{2+} and it seems to be related to a CRAC mechanism.

Efforts failed to detect a Ca^{2+}-selective entry channel, as recently described in mast cells and EN cells (Hoth and Penner, 1992; Lückhoff and Clapham, 1992). We argue that in EN cells CRAC have some properties in common with Ca^{2+}-permeable nonselective cation channels.

Leak Currents in Endothelial Cells

Activation of hyperpolarization inducing ion channels like Ca^{2+}-dependent K^+-channels would increase the driving force for a Ca^{2+}-influx in EN cells. However, sometimes the values reported for resting potentials in cultured endothelial cells are far from the potassium equilibrium potential. They are shifted towards the expected potassium equilibrium potential after application of various agonists such as ATP or bradykinin. This shift in the resting membrane potential increases the driving force for calcium. For such an intriguing mechanism a pathway for Ca^{2+} must be available.

We have tried to directly test this hypothesis of the existence of a "leak" Ca^{2+}-entry pathway by using a method of simultaneous voltage

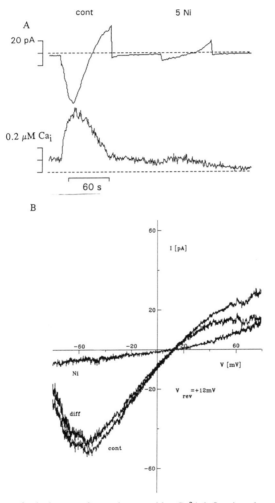

Figure 5. Existence of a leakage pathway that provides Ca^{2+}-influx into human EN cells. A: in voltage clamped cells (perforated patch-clamp technique, Korn and Horn, 1989) application of voltage ramps induced a Ca^{2+}-transient that follows the driving force. Application of 5 mM Ni^{2+} to the same cell completely abolished the Ca^{2+}-transient. B: Application of voltage ramps from -80 to $+80$ mV ($500\,\mu s$ sampling interval, 500 Hz filter) allows to construct current-voltage relations of the Ni^{2+}-blockable current. This currents again reverses near 0 mV.

clamp and intracellular Ca^{2+}-measurements. To avoid Ca^{2+} entry through a high resistance pathway between the rim of the pipette and the cell surface, we used the perforated patch-clamp technique. Pores formed by nystatin are impermeable for Ca^{2+} (Horn and Marty, 1988; Korn and Horn, 1989). In a series of experiments, we demonstrated that a high concentration of Ca^{2+} (2.5 mM) in the patch pipette did not affect intracellular Ca^{2+}. The changes in $[Ca^{2+}]_i$ by variation of the driving force disappeared in Ca^{2+}-free solutions. Therefore, changes in intracellular Ca^{2+} by variation of the membrane potential must be mediated by transmembrane pathways.

Figure 5A shows such an experiment: in a non-stimulated EN cell application of a voltage ramp from -80 to $+80$ mV induced an inward current simultaneously with an increase in Ca^{2+}_i. A significant correlation can be obtained between Ca^{2+}_i and the driving force for Ca^{2+}. Nickel ions (5 mM) completely and reversibly blocked the Ca^{2+}-transients and the measured currents.

Thus, the Ca^{2+}-influx pathway should be considered as a passive leak. We still do not know the nature of this pathway. By calculating the difference between the currents measured before and after application of Ni^{2+}, we again obtained reversal potentials near 0 mV. In four cells a reversal potential of 9 ± 6 mV for the current blocked by nickel was measured. Such a leak could be a candidate for providing an increased Ca^{2+} influx by cell hyperpolarization. It is intriguing to speculate whether already depleted stores in EN cells are responsible for generating a sustained leak through which Ca^{2+} could permeate, and only controlled by the electrochemically driving force. Block of this pathway by nickel – which also inhibits CRAC induced Ca^{2+}-entry – may suggest such a possibility (Jacob, 1990).

Conclusions

Ion channels are involved in the control of Ca^{2+}-signaling in almost each cell. In EN cells, however, we still have little information on their existence, properties, modulation, functional impact of ion channels and electrogenic transporters. At least four different mechanisms that all include ion transport through not very selective pathways have to be considered. Mechanically gated ion channels may provide a Ca^{2+}-entry pathway that is not highly selective for Ca^{2+} but seems to be more permeable for divalents than for monovalents. Presently, no details are available to further characterize these important mechano-sensing properties of EN cells. Nonselective ion channels gated by agonists that bind to surface receptors (R) seem to be – at least when activated by histamine – permeable for Ca^{2+}. These channels may provide a Ca^{2+}-entry route after stimulation of EN cells by agonists. G-protein activation

POSSIBLE NONSELECTIVE PATHWAYS

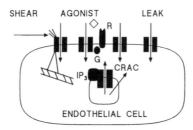

Figure 6. Synopsis of the nonselective pathways discussed in the test (see conclusions).

(G) might be involved in gating of these channels. A down regulation by PKC provides a negative feedback mechanism for Ca^{2+}-signaling. An intriguing mechanism indicates gating of Ca^{2+}-permeable ion channels by Ca^{2+}-release from Ca^{2+} stores in the endoplasmic reticulum (ER) via binding of Ins(1, 4, 5)P_3 to the Ins(1, 4, 5)P_3-receptor Ca^{2+}-release channel in the ER membrane. By an unknown mechanism, the ER store signals to a Ca^{2+}-permeable membrane channel in the plasma membrane to open (Ca^{2+}-release activated Ca^{2+} channels, CRAC). We have some indication that in EN cells this pathway also involves a nonselective cation channel.

Another pathway seems to be associated with a nickel-sensitive leak that allows Ca^{2+} ions to enter an EN cell following their driving force. It is intriguing to speculate whether the agonist and leak pathway are related to CRAC mechanisms. A synopsis of the described nonselective pathways is given in Figure 6.

References

Horn R, Marty A (1988). Muscarinic activation of ionic currents measured by a new whole cell recording method. J. Gen. Physiol. 92:145–159.

Hoth M, Penner R (1992). Depletion of intracellular calcium stores activates a calcium current in mast cells. Nature 355:353–356.

Jacob (1990). Agonist-stimulated divalent cation entry into single cultured human umbilical vein endothelial cells. J. Physiol. 421:55–77.

Jaffe EA, Nachman RL, Becker CG, Minick CR (1973). Culture of human endothelial cells derived from umbilical vein. J. Clin. Invest. 52:2745–2756.

Korn SJ, Horn R (1989). Influence of sodium-calcium exchange on calcium current rundown and the duration of calcium-dependent chloride currents in pituitary cells, studied with whole cell and perforated patch recording. J. Gen. Physiol. 94:789–812.

Lückhoff A, Clapham DE (1992). Inositol 1,3,4,5-tetrakisphosphate activates an endothelial Ca^{2+}-permeable channel. Nature 355:356–358.

Meldolesi J, Clementi E, Fasolato C, Zacchetti D, Pozzan T (1991). Ca^{2+} influx following receptor activation. Trends Pharmacol. Sci. 12:289–292.

Mendelowitz D, Bacal K, Kunze DL (1992). Bradykinin-activated calcium influx pathway in bovine aortic endothelial cells. Am. J. Physiol. 262:H942–H948.

Neher E (1989). Combined Fura-2 and patch clamp measurements in rat peritoneal mast cells. In: Neuromuscular Junction. Sellin LC, Libelius R, Thesleff S, editors. Amsterdam: Elsevier, pp. 65–76.

Newby A, Henderson AH (1990). Stimulus-secretion coupling in vascular endothelial cells. Ann. Rev. Physiol. 52:661–674.

Nilius B (1990). Permeation properties of a nonselective cation channel in human vascular endothelial cells. Pflügers Arch. 416:609–611.

Nilius B (1991a). Regulation of transmembrane calcium fluxes in endothelium. News Physiol. Sci. 6:110–114.

Nilius B (1991b). Ion channels and regulation of transmembrane Ca^{++} influx in endothelium. In: Electrophysiology and Ion Channels of Vascular Smooth Muscle and Endothelial Cells. Sperelakis N, Kuriyama H, editors. New York: Elsevier, pp. 317–325.

Nilius B, Riemann D (1990). Ion channels in human endothelial cells. Gen. Physiol. Biophys. 9:89–112.

Nollert MU, Diamond SL, McIntire LV (1991). Hydrodynamic shear stress and mass transport modulation of endothelial cell metabolism. Biotechnology and Bioengineering 38:588–602.

Takemura H, Hughes AR, Thastrup O, Putney JW Jr (1989). Activation of calcium entry by the tumor promotor Thapsigargin in parotid acinar cells. J. Biol. Chem. 264:12266–12271.

Thastrup O, Cullen PJ, Drobak BK, Hanley MR, Dawson AP (1991). Thapsigargin, a tumor promotor, discharges intracellular Ca^{2+} stores by specific inhibition of the endoplasmic reticulum Ca^{2+}-ATPase. Proc. Nat. Acad. Sci. USA 87:2466–2470.

Putney JW Jr (1990). Capacitative calcium entry revisited. Cell Calcium 11:611–624.

Schwarz G, Droogmans G, Callewaert G, Nilius B (1992a). Shear stress induced calcium transients in human endothelial cells from umbilical cord veins. J. Physiol. (London) 458:527–538.

Schwarz G, Droogmans G, Nilius B (1992b). Shear stress induced membrane currents in human vascular endothelial cells. Pflügers Arch. 421:394–396.

Yamamoto Y, Chen G, Miwa K, Suzuki H (1992). Permeability and Mg^{++} blockade of histamine-operated cation channel in endothelial cells from rat intrapulmonary artery. J. Physiol. (London) 450:395–408.

Weber A, Siemen D (1989). Permeability of the nonselective channel in brown adipocytes to small cations. Pflügers Arch. 414:564–570.

Nonselective Cation Channels: Pharmacology, Physiology and Biophysics
ed. by D. Siemen & J. Hescheler
© 1993 Birkhäuser Verlag Basel/Switzerland

Properties and Regulation of Human Platelet Cation Channels

Jörg Geiger and Ulrich Walter

Medizinische Universitätsklinik, Klinische Forschergruppe, D-97078 Würzburg, FRG

Summary
The stimulation of calcium influx by various human platelet agonists which differ in their activation pathways was investigated. ADP activates a receptor-operated cation channel (ROC) and stimulates a phospholipase C (PLC)/inositoltrisphosphate (IP_3)-mediated calcium mobilization associated with a secondary calcium influx. Thrombin only stimulates the PLC/IP_3-mediated calcium mobilization and associated calcium influx, perhaps followed by an additional phase of calcium influx. The platelet calcium response after incubation with the thromboxane A_2 mimetic U 46619 is similar but more transient compared to that after thrombin stimulation.

Tert-butylhydroquinone (an inhibitor of endoplasmatic reticulum Ca^{2+}-ATPases and cyclooxgenase) elevates cytosolic calcium levels by emptying intracellular calcium stores and stimulates a biphasic calcium influx. Activation of platelet cAMP- and cGMP-dependent protein kinases inhibits the ADP- and thrombin-evoked, calcium store-associated cation influx, but not the fast receptor operated cation influx induced by ADP. Experiments with various ADP-analogs, ATP and ATP-γ-S suggest that two different ADP-receptors may mediate the calcium responses in human platelets.

Introduction

Numerous hormones, drugs and vasoactive substances stimulate or inhibit the activation of human platelets. Activators such as thrombin, thromboxane A_2, vasopressin, platelet activating factor, ADP and collagen cause platelet shape change, adhesion, aggregation and degranulation. Most platelet agonists elevate cytosolic free calcium via activation of phospholipase C (PLC)/inositoltrisphosphate (IP_3)-induced calcium-release from intracellular stores and an associated store-related calcium influx followed by an activation of several protein kinases including myosin light chain kinase (MLCK), protein kinase C (PKC) and protein tyrosine kinases (Siess, 1989; Rink and Sage, 1990; Halbrügge and Walter, 1992). ADP is the only platelet agonist for which an additional rapid calcium influx mediated by a receptor-operated cation channel (ROC) has been established (Rink and Sage, 1990). Some platelet agonists also inhibit platelet adenylyl cyclase by a G_i-protein-dependent mechanism (Siess, 1989). In contrast, agents which stimulate either platelet cAMP/cAMP-dependent protein kinase (e.g., PG-I$_2$, PG-E$_1$, adenosine) or cGMP/cGMP-dependent protein kinase (e.g., endothelium-derived relaxing factor, NO, and other nitrovasodilators) are powerful inhibitors of platelet activation (Siess, 1989; Halbrügge and Walter, 1992). In this chapter we will discuss some properties and the regulation of cation (calcium) channels which are activated by platelet agonists, in particular, by ADP, thrombin, and thromboxane A_2.

Properties of Platelet Cation Channels

Platelet agonists differ in their regulation of calcium influx (Rink and Sage, 1990). ADP induces a fast calcium influx which is due to an ADP-receptor-operated cation channel. ADP-evoked calcium rise commences without a measurable delay (< 10 ms) and peaks within 200 ms (Sage et al., 1990; Sage and Rink, 1987). "Cell attached" patch-clamp studies (Mahaut-Smith et al., 1990) and "whole-cell" patch technique (Mahaut-Smith et al., 1992) provided evidence that this channel is receptor-operated and nonselective. Additional experiments with fluorescent indicators and "cell attached" patch-clamp studies demonstrated that this channel is permeable to Ca^{2+}, Na^+, Mn^{2+}, Ba^{2+}, and Sr^{2+} (Sage et al., 1991). The fast ADP-evoked cation influx is followed by a second phase of calcium influx with a delay of 200 ms (Sage et al., 1990). This entry mechanism appears to be more selective than the first phase of ADP-evoked calcium influx as it is only permeable to Ca^{2+}, Ba^{2+}, Sr^{2+}, and Mn^{2+} (Ozaki et al., 1992). In contrast to the first influx phase, the delay between ADP addition and cation influx of this secondary phase is temperature-sensitive.

Thrombin-evoked calcium influx commences after a delay of 400 ms and peaks within 1.5 s (Sage and Rink, 1987). Recent "whole-cell" patch-clamp studies indicated that this influx occurs through small conductance channels or by an electroneutral pathway (Mahaut-Smith, 1992). This channel appears to be identical with the channel responsible for the second phase of calcium influx after ADP stimulation. Our experiments indicate that there may be a second calcium influx which keeps the intracellular calcium concentration after stimulation at a sustained level, while the intracellular calcium concentration after ADP stimulation returns within 80–100 s to the basal level.

Stimulation of platelets with the thromboxane A_2 mimetic U 46619 induces a calcium influx which is similar to the thrombin-evoked calcium influx (Figure 1). However, the U 46619-induced calcium influx is of short duration, and the intracellular calcium level rapidly declines within 60 s.

A rise in platelet intracellular calcium can also be induced with the endoplasmatic reticulum (ER) Ca^{2+}-ATPase inhibitors tert-butylhydroquinone (t-BHQ) and thapsigargin. T-BHQ does not only inhibit ER Ca^{2+}-ATPase, but also the platelet cyclooxygenase which is responsible

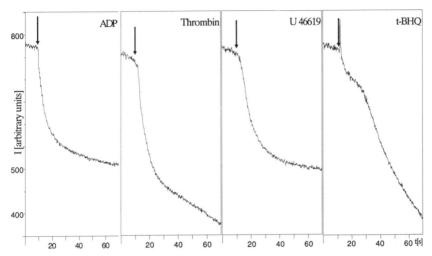

Figure 1. Calcium influx in human platelets induced by ADP, thrombin, U 46619, and tert-butylhydroquinone (t-BHQ) determined by manganese quenching of fura-2 fluorescence. Human platelets loaded with fura-2 were stimulated with either 40 μM ADP, 0.5 u/ml thrombin, 5 μM U 46619 (a stable thromboxane A_2-analog) or 200 μM tert-butylhydroquinone. The buffer solution contained 0.5 mM $MnCl_2$ and 0.5 mM $CaCl_2$. The fluorescence signal was collected at 360 nm for the time indicated in arbitrary units (I).

for the eicosanoid pathway in platelets (Brüne and Ullrich, 1991). Therefore, the main effect of t-BHQ is a fast rise of intracellular calcium level due to the mobilization of calcium from intracellular stores. Influx experiments (Figure 1) showed that t-BHQ can induce a biphasic calcium influx. The whole calcium signal induced by t-BHQ resembles that obtained with thrombin although the first phase of calcium influx is less strong.

Regulation of Platelet Cation Channels

Stimulators of platelet cAMP- and cGMP-dependent protein kinase are potent inhibitors of agonist-induced shape change, secretion, and aggregation. As these processes are coupled to a rise of intracellular calcium level, a crosstalk between the cGMP- and cAMP-second messenger systems and platelet calcium regulation is necessary. The action of some cAMP-dependent protein kinase activators has already been investigated (Sage and Rink, 1985). To exclude possible side-effects unrelated to protein kinase action, we incubated platelets with the highly specific and membrane permeant cGMP-analog 8-para-chlorophenylthio-cGMP (8-pCPT-cGMP) (Butt et al., 1992) and the cAMP-analog Sp-5,6-dichlorobenzimidazolriboside-3,5′-monophosphorothioate (5,6-DiCl-BIMPS) (Sandberg et al., 1991).

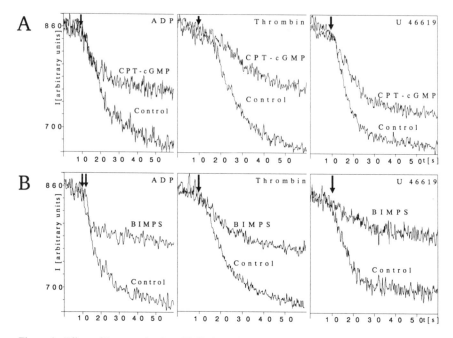

Figure 2. Effect of human platelet cGMP-dependent protein kinase (cGMP-PK) (upper row "A") and cAMP-dependent protein kinase (cAMP-PK) stimulation (lower row "B") on ADP-, thrombin-, and U 46619-evoked calcium influx. Fura-2 loaded human platelets were incubated for 10 min at 37°C with either 500 μM 8-p-CPT-cGMP (a selective, membrane permeant analog of cGMP) or 100 μM 5,6-Di-Cl-cBIMPS (a selective membrane permeant analog of cAMP) and thereafter stimulated with the agonists (indicated by arrows) in buffer solution containing 0.5 mM MnCl$_2$ and 0.5 mM CaCl$_2$. Fluorescence was measured at 360 nm.

Stimulation of either cAMP- or cGMP-dependent protein kinase inhibited mobilization of calcium from intracellular stores (Geiger et al., 1992). Furthermore, the rapid ADP-evoked calcium influx was not affected while the secondary store-associated calcium influx was inhibited (Figure 2). Although the thrombin-stimulated calcium mobilization was completely inhibited (data not shown), a small calcium influx was still observed (Figure 2). This is further evidence for another yet undefined calcium influx. Calcium influx induced by U 46619 was also inhibited by cAMP-/cGMP-protein kinase activation.

Ligand Specificity of ADP-Regulated Cation Channels

Three ADP-induced signaling pathways have been observed in human platelets: ADP-evoked calcium influx through a receptor-operated cation channel (Rink and Sage, 1990), activation of phospholipase C and subsequent calcium mobilization from intracellular stores, and

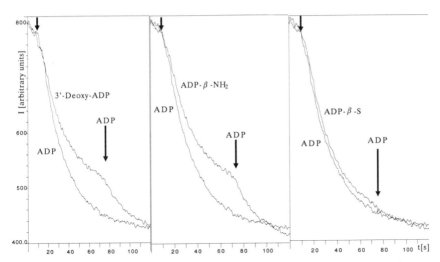

Figure 3. Effects of the ADP analogs 3′-desoxy-ADP, ADP-β-NH$_2$ and ADP-β-S on human platelet calcium influx. Fura-2 loaded human platelets were first stimulated with ADP (40 μM) or the ADP analog (40 μM) as indicated (first arrow) and then again in both cases with 40 μM ADP (second arrow).

inhibition of adenylyl cyclase via a G$_i$-protein mediated pathway (Sage and Heemskerk, 1992).

The number, type, and specificity of these receptors have been investigated by functional assays (e.g., by determination of aggregation and shape change) and by binding assays with platelet membranes or fixed platelets (Jefferson et al., 1988; Greco et al., 1992). These experiments showed that ATP (normally a potent activator of purinergic receptors) acts as an inhibitor of human platelets. ADP analogs substituted at the C-2 position of the purine ring were found to be more potent activators than ADP itself (Cusack and Hourani, 1982). We investigated the effect of ADP analogs on platelet calcium influx, calcium mobilization, and cAMP levels. Our results indicate that the platelet ADP receptor is very specific. ATP and ATP-γ-S did not activate the platelet calcium response but inhibited the fast ADP-evoked calcium influx and calcium mobilization, while ADP-induced secondary calcium influx was still observed. ADP-derivatives substituted at the phosphate moiety were weak platelet agonists with the exception of α,β-methylene-ADP which did not cause a calcium response. Interestingly, ADP-β-S was able to activate the ADP receptor-operated cation channel while ADP-β-NH$_2$ was less efficient (Figure 3). ADP-β-S, a weak activator of calcium mobilization (data not shown), caused calcium influx similar to that observed with ADP (Figure 3). Also, changes at the ribose ring did not cause a complete loss of calcium elevating capacity: the 3′-desoxy derivative of ADP-stimulated calcium mobilization and store-dependent secondary calcium influx, but had little effect on the receptor-operated cation channel. Although all three ADP analogs caused a receptor-mediated calcium response, the calcium influx could be further stimulated by ADP added later (in the case of 3'-desoxy ADP and ADP-β-NH$_2$), while ADP/ADP-β-S-stimulated platelets did not respond with additional calcium influx after a second ADP addition (Figure 3). ADP substituted at position 8 of the purine ring was completely inactive. Other nucleotide diphosphates such as GDP, CDP or IDP did not stimulate or inhibit platelet calcium regulation. All these results suggest that the platelet calcium responses are regulated by two different ADP-receptors with distinct properties.

Role of Cation Channels in Platelet Regulation

Platelet agonists such as ADP, thrombin or thromboxan A$_2$ activate phospholipase C (PLC), elevate cytosolic calcium by an inositol 1,4,5-trisphosphate-dependent release of calcium from intracellular stores, and stimulate the entry of extracellular Ca^{2+} (Figure 4). ADP is the only platelet agonist known to cause a fast Ca^{2+} entry mediated by a receptor-operated cation channel. This rapid influx precedes the ADP-evoked release of calcium from intracellular stores and a second phase of store-related calcium entry. The rise of cytosolic free Ca^{2+} stimulates the activity of myosin light chain kinase (MLCK).

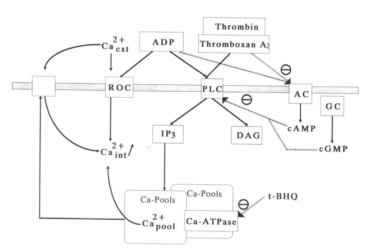

Figure 4. Calcium regulation in human platelets. The activation or inhibition (−) of receptor-operated cation channel, phospholipase C (PLC) and adenylyl cyclase (AC) by platelet agonists is indicated. Other abbreviations: IP$_3$, 1,4,5-inositoltrisphosphate; DAG, diacylglycerol; GC guanylyl cyclase; t-BHQ, tert-butylhydroquinone. Intracellular calcium (Ca^{2+}$_{int}$) is elevated by mobilization from intracellular calcium pools (Ca^{2+}$_{pool}$) and by influx through receptor-operator (ROC) and yet undefined cation channels.

Activation of MLCK by calcium together with the activation of protein kinase C (PKC) by diacylglycerol (DAG) ultimately causes platelet shape change, adhesion, aggregation, and degranulation. Platelet inhibitors which elevate cAMP or cGMP block the PLC/IP$_3$-mediated calcium mobilization from intracellular stores and associated store-related cation influx without major effects on the ADP-operated cation channel.

Acknowledgements

This work was supported by the Deutsche Forschungsgemeinschaft (Ko 210/11-3). The authors thank S. Ebert for invaluable help in preparing this paper.

References

Brüne B, Ullrich V (1991). Calcium mobilization in human platelets by receptor agonists and calcium-ATPase inhibitors. FEBS Lett. 284:1–4.

Butt E, Nolte C, Schulz S, Beltman J, Beavo JA, Jastorff B, Walter U (1992). Analysis of the functional role of cGMP-dependent protein kinase in intact human platelets using a specific activator 8-para-chlorophenylthio-cGMP. Biochem. Pharmacol. 43:2591–2600.

Cusack NJ, Hourani SMO (1982). Competitive inhibition by adenosine 5'-triphosphate of the actions on human platelets of 2-chloroadenosine 5'-diphosphate, 2-azidoadenosine 5'-diphosphate and 2-methylthioadenosine 5'-diphosphate. Br. J. Pharmac. 77:329–333.

Geiger J, Nolte C, Butt E, Sage SO, Walter U (1992). Role of cGMP and cGMP-dependent protein kinase in nitrovasodilator inhibition of agonist-evoked calcium elevation in human platelets. Proc. Natl. Acad. Sci. USA 89:1031–1035.

Greco NJ, Tandon NN, Jackson BW, Jamieson GA (1992). Low structural specificity for nucleotide triphosphates as antagonists of ADP-induced platelet activation. J. Biol. Chem. 267:2966–2970.

Halbrügge M, Walter U (1993). The regulation of platelet function by protein kinases. In: Protein Kinases in Blood Cell Function. Huang C-K, Sha'fi RI, editors. Boca Raton: CRC Press: pp 245–298.

Jefferson JR, Harmon JT, Jamieson GA (1988). Identification of High-Affinity (K$_d$ 0.35 μmol/L) and Low-Affinity (K$_d$ 7.9 μmol/L) Platelet Binding Sites for ADP and Competition by ADP Analogues. Blood 71:110–116.

Mahaut-Smith MP, Sage SO, Rink TJ (1990). Receptor-activated single channels in intact human platelets. J. Biol. Chem. 265:10479–10483.

Mahaut-Smith MP, Sage SO, Rink TJ (1992). Rapid ADP-evoked currents in human platelets recorded with the nystatin permeabilized patch technique. J. Biol. Chem. 267:3060–3065.

Ozaki Y, Yatomi Y, Kume S (1992). Evaluation of platelet calcium ion mobilization by the use of various divalent ions. Cell Calcium 13:19–27.

Rink TJ, Sage SO (1990). Calcium signaling in human platelets. Ann. Rev. Physiol. 52:431–449.

Sage SO, Heemskerk JWM (1992). Thromboxane receptor stimulation inhibits adenylate cyclase and reduces cyclic AMP-mediated inhibition of ADP-evoked responses in fura-2-loaded human platelets. FEBS Lett. 298:199–202.

Sage SO, Reast R, Rink TJ (1990). ADP evokes biphasic Ca^{2+} influx in fura-2-loaded human platelets. Biochem. J. 265:675–680.

Sage SO, Rink TJ (1987). Stimulated calcium efflux from fura-2-loaded human platelets. J. Physiol. 393:513–524.

Sage SO, Rink TJ (1985). Inhibition by forskolin of cytosolic calcium rise, shape change and aggregation in quin-2-loaded human platelets. FEBS Lett. 188:135–140.

Sage SO, Rink TJ (1987). The kinetics of changes in intracellular calcium concentration in fura-2-loaded human platelets. J. Biol. Chem. 262:16364–16369.

288

Sage SO, Rink TJ, Mahaut-Smith MP (1991). Resting and ADP-evoked changes in cytosolic free sodium concentration in human platelets loaded with the indicator SBFI. J. Physiol. 441:559–573.

Sandberg M, Butt E, Nolte C, Fischer L, Halbrügge M, Jahnsen T, Genieser H-G, Jastorff B, Walter U (1991). Characterization of Sp-5,6-dichlorobenzimidazolriboside-3′,5′-monophosphorothioate (Sp-5,6-DCI-cBIMPS) as potent and specific activator of cAMP-dependent protein kinase in cell extracts and intact cells. Biochem. J. 279:521–527.

Siess W (1989). Molecular mechanism of platelet activation. Physiol. Rev. 69:58–178.

Nonselective Cation Channels: Pharmacology, Physiology and Biophysics
ed. by D. Siemen & J. Hescheler
© 1993 Birkhäuser Verlag Basel/Switzerland

The Role of a PDGF-Activated Nonselective Cation Channel in the Proliferative Response

J. Jay Gargus*, A. M. Frace, and F. Jung

Department of Physiology and Section of Medical Genetics, Emory University School of Medicine, Atlanta, Georgia 30322, USA

Summary
Murine fibroblasts have a 28 pS calcium- and voltage-insensitive NSC that becomes quiescent at G_0 arrest and is rapidly and specifically activated by PDGF. Activation is produced by the discrete loss of long channel closures. The NSC can be rapidly and reversibly blocked with the NSAID flufenamic acid, through a prostaglandin-independent mechanism. The cell cycle (not viability) is blocked concomitantly with NSC block. A somatic cell mutant with altered NSC conductance has been isolated and used to clone the genomic locus of the channel. The mutant growth phenotype adds further support to the participation of NSC conductance in cell cycle control.

Introduction

In serum-deprived G_0-arrested fibroblasts, the addition of serum or growth factors initiates a cascade of events which culminate in DNA synthesis and mitosis. Early events in this response occur at the cell membrane and fast changes in ion flux have long been reported (Reuss et al., 1987). We have previously demonstrated a PDGF-activated nonselective cation channel (the NS channel) to participate in this response (Frace and Gargus, 1989; Jung et al., 1992). Although prior to our studies there had been no direct demonstration that NSC channels play a role in growth factor response, the older descriptions of serum and growth factor responses include loss of membrane potential and nonselective increases in membrane conductance. Activation of NS channels would be expected to induce such behavior and may offer a molecular explanation for these reported currents. The strongest links yet forged between a specific ion channel's activation and mitogenesis is presented in the lymphocyte mitogenic response to lectins, achieved by quantitatively correlating blocker effects on channel activity and proliferation (DeCoursey et al., 1984; Lewis and Cahalan, 1990).

*Address correspondence to current address: Department of Biophysics and Physiology, University of California at Irvine, Irvine, CA 92717, USA

Occurrence

In exponentially growing murine LMTK⁻ fibroblasts the NS channel is the predominant channel in the plasma membrane. It can be found in over half the membrane patches, typically with multiple channels per patch. Its open probability is ~ 0.3. The channel becomes quiescent at G_0 arrest induced by serum deprivation, and becomes rapidly active within seconds upon adding serum or pure PDGF at mitogenic concentrations (Frace and Gargus, 1989). The NS channel can also be studied in L-cell plasma membrane vesicles reconstituted into a planar lipid bilayer (Gargus and Coronado, 1985).

Electrophysiology

In physiological solutions (Frace and Gargus, 1989), the current-voltage relationship for the channel displays a slope conductance of 28 pS in cell-attached or inside-out patch conformation, and the reversal potential of channels in inside-out patches is at 0 mV regardless of whether K^+ or Na^+ is the major internal cation in solution. When the external solution is supplemented to contain 20 mM calcium or is replaced with an isotonic barium chloride solution (90 mM), no inward currents are detectable, suggesting minimal divalent cation permeation through the channel. Replacement of chloride with glutamate produces no effect on channel properties, and replacement studies with isotonic tetraethylammonium chloride ($TEA^+ Cl^-$) solutions show neither TEA^+ nor Cl^- to be permeant, pointing out the absence of anion conduction through the channel. The channel is not voltage-sensitive, with channel-open probability (cell-attached) being independent of membrane holding potential over the range of $+100$ to -100 mV. The openings of the NS channel are burstlike and exhibit complicated kinetics with at least two exponentials required to fit open time histograms. The closures of the channel contain two resolvable components of 0.48 and 15.9 mS and a third extremely fast component or flicker closure too rapid to resolve properly in our recording system. We are unable to demonstrate any correlation of channel open probability with internal calcium concentration (0.5 nM to $> 100\ \mu$M). At pCa ~ 9, activity rapidly disappears. This is irreversible, although noise of aberrant channel activity can occasionally be seen upon rapid reperfusion with increased calcium. In this respect, the behavior of the NS channel is similar to that of the calcium-dependent calmodulin-activated sodium channels of Paramecium (Saimi and Ling, 1990), where calcium-dependent behavior is accounted for by the dissociation, in the absence of calcium, of a channel-activating factor.

Pharmacology

The only useful inhibitors of the L-cell NS channel we have uncovered are the antiinflammatory agents flufenamic acid (FFA) and mefenamic acid (MA) (Jung et al., 1992), findings comparable to those of Gögelein et al. on the pancreatic NSC (1990). With FFA between 0 and 100 μM, the channel open probability decreases monotonically with increasing FFA, the data being well fit by a complement of the Langmuir adsorption isotherm, with a K_i of about 10 μM. While channel block is maintained through prolonged exposure to FFA, the inhibition is rapidly reversible upon perfusing the bath with FFA-free solution. For a given patch, this sequence of inhibition with FFA and recovery with FFA-free medium can be repeated a number of times, although the recovery is usually not complete. The efficacy of inhibition is independent of the calcium concentration in the bath. Results qualitatively similar to those obtained with FFA were obtained with MA.

As a prominent pharmacological effect of FFA is inhibition of prostaglandin biosynthesis, we investigated the role this pathway might play in channel activity (Jung et al., 1992). Indomethacin, another potent inhibitor of prostaglandin biosynthesis that is structurally dissimilar to FFA, has no effect on NS channel activity at concentrations up to 100 μM. Therefore, it is most likely that the channel blocking effects of FFA and MA are due to their common structure, and not their ability to inhibit prostaglandin biosynthesis.

FFA can be seen to inhibit a distinct component of unidirectional K^+ efflux from L cells (Jung et al., 1992). The behavior of the FFA-sensitive component of K^+ efflux, $^{\circ}M_K a$, is consistent with the notion that it reflects K^+ efflux through the NS channel: the ensemble average of the microscopic properties of the NS channel quantitatively reproduce the flux, and their pharmacology is identical. This assignment of $^{\circ}M_K a$ to the NS channel is further supported by the observation that the LTK-1 mutant (see below) is dramatically reduced specifically in $^{\circ}M_K a$, consistent with its decreased single-channel conductance.

Tetraethylammonium, 4-aminopyridine, stilbenes, amiloride, bumetanide, verapamil, diltiazam, nickel, cobalt and gadolinium are without effect on the NS channel.

Since in the cell-attached configuration the receptors exposed to the channel activating growth factor are isolated from the membrane patch that contains the studied channel, the coupling of PDGF receptor to NS channel must be through some mobile second messenger. However, the rapidity with which the channel responds (~ 10 s) designates it as one of the most proximal physiological correlates yet described of PDGF receptor activation (Williams, 1989), faster than previously described membrane events such as Na/H antiporter activation or calcium transients, all being far faster than nuclear correlates of activation such as immediate early oncogene expression or DNA synthesis. The nature of the activating intermediate remains to be determined for the NS channel. It cannot simply be free intracellular Ca^{2+} ions interacting directly with the channel, as the NS channel appears Ca-insensitive. Pharmacological studies also tend to rule out cAMP, cGMP, ATP, PKA, PKC, and calmodulin.

Molecular Genetics

We have isolated a mutant L-cell with a functionally altered NS channel (Gargus et al., 1978; Gargus, 1987b). The mutant, LTK-1, was isolated by a single-step selection from mutagenized LMTK$^-$ cells by its ability to proliferate at a low (0.2 mM) K_o^+ (the LK^R phenotype), a selective condition that yields no viable clone from a plating of 10^8 WT cells. It maintains an elevated steady-state K_i^+ concentration in medium containing 0.2 mM K_o^+ (whereas the parent LMTK$^-$ cell cannot), it has a unidirectional $^{42}K^+$ influx indentical to that of the parent, but has a three-fold increase in the half-time of the major component of K^+ efflux, nearly all assignable to $^oM_K a$, the component attributed to the NS channel (see above).

Single-channel studies show the NS channel to be qualitatively altered in the LTK-1 mutant (Gargus and Coronado, 1985). This is the only K^+ transport mechanism found to be altered and this alteration serves to explain the altered macroscopic flux and growth phenotypes. Patch electrode studies show the channel is not notably less prevalent in the mutant, it is unaltered in its voltage insensitivity, its responsiveness to PDGF, or in its open-time and closed-time kinetics, but it is clearly altered in its single-channel conductance properties. Combining the patch electrode data on the NS channel with that obtained by reconstituting the channel in a planar lipid bilayer (in collaboration with Dr. R. Coronado), we can greatly extend our interpretation of just *how* NS channel activity has been altered by the LK^R mutation in LTK-1. For the altered NS channel behavior to continue to be seen in a completely reconstituted system argues that the mutation has changed a protein so inherent to the ion conductance pathway that they stay together through reconstitution: most likely it is one of the proteins of the channel itself. In the high ionic strength buffers required in the reconstitution experiments (0.5 M NaCl/KCl), the LMTK$^-$ channel has a conductance of 50 pS, a permeability ratio P_K/P_{Na} of 1, shows voltage insensitivity, and slow bursting kinetics. Because of all of the features shared in common with the NS channel seen in patch electrode studies, it is most likely that this channel reflects its reconstitution, the only differences detected being reasonably a result of the high ionic strength of the reconstitution buffers.

When reconstitution studies identical to those performed with the WT parent were performed using the LTK-1 mutant, identical channels were *not* seen. The channel seen had the same voltage insensitivity and bursting kinetics of the WT channel, but had less than half the unitary conductance (Gargus and Coronado, 1985).

It is most parsimonious to reason that the nonselective, slow-bursting, voltage-insensitive channels seen in reconstitution reflect the very common channels which happen to have very similar characteristics seen in patch electrode recording, and that the LTK-1 mutation has altered this *one* mechanism, whether it be observed in reconstitution or patch or flux, through an alteration in the coding sequence for a region of this channel which affects its unitary conductance, while leaving unaltered channel kinetics, voltage insensitivity, and number.

The LTK-1 mutation was shown to be dominant to the WT allele in somatic cell hybrids (Gargus, 1987a) and the LTK-1 mutant gene and phenotype were shown to be transferable via DNA-mediated gene transfer from either high molecular weight genomic DNA or cosmid genomic library DNA (Mitas et al., 1988). Transformants have been shown to express the growth and transport phenotype of the DNA-donating mutant. Interestingly, they also manifest a prolonged cell cycle (Mitas et al., 1988) and an unusual sensitivity to the Rb^+ ion (Gargus et al., 1992), both like the LTK-1 mutant (see below).

A cosmid library transformant, 1–4.1, has only a single cosmid integrant by dot-blot and southern analysis and has a K^+ flux like the LTK-1 mutant. High molecular weight DNA from it can be used to produce secondary then tertiary transformants.

A cosmid library was prepared from 1–4.1 DNA in Supercos cosmid vector (Tuscan and Gargus, unpublished observations). This library was then screened by hybridization with the HSV-TK cassette of the original library vector. Five overlapping cosmid clones were isolated and restriction-mapped, defining the LK^R-containing transgene locus from 1–4.1. Single copy sequence is now being used to probe zooblots to define those useful for cDNA library screening.

Function

An initial observation suggesting that the NS channel may be regulated by growth factors is that after cells are rendered quiescent by incubation in serum-free media for 24 h, NS channel activity is very rare, with fractional open time being virtually zero for prolonged periods (Frace and Gargus, 1989). Further, if mitogenesis is initiated by addition of serum in solutions bathing the cells, while maintaining cell-attached recording, there is a rapid activation of the channel. After activation, the channels maintain a high percent open time (> 75%) for several minutes, then gradually become much less active even in the continued presence of serum. Occasionally, they can be followed into another burst of activity a few minutes later, as if the channels are slowly oscillating between states.

We were able to identify PDGF (BB) and PDGF alone to be responsible for channel activation by the application of pure recombinant growth factor at ng/ml concentrations (Frace and Gargus, 1989). Fibroblast-derived growth factor, epidermal growth factor, vasopressin, and bombesin (each known to raise intracellular calcium in fibroblasts) insulin, transferrin, selenium, and alpha-thrombin are all incapable of activating the NS channel, either alone or in combination.

Two properties of the channel, unit conductance and open time kinetics, appear unaltered by PDGF activation, but stimulation does cause the discrete loss of the population of long channel closures, thereby producing the net effect of channel activation. The discovery that PDGF is capable of activating the NS channel provided the first molecular evidence of a specific channel's participation in a growth factor response.

A hint to the functional importance of NS channel conductance to cell cycle was provided by the long cycle duration observed for all independent mutants and transformants of the LTK-1 class: apparently mutations that decreased NS conductance prolong cell cycle (Gargus, 1987b). To address directly the physiological significance of ion flux through the NS channel to cell proliferation, we took advantage of the inhibitory effect of FFA on the channel. Two separate estimates of cell proliferation, cell growth constant and cloning efficiency, were calculated and plotted as a function of [FFA]. For LMTK$^-$ and LTK-1 cells an inhibitory dose, ID_{50}, of $50-100$ μM was obtained for both measures of growth (Jung et al., 1992). MA, while slightly less potent than FFA, produced effects that were qualitatively the same (while indomethacin at 100 μM had no effect).

These two measures show the cell cycle to be affected by FFA and MA in the same potency sequence (FFA better than MA), and over the same range of concentrations that produce channel block. The fitted ID_{50} for FFA in LMTK$^-$ and LTK-1 is roughly five times the K_i fitted for the single-channel current or macroscopic K^+ efflux, implying that not until approximately 80% of the NS channels are blocked is there a half-inhibition of growth, comparable to results obtained with channel blockers on lectin-induced lymphocyte proliferation (Chandy et al., 1984).

References

Chandy KG, DeCoursey TE, Cahalan MD, McLaughlin C, Gupta S (1984). Voltage-gated potassium channels are required for human T lymphocyte activation. J. Exp. Med. 160:369–385.

DeCoursey TE, Chandy KG, Gupta S, Cahalan MD (1987). Mitogen induction of ion channels in murine T lymphocytes. J. Gen. Physiol. 89:405–420.

Frace AM, Gargus JJ (1989). Activation of single-channel currents in mouse fibroblasts by platelet-derived growth factor. Proc. Natl. Acad. Sci. USA 86:2511–2515.

Gargus JJ (1987a). Selectable mutations altering two mechanisms of mammalian K^+ transport are dominant. Am. J. Physiol. 252:C515–C522.

Gargus JJ (1987b). Mutant isolation and gene transfer as tools in the study of transport proteins. Am. J. Physiol. 252:C457–C467.

Gargus JJ, Coronado R (1985). A selectable mutation alters the conductance of a mammalian K^+ channel. Fed. Proc. 44:190.

Gargus JJ, Miller IL, Slayman CW, Adelberg EA (1978). Genetic alterations in potassium transport in L cells. Proc. Natl. Acad. Sci. USA 75:5589–5593.

Gargus JJ, Mitas M, Selvaraj S (1992). Rubidium sensitivity: new phenotype of mutants in PDGF-activated cation channel. FASEB J. 6:A1897.

Gögelein H, Dahlem D, Englert HC, Lang HJ (1990). Flufenamic acid, mefenamic acid and niflumic acid inhibit single nonselective cation channels in the rat exocrine pancreas. FEBS Lett. 268:79–82.

Jung F, Selvaraj S, Gargus JJ (1992). Blockers of PDGF-activated nonselective cation channel inhibits cell proliferation. Am. J. Physiol. 262:C1464–C1470.

Lewis RS, Cahalan MD (1990). Ion channels and signal transduction in lymphocytes. Ann. Rev. Physiol. 52:415–430.

Mitas M, Coogan C, Gargus JJ (1988). Gene transfer of a putative mammalian K^+ channel gene from genomic and cosmid DNA. Am. J. Physiol. 255:C12–C18.

Reuss L, Cassel D, Rothenberg P, Whitely B, Mancuss D, Glaser L (1986). Mitogens and ion fluxes. Curr. Top. Membr. Trans. 27:3–54.

Saimi Y, Ling K-Y (1990). Calmodulin activation of calcium-dependent sodium channels in excised membrane patches of Paramecium. Science 249:1441–1444.

Williams LT (1989). Signal transduction by the platelet-derived growth factor receptor. Science 243:1564–1570.

Nonselective Cation Channels: Pharmacology, Physiology and Biophysics
ed. by D. Siemen & J. Hescheler
© 1993 Birkhäuser Verlag Basel/Switzerland

Cation Channels in Oocytes and Early States of Development: A Novel Type of Nonselective Cation Channel Activated by Adrenaline in a Clonal Mesoderm-Like Cell Line (MES-1)

Thomas Kleppisch[1], Anna M. Wobus[2], Jürgen Hescheler[3]

[1]Institut für Physiologie, Humboldt-Universität zu Berlin, D-10115 Berlin, FRG;
[2]Institut für Pflanzengenetik und Kulturpflanzenforschung, D-06466 Gatersleben, FRG;
[3]Institut für Pharmakologie, Freie Universität Berlin, D-14195 Berlin, FRG

Summary
The expression of receptors and ion channels alters during growth, maturation, and after fertilization of oocytes reflecting functional changes. Besides voltage-dependent ion channels, oocyte membranes possess an IP_3-activated cation channel mediating a prolonged Ca^{2+} influx. The Ca^{2+} is thought to be involved in maturation and fertilization. Alternatively, mono- and divalent cations can enter oocytes via stretch-activated channels. The oocyte channel population is further modified during subsequent embryogenesis, suggesting that ionic channels obviously become expressed at specific states of embryological differentiation and in tissue-specific manner. The resulting differences in functional ion channel populations of adult cells underlie the large diversity of cells and their function. Conversely, differentiation and cell proliferation themselves depend on ion transport. Ca^{2+} ions have been shown to play a pivotal role in these processes. Nonselective cation channels represent one possible pathway for Ca^{2+} entry into the cell and, therefore, might be involved in the regulation of embryological development. Undifferentiated embryonal carcinoma cells (P19), visceral endoderm-like cells (END-2), epithelioid ectoderm-like cells (EPI-7), mesoderm-like cells (MES-1), and parietal yolk sac cells (PYS-2) have been used as a model to study the expression of ionic channels during early development. In MES-1 cells a nonselective cation current was activated by adrenaline. Interestingly, the intracellular pathway for activation of these channels involved the cascade of activation of the cAMP-dependent protein kinase (PKA) resulting in protein phosphorylation. This mechanism is well known for Ca^{2+} channel stimulation in cardiac and skeletal muscle both originating from the mesoderm.

Introduction

Maturation, proliferation, and embryological differentiation require many complex regulatory processes which are not yet completely understood. Obviously, intracellular Ca^{2+} is an important factor in the control of many cellular functions including developmental changes. This view is supported by the finding of transcription factors regulated by Ca^{2+}, e.g., via Ca^{2+}-dependent protein kinases (Morgan and Curran, 1988; Kapiloff et al., 1991). Hence, pathways controlling the Ca^{2+}

Address for correspondence: J. Hescheler, Pharmakologisches Institut, Freie Universität Berlin, Thielallee 69–73, D-14195 Berlin, FRG.

entry into cells at distinct stages of development may be involved in the regulation of differentiation. In oocytes, besides voltage-dependent Ca^{2+} channels (Okamoto et al., 1977; Hagiwara and Jaffe, 1979; Moody, 1985), an $InsP_3/InsP_4$ sensitive Ca^{2+} permeable cation channel was detected (Snyder et al., 1988; Mahlmann et al., 1989; DeLisle et al., 1992). Injection of $InsP_3$ isomers or stimulation of membrane receptors regulating phosphoinositide turnover activate a prolonged Ca^{2+} influx. This Ca^{2+} influx was revealed by recording Ca^{2+}-activated Cl^- currents. Ca^{2+}-activated Cl^- currents show an initial transient component due to Ca^{2+} release from intracellular stores which is followed by a more prolonged component. The sustained component is based on a Ca^{2+} influx through Ca^{2+} permeable membrane channels, since it is abolished in Ca^{2+} free extracellular medium (Snyder et al., 1988). The inositol phosphates, $InsP_3$ and $InsP_4$, have been shown to act in synergy to stimulate the influx of extracellular Ca^{2+} into oocytes (Snyder et al., 1988). This $InsP_3$-induced Ca^{2+} influx may represent a mechanism for a long-lasting cell activation, since, in contrast to voltage-dependent Ca^{2+} channels, it does not show significant inactivation upon stimulation. Mechanosensitive channels permeable for both mono- and divalent cations represent an alternative pathway for Ca^{2+} entry. They are consistently expressed in oocytes (Methfessel et al., 1986; Taglietti et al., 1988; Moody and Bosma, 1989; Yang and Sachs, 1989). These channels closely resemble stretch-activated channels found in other preparations (reviewed by Yang and Sachs, this volume). They are activated in cell-attached patches when moderate suction or pressure is applied to the pipette. Energy changes resulting from deformation of the cytoskeleton are thought to cause directly or indirectly (*via* a G-protein activation) channel opening. Characteristically, the activity of stretch-activated channels in oocytes increases with membrane depolarization. Gadolinium and amiloride effectively block these channels. Amiloride-sensitive ion influx is involved in regulation of fertilization, cell proliferation, and differentiation. This might be related, at least partly, to the function of stretch-activated channels.

The expression of receptors and ion channels changes during growth, maturation, and after fertilization of oocytes (Eusebi et al., 1979; Kusano et al., 1982; Taglietti et al., 1984; Moody 1985) and during embryological differentiation (Ebihara and Speers, 1984; Simonneau et al., 1985; Hirano and Takahashi, 1987; Kubo, 1989). Electrophysiological characteristics of differentiation states of the three primary germ layers in mammalian embryos are poorly examined. Due to very small cell numbers and rapid changes of developmental states in early embryos, committed cells of the primary germ layers are difficult to obtain for patch-clamp studies. Therefore, we used the well established cellular model of pluripotent embryonic carcinoma P19 cells (McBurney and Rogers 1982) and their differentiated derivatives END-2, EPI-7, MES-1, and PYS-2 characterized as endoderm-like, epithelioid ectoderm-like, mesoderm-like, and parietal yolk sac cells, respectively (Mummery et al., 1985; Mummery et al., 1986; Adamson et al., 1977). Using these cell lines, we examined the effects of various receptor agonists.

Occurrence

A novel type of nonselective cation channel is induced in about 65% of MES-1 cells by adrenaline, isoproterenol, forskolin as well as by the specific activator of the cAMP-dependent protein kinase (PKA) cBIMPS. Other receptor agonists including carbachol, bradykinine, serotonin, ATP, and somatostatin did not stimulate cation currents. In undifferentiated P19, END-2, PYS-2 as well as EPI-7 cells, none of the hormones tested was found to induce cation currents. The MES-1 cell line was selected after DMSO treatment of EC P19 cells (Mummery et al., 1986) known to induce mesodermal differentiation (Edwards et al., 1983).

Electrophysiology

The adrenaline-induced current found in MES-1 cells shows a linear current-voltage (I-V) relationship with a reversal potential close to 0 mV under physiological conditions. This suggests that this current might be a nonselective cation current. The two findings that i) reduction of Cl^- in the extracellular solution did not affect the adrenaline-induced current and ii) removal of extracellular Na^+ suppressed the inward current strongly support that the adrenaline-induced current in MES-1 cells is a nonselective cation current. The intracellular Ca^{2+} concentration could be increased by a Ca^{2+} influx *via* NSC. Noteworthy is that, voltage-dependent Ca^{2+} channel currents in MES-1 cells are largely reduced during activation of the NSC. This might be due to Ca^{2+}-dependent Ca^{2+} channel inactivation.

The onset of the adrenaline-induced nonselective cation current in MES-1 cells is rather slow. Maximal activation of the inward current at -60 mV was reached with a delay of approximately 30 s and the effect also slowly reversed following wash-out. Similar to adrenaline, the β-adrenoceptor agonist isoprenaline (ISO, 1 μM) reversibly stimulated the linear nonselective current. In cardiac cells originating from mesoderm, β-adrenoceptors are coupled to an intracellular cascade of enzymatic reactions resulting in cAMP-dependent protein phosphorylation (Trautwein and Hescheler, 1990). Both the adenylyl cyclase activator, forskolin (10 μM) and the direct activator of the cAMP-dependent protein kinase, Sp-5,6-dichloro-1-b-D-ribofuranosyl-benzimidazole-3′-

300

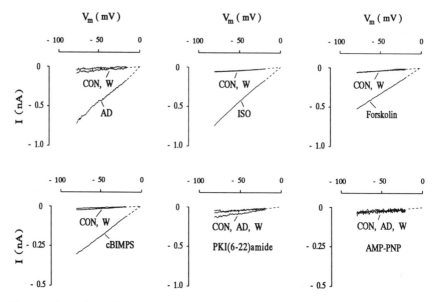

Figure 1. A novel nonselective cation current in MES-1 cells is activated by cAMP-dependent phosphorylation. Currents were recorded during voltage-clamp ramps from -100 to 0 mV; the holding potential was -80 mV. The extracellular solution contained (in mM): NaCl 125, $BaCl_2$ 10.8, CsCl 5.4, $MgCl_2$ 1.0, glucose 10, HEPES 10 (pH 7.4 at 37°C), while the standard pipette solution had the following composition (in mM): CsCl 120, $MgCl_2$ 3, EGTA 10, HEPES 5/CsOH (pH 7.4 at 37°C). Shown are current traces recorded before (CON) and during superfusion of the cell with adrenaline (AD, 10 μM), isoprenaline (ISO, 0.1 μM), forskolin (10 μM), and the protein kinase A activator cBIMPS (200 μM), respectively, and after wash-out (W). Two experiments are shown, where the intracellular solution was supplemented with the protein kinase A inhibitor PKI(6-22)amide (2 μM), or ATP was replaced by AMP-PNP (5 mM), respectively (as indicated in the corresponding panels). Current traces were recorded after 10 min infusion of the cell. No nonselective cation currents were activated under these experimental conditions.

5′-monophosphorothioate (cBIMPS, 200 μM) induced the linear nonselective current in MES-1 cells (Figure 1). A possible direct activation of NSC by cAMP was excluded by intracellular infusion of the specific inhibitor of protein kinase A (PKI(6-22)amide, 2 μM), or by exchanging the intracellular ATP by 5′-adenylimidodiphosphate (AMP-PNP, 5 mM), respectively. These approaches should suppress phosphorylation without affecting increase in cAMP. The intracellular concentration of compounds diffusing from the pipette into the cell is supposed to reach about 50% of the concentration in the pipette solution within 2–5 min after disruption of the membrane patch (Pusch and Neher, 1988). Infusion of both compounds fully suppressed the adrenaline-induced NSC in MES-1 cells within 10 min (Figure 1). These observations clearly demonstrate that a phosphorylation step is involved in the activation of NSC in MES-1 cells, rather than a direct activation by cAMP (see scheme Figure 2). No nonselective cation currents were

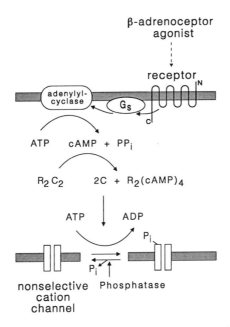

Figure 2. The cascade of enzymatic reactions involved in cAMP-dependent phosphorylation resulting in activation of NSC in MES-1 cells. Adenylyl cyclase is activated via the β-adrenoceptor (mediated by the G-protein G_s). The subsequent rise of intracellular cAMP causes the dissociation of the cAMP-dependent protein kinase (PKA) into the active form and regulatory subunits. The activated PKA phosphorylates the channel itself or a protein involved in activation of NSC. Dephosphorylation is catalyzed by phosphatases.

activated following application of other tested receptor agonists including carbachol, bradykinin, serotonin, ATP, and somatostatin.

Function

The receptor-induced opening of cation channels may represent an important mechanism of cell activation. In fibroblasts, NSC have been shown to mediate the mitogenic response to growth factors (see Gargus et al., this volume). Their cell cycle could be blocked by inhibiting these NSC. In mesodermal cells, it is difficult to establish a clear correlation between NSC and physiological functions, since so far a selective blocker is not available. However, one might assume that NSC contributing to the Ca^{2+} influx are involved in the regulation of embryological development, differentiation, and proliferation. This is supported by the recent findings of Ca^{2+} dependent transcription factors. The fact that adrenaline-activated NSC were found exclusively in MES-1 mesoderm-like cells might indicate a specific role in the regulation of early gene expression leading to excitable muscle tissue. The role of β-adreno-ceptors in the early embryological development is not yet clear. Our data suggest that adrenaline may represent one of the first hormones modulating cellular functions, especially of the mesodermal lineage, by stimulating β-adrenoceptors.

Acknowledgements
We thank Dr. C. Mummery (Utrecht, The Netherlands) for providing us with END-2, EPI-7, and MES-1 cell lines, Dr. P. Gruss (Göttingen, FRG) for P19 cells, and Dr. J. Forejt (Praha, Czech Republic) for PYS-2 cells. We are also grateful to M. Bigalke, I. Reinsch, and S. Sommerfeld for expert technical assistance, W. Stamm for skillful engineering, and Dr. Geniesser (Biolog, Bremen, FRG) for the kind gift of Sp-5,6-DCl-cBIMPS. These studies were supported by funds of the Deutsche Forschungsgemeinschaft to A.W. and J.H.

References

Adamson ED, Evans MJ, Magrane GG (1977). Biochemical markers of the progress of differentiation in cloned teratocarcinoma cell lines. Eur. J. Biochem. 79:607–615

DeLisle S, Pittet D, Potter BVL, Lew PD, Welsh MJ (1992). InsP$_3$ and Ins(1,3,4,5)P$_4$ act in synergy to stimulate influx of extracellular Ca^{2+} in *Xenopus* oocytes. Am. J. Physiol. 262:C1456–1463.

Ebihara L, Speers WC (1984). Ionic channels in a line of embryonal carcinoma cells induced to undergo neuronal differentiation. Biophys. J. 46:827–830.

Edwards MK, McBurney MW (1983). The concentration of retinoic acid determines the differential cell types formed by teratocarcinoma cell line. Dev. Biol. 98:187–191.

Eusebi F, Magina F, Alfei L (1979). Acetylcholine-elicited responses in primary and secondary mammalian oocytes disappear after fertilization. Nature 277:651–653.

Hagiwara S, Jaffe LA (1976). Electrical properties of egg membranes. Ann. Rev. Biophys. Bioeng. 8:385–416.

Hirano T, Takahashi K (1987). Development of ionic channels and cell-surface antigens in the cleavage-arrested one-cell embryo of ascidian. J. Physiol. 386:113–133.

Kapiloff MS, Mathis JM, Nelson CS, Lin CR, Rosenfeld MG (1991). Calcium/calmodulin-dependent protein kinase mediates a pathway for transcriptional regulation. Proc. Natl. Acad. Sci. USA 88:3710–3701.

Kubo (1989). Development of ion channels and neurofilament during neuronal differentiation of mouse embryonal carcinoma cell lines. J. Physiol. 409:497–523.

Kusano K, Miledi R, Stinnaker J (1982). Cholinergic and catecholaminergic receptors in the *Xenopus* oocyte membrane. J. Physiol. 328:143–170.

Mahlmann S, Meyerhof W, Schwarz JR (1989). Different roles if IP$_4$ and IP$_3$ in the signal pathway coupled to the TRH receptor in microinjected *Xenopus* oocytes. FEBS Lett. 249:108–112.

McBurney MW, Rogers BJ (1982). Isolation of male embryonal carcinoma cells and their chromosome replication patterns. Dev. Biol. 89:503–508.

Methfessel C, Witzemann V, Takahashi T, Mishima M, Numa S, Sakman B (1986). Patch clamp measurements on *Xenopus laevis* oocytes: currents through endogenous channels and implanted acetylcholine receptor and sodium channels. Pflügers Arch. 407:577–588.

Moody WJ (1985). The development of calcium and potassium currents during oogenesis in the starfish, *Leptasterias hexactis*. Dev. Biol. 112:405–413.

Moody WJ, Bosma MM (1989). A nonselective cation channel activated by membrane deformation in oocytes of the ascidian *Boltenia villosa*. J. Membr. Biol. 107:179–188.

Morgan JI, Curran, T (1988). Calcium as a modulator of the immediate-early gene cascade in neurons. Cell Calcium 9:303–311.

Mummery CL, Feijen A, van der Saag PT, van den Brink CE, de Laat SW (1985). Clonal variants of differentiated P19 embryonal carcinoma cells exhibit epidermal growth factor receptor kinase activity. Dev. Biol. 109:402–410.

Mummery CL, Feijen A, Moolenaar WH, van edn Brink CE, de Laat SW (1986). Establishment of a differentiated mesodermal line from P19 EC cells expressing functional PDGF and EGF receptors. Exp. Cell. Res. 165:229–241.

Okamoto H, Takahashi K, Yamashita N (1977). Ionic currents through the membrane of mammalian oocyte and their comparison with those in the tunicate and sea urchin. J. Physiol. 267:465-495.

Pusch M, Neher E (1988). Rates of diffusional exchange between small cells and a measuring patch pipette. Pflügers Arch. 411:204–211.

Simonneau M, Distasi M, Tauc L, Poujeol C (1985). Development of ionic channels during mouse neuronal differentiation. Journal de physiologie 80:312–320.

Snyder PM, Krause KH, Welsh MJ (1988). Inositol trisphosphate isomers but not inositol 1,3,4,5,-tetrakisphosphate, induce calcium influx in *Xenopus laevis* oocytes. J. Biol. Chem. 263:11048–11051.

Taglietti V, Tanzi F, Romero R, Simoncini L (1984). Maturation involves suppression of voltage-gated currents in the frog oocyte. J. Cell. Physiol. 121:576–588.

Taglietti V, Toselli M (1988). A study of stretch-activated channels in the menbrane of frog oocytes with Ca^{2+} ions. J. Physiol. 407:311–328.

Trautwein W, Hescheler J (1990). Regulation of cardiac L-type calcium current by phosphorylation and G proteins. Annu. Rev. Physiol. 52:257–274.

Yang XC, Sachs F (1989). Block of stretch-activated ion channels in *Xenopus* oocytes by gadolinium and calcium ions. Science 243:1068–1071.

Nonselective Cation Channels: Pharmacology, Physiology and Biophysics
ed. by D. Siemen & J. Hescheler
© 1993 Birkhäuser Verlag Basel/Switzerland

Slowly-Activating Cation Channels in the Vacuolar Membrane of Plants

Thomas Weiser

Boehringer Ingelheim KG, ZNS-Pharmakologie, Bingerstrasse, D-55216 Ingelheim, FRG

Summary
Among other ion channels and transport proteins, the membrane of plant vacuoles contains a voltage- and calcium-dependent cation channel with activation kinetics in the range of seconds. This SV(= slow vacuolar)-channel has a unit conductance of 60 to 80 pS (in symmetrical 100 mM cation solution) and is strictly inward rectifying. Investigations on the pharmacology of this protein revealed reasonable similarities to calcium-dependent potassium channels of large conductance.

Introduction

Almost every plant cell possesses a large central vacuole which can make up more than 90% of the cell volume. This organelle plays an important role in the life of the organism and is the main compartment for the storage of solutes and metabolical products.

Thus, the vacuolar membrane, the tonoplast, contains a great variety of transport systems for such compounds. The application of the patch-clamp technique to this organelle made it possible for the first time to study the electrophysiological properties of the tonoplast, revealing the predominance of an ubiquitous genus of cation channels. In this paper a short review concerning some aspects of these channels is presented.

Occurrence and Appearance

In 1987, Hedrich and Neher found a cation-channel in the vacuolar membrane of sugar beet tap roots which exhibited activation kinetics in the range of seconds, and was therefore named "slow-vacuolar" (SV)-channel. It opens only upon hyperpolarization of the vacuole interior; additionally, the activity is dependent on the calcium-concentration on the cytoplasmatic side. The selectivity is equal for potassium and sodium, but also anions can permeate (i.e., malate and chloride, $gK/gCl = 6$). The single-channel conductance is in the range of 60 to 80 pS (with symmetrical 100 mM potassium on either side of the

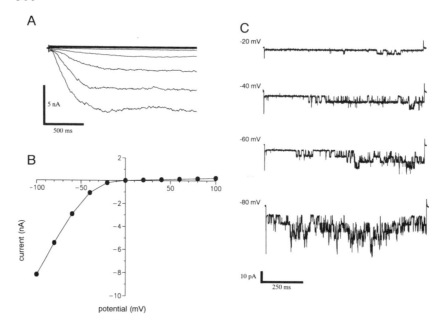

Figure 1. Electrophysiological properties of the SV-channel in *Allium cepa* bulb scale vacuoles. In (A), the response of a vacuole to voltage pulses from −100 to +100 mV is given (20 mV increment; experiment in the "whole vacuolar" mode of the patch-clamp technique). The SV-channel-mediated conductance exhibits slow activation kinetics and is strongly rectifying (B). Single-channel experiments at various holding potentials demonstrate the slow activation and the voltage dependence of open probability. (C; experiment in the "outside-out" mode). Media contained (in mM): KCl 100, $MgCl_2$ 2, $CaCl_2$ 0.1, Mannitol 300, Tris/Mes 5, pH 7.2 on either side of the membrane. Potentials refer to the vacuolar interior (= recording pipette). (Data from T. Weiser, unpublished.)

tonoplast). Examinations of vacuoles from other plants and tissues revealed an abundance of this type of channel in almost every vascular plant cell (Hedrich et al., 1988). In Figure 1 some data about the electrophysiological appearance of the SV-channel in onion (*Allium cepa*) bulb scales are presented.

Calcium Dependence

One characteristic property of SV-channels is their strict dependence on cytosolic calcium. In most preparations, the channel exhibits its maximum open probability at calcium concentrations of about 10^{-4} to 10^{-3} M (cf. Tanifuji et al., 1988; Colombo et al., 1989). This seems puzzling at first glance, since the cytoplasmatic calcium concentration in an intact plant cell is three to four orders of magnitude smaller. This contradiction may be circumvented if one takes into account that experiments with isolated vacuoles expose this organelle to far from physiological conditions. In some animal cells, the wash-out of cytoplasmatic factors is responsible for the shift of calcium-dependence, e.g. in a cation channel in mouse pancreatic acinar cells (Maruyama and Petersen, 1984). Bertl and Slayman (1990) demonstrated for yeast vacuoles that the sensitivity of a cation channel to calcium can be enhanced by the presence of antioxidant compounds in the surrounding medium. Thus, these high concentrations for a reasonable channel activity may be due to the experimental conditions.

Similar to other calcium-dependent ion-channels (i.e., Saimi and Ling, 1990; Preston et al., 1990; Onozuka et al., 1987) the calcium-dependence of the SV-channel is mediated by calmodulin or a calmodulin-like domain of the channel protein. Weiser et al. (1990) found a concentration-dependent decrease of the SV-channel activity of *Chenopodium rubrum* in the presence of the calmodulin antagonist N-(6-aminohexyl)-5-chloro-1-naphten-sulfonamid (W-7) and its chlorine-deficient derivative (W-5). This effect can be partly reversed by synchronous application of plant calmodulin. Surprisingly, calmodulin from bovine brain lacks this ability, suggesting a reasonable selectivity of the channel protein for calmodulin from plant sources.

Apart from the dependence on calcium and voltage, also other regulatory mechanisms were found. Pantoja et al. (1992) report the dependence of an SV-channel in sugar beet suspension cells on cytoplasmatic chloride. Reduction of the chloride concentration causes a decrease of channel open probability and the appearance of an additional closed state. Moreover, in some preparations channel activity is regulated from the interior of the vacuole: vacuoles from young *Vignia unguiculata* seedlings exhibit almost no channel activity when studied in the "attached"-mode of the patch-clamp technique (in contrast to experiments performed in the "outside-out"-mode). After establishing the "whole-vacuolar"-mode, the voltage-activated conductance requires about 10 min to reach saturation. This effect can be explained by the diffusion of an intra-vacuolar inhibiting factor with a molecular weight between 20 and 200 kD out of this compartment into the recording pipette (Maathuis and Prins, 1991).

Pharmacology

Up to now, only a few studies concerning the pharmacology of SV-channels have been published. Hedrich and Kurkdjian (1988) found the inhibition of the SV-channel of beta-taproots by micromolar concentrations of the anion transport blockers zinc and the stilbene-derivatives DIDS and SITS. Pantoja et al. (1990) report the channel in beta-suspension cultures to be sensitive to submicromolar amounts of amiloride, an inhibitor of Na/H antiporters.

More detailed studies were performed on suspension-cultured cells of *Chenopodium rubrum* (Weiser and Bentrup, 1990; Weiser and Bentrup, 1991; Weiser and Bentrup, 1993). Here, at $-100\,mV$ transtonoplast potential, SV-channels were half-maximally inhibited by 300 nM quinacrine, $6\,\mu M$ (+)-tubocurarine, $7\,\mu M$ 9-aminoacridine, and $35\,\mu M$ quinine, respectively. All these compounds are known to be inhibitors of calcium-dependent potassium-channels of large conductance (Maxi-K-channels). The most convincing evidence for great similarity to this class of ion-channels arises form the fact that this SV-channel is inhibited by nanomolar concentrations of charybdotoxin, but is insensitive to apamin. These compounds are proposed to specifically distinguish between different species of K(CA)-channels (Moczydlowski et al., 1988). Moreover, the mechanism of action on single-channel performance, namely, a "slow" block (Hille, 1984), is very similar for both SV- and Maxi-K-channels.

Function

In contrast to the large number of studies on the SV-channels and the abundance in almost any plant vacuole preparation, the knowledge about their physiological role is only limited. Coyaud et al. (1987) and Iwasaki et al. (1992) discuss a function in the accumulation of malate into the vacuole (especially in CAM-plants) due to the fact that SV-channels are, to some extent, also permeable to anions. Maathuis and Prins (1990) report that the growing conditions have an influence on the channel's performance: SV-channels in the vacuoles of *Plantago* root-cells exhibit a drastically reduced open probability when the plants are grown on media containing high concentrations of sodium salts.

These two examples demonstrate that SV-channels of similar properties may play completely different roles in plant cells of different origin, but there are other reasons that make it difficult to reveal the physiological functions of this type of channel.

Although the preparation of vacuoles from living cells can be performed within a few minutes, the experimental conditions are far from physiological. Thus, the vacuole is no longer embedded in protoplasm, but faces the "dead" environment of an artificial electrolyte-solution, free of physiological stimuli (like changes in concentration of solutes, membrane potential, osmotic pressure, etc.). Moreover, regulating factors of the cytoplasm may be washed out. This is probably one reason why, under physiological calcium-concentrations and voltage gradients (of some 10 millivolts) the SV-channel in most preparations remains silent. To activate it to a measurable extent, stimuli in unphysiological ranges have to be applied.

Up to now, it has not been possible to study the performance of this channel in the most efficient way, which would be in its natural environment, in an intact vacuole of a living plant cell.

Since the first studies on the SV-channel, about eight years ago, much fundamental work on the properties and occurrence of different SV-channels has been performed. Now, knowing part of its pharmacology and regulation, it seems promising to study this transport protein with biochemical and molecular biological techniques to gain more information about its position in the large family of ion channels.

References

Bertl A, Slayman CL (1990). Cation-selective channels in the vacuolar membrane of *Saccharomyces*: Dependence on calcium, redox state and voltage. Proc. Natl. Acad. Sci. USA 87:7824–7828.

Colombo R, Cerana R, Lado P, Peres A (1989). Regulation by calcium of voltage-dependent tonoplast K+ channels. Plant Physiol. Biochem. 27(4):557–562.

Coyaud L, Kurkdjian A, Kado R, Hedrich R (1987). Ion channels and ATP-driven pumps involved in ion transport across the tonoplast of sugarbeet vacuoles. Biochimica et Biophysica Acta 902:263–268.

Hedrich R, Barbier-Brygoo H, Felle H, Flügge UI, Lüttge U, Maathuis FJM, Mar S, Prins HBA, Raschke K, Schnabl H, Schröder JI, Struve I, Taiz L, Ziegler P (1988). General mechanisms for solute transport across the tonoplast of plant vacuoles: a patch-clamp, survey of ion channels and proton pumps. Botanica Acta 101:7–13.

Hedrich R, Kurkdjian A (1988). Characterization of an anion-permeable channel from sugar beet vacuoles: effect of inhibitors. EMBO Journal 7:3661–3666.

Hedrich R, Neher E (1987). Cytoplasmic calcium regulates voltage-dependent ion channels in plant vacuoles. Nature 329:833–835.

Hille B (1984). Ionic Channels of Excitable Membranes. Sunderland, Mass.: Sinauer Associates.

Iwasaki I, Arata H, Kijima H, Nishimura M (1992). Two types of channels involved in the malate ion transport across the tonoplast of a crassulacean acid metabolism plant. Plant Physiol. 98:1491–1497.

Maathuis FJM, Prins HBA (1990). Patch clamp studies on root cell vacuoles of a salt-tolerant and a salt-sensitive *Plantago* species. Plant Physiol. 92:23–28.

Maathuis FJM, Prins HBA (1991). Inibition of inward rectifying tonoplast channels by a vacuolar factor: physiological and kinetic implications. J. Membrane Biol. 122:251–258.

Maruyama Y, Petersen OH (1984). Single calcium-dependent cation channels in mouse pancreatic acinar cells. J. Membr. Biol. 81:83–87.

Moczydlowski E, Lucchesi K, Ravindran A (1988). An emerging pharmacology of peptide toxins targeted against potassium channels. J. Membrane Biol. 105:95–111.

Onozuka M, Furuichi H, Kishii K, Imai S (1987). Calmodulin in the activation process of calcium-dependent potassium channel in *Euhadra* neurones. Comp. Biochem. Physiol. 86A:589–593.

Pantoja O, Dainty J, Blumwald E (1990). Tonoplast ion channels from sugar beet cell suspensions. Plant Physiol. 94:1788–1794.

Pantoja O, Dainty J, Blumwald E (1992). Cytoplasmic chloride regulates cation channels in the vacuolar membrane of plant cells. J. Membrane Biol. 125:219–229.

Preston RR, Wallen-Friedmann MA, Saimi Y, Kung C (1990). Calmodulin defects cause the loss of Ca^{2+}-dependent K^+ currents in two pantophobiac mutants of *Paramecium tetraurelia*. J. Membrane Biol. 115:51–60.

Saimi Y, Ling K-Y (1990). Calmodulin activation of calcium-dependent sodium channels in excised membrane patches of *Paramecium*. Science 249(4975):1441–1444.

Tanifuji M, Sato M, Wada Y, Anraku Y, Kasai M (1988). Gating behaviors of a voltage-dependent and Ca^{2+}-activated cation channel of yeast vacuolar membrane incorporated into planar lipid bilayer. J. Membrane Biol. 106:47–55.

Weiser T, Bentrup F-W (1990). (+)-tubocurarine is a potent inhibitor of cation channels in the vacuolar membrane of *Chenopodium rubrum* L. FEBS Lett. 277:220–222.

Weiser T, Bentrup F-W (1991). Charybdotoxin blocks cation-channels in the vacuolar membrane of suspension cells of *Chenopodium rubrum* L. Biochimica et Biophysica Acta 1066:109–110.

Weiser T, Bentrup F-W (1993). Pharmacology of the SV-channel in the vacuolar membrane of *Chenopodium rubrum* L. suspension cells. J. Membrane Biol., in press.

Index

In the index f (ff) means that the indexed word is also referred to on the following page(s).

BIRKHÄUSER
LIFE SCIENCES

Experientia Supplementum

Comparative Molecular Neurobiology

Edited by
Y. Pichon, *Université de Rennes, France*

1993. 434 pages. Hardcover. ISBN 3-7643-2785-5 (EXS 63)

Most comparative studies of the physiological and pharmacological properties of the receptors and ionic channels of various animal species have so far stressed the differences. More recent studies based on the knowledge of the primary structure of these proteins as obtained using molecular cloning techniques emphasize the common features and have led to the concept of superfamilies. These superfamilies are believed to be derived from common ancestors through evolution. To understand how this happened, it is necessary to compare the sequences and the properties of the receptors in species sufficiently distant in the evolutionary tree. Until recently, this kind of information was lacking. In the present volume, specialists in the field of comparative molecular neurobiology, most of them working on both vertebrate and invertebrate species, report their recent findings concerning the three most important superfamilies: the Ligand-Gated Ion Channels superfamily (n-ACh, $GABA_A$, glycine), the Second-Messenger Linked receptor superfamily (m-ACh, catecholamines, peptides) and the Voltage-Gated Ion Channels (Na^+, K^+ and Ca^{2+}) superfamily.

Please order through your bookseller or directly from:
Birkhäuser Verlag AG, P.O. Box 133,
CH-4010 Basel / Switzerland (Fax ++41 / 61 / 721 7950)
Orders from the USA or Canada should be sent to:
Birkhäuser Boston
44 Hartz Way, Secaucus, NJ 07096-2491 / USA
Call Toll-Free 1-800-777-4643

Birkhäuser

Birkhäuser Verlag AG
Basel · Boston · Berlin

Prices are subject to change without notice. 9/93

Molecular Biology of G-Protein-Coupled Receptors

Edited by
M.R. Brann, Univ. of Vermont, USA, (Ed.)

1992. 326 pages. Hardcover. ISBN 3-7643-3465-7 (AMGP)

Recent discoveries have demonstrated that a wide array of biologically active substances, including such neuroactive substances as neurotransmitters, hormones, neuropeptides, etc., produce their effects by interacting with a major class of receptors, namely, those that couple with G-proteins. This new area is systematically reviewed in this volume by neuroscience leaders neurochemists, pharmacologists, molecular neurobiologists – surveying the new findings and their relevance to basic and clinical issues in brain research. Discussed are ACh-muscarinic, adrenergic, dopamine, serotonin, and neuropeptide receptors and their G-protein involvement. Olfactory receptors and G-protein receptors are also covered. The fundamental G-protein to receptor cascade

underlying neuronal interaction, with its direct application to cell biology in general, is finally being elucidated by newly developed genetic engineering techniques.

This book, combining high-level review papers and new results by the world's leading research workers, will help experts stay abreast of new developments and allow others entrance to the field to find their way through the relevant literature. Readers will include biochemists, pharmacologists, neuroscientists, and others interested in the mechanisms underlying intercellular communication and the development of pharmaceutical agents to treat disease states that result from cell receptor abnormalities.

Please order through your bookseller or directly from:
Birkhäuser Verlag AG, P.O. Box 133,
CH-4010 Basel / Switzerland (Fax ++41 / 61 / 721 7950)
Orders from the USA or Canada should be sent to:
Birkhäuser Boston
44 Hartz Way, Secaucus, NJ 07096-2491 / USA
Call Toll-Free 1-800-777-4643

Birkhäuser

Birkhäuser Verlag AG
Basel · Boston · Berlin

Prices are subject to change without notice. 9/93

BIRKHÄUSER
LIFE SCIENCES

Information at First Hand

Formation and Regeneration of Nerve Connections

Edited by
S. C. Sharma, *The New York Medical College, NY, USA*
J. W. Fawcett, *The Physiological Laboratory, Cambridge, UK*

1993. 260 pages. Hardcover. ISBN 3-7643-3563-7

What causes brain cells to develop into functional systems? How are neuronal specificity and differentiation accomplished in the developing brain? What factors will bring about regeneration of injured nerves? World leaders in the fields of developmental cell biology and neuronal specificity discuss these and other issues in this landmark book about the mechanisms of development and regeneration of the nervous system, in particular as they involve visual system development. Authors of the chapters in this carefully edited book are themselves responsible for most of the discoveries and advances in the field over the past 25 years.

Neuroscientists, vision researchers, and those interested in understanding the basic principles of developmental cell biology will appreciate this important view of neuronal specificity and differentiation and its implications for the regeneration of injured nerves.

Contents:

Please order through your bookseller or directly from:
Birkhäuser Verlag AG, P.O. Box 133,
CH-4010 Basel / Switzerland (Fax ++41 / 61 / 721 7950)
Orders from the USA or Canada should be sent to:
Birkhäuser Boston
44 Hartz Way, Secaucus, NJ 07096-2491 / USA
Call Toll-Free 1-800-777-4643

Birkhäuser

Birkhäuser Verlag AG
Basel · Boston · Berlin

Prices are subject to change without notice. 9/93